Advanced Technologies for the Optimization of Internal Combustion Engines

Advanced Technologies for the Optimization of Internal Combustion Engines

Editors

Cinzia Tornatore
Luca Marchitto

MDPI • Basel • Beijing • Wuhan • Barcelona • Belgrade • Manchester • Tokyo • Cluj • Tianjin

Editors
Cinzia Tornatore
Italian National Research
Council
Italy

Luca Marchitto
Italian National Research
Council
Italy

Editorial Office
MDPI
St. Alban-Anlage 66
4052 Basel, Switzerland

This is a reprint of articles from the Special Issue published online in the open access journal *Applied Sciences* (ISSN 2076-3417) (available at: https://www.mdpi.com/journal/applsci/special_issues/ICE_Optimization).

For citation purposes, cite each article independently as indicated on the article page online and as indicated below:

LastName, A.A.; LastName, B.B.; LastName, C.C. Article Title. *Journal Name* **Year**, *Volume Number*, Page Range.

ISBN 978-3-0365-2568-6 (Hbk)
ISBN 978-3-0365-2569-3 (PDF)

Contents

About the Editors

Cinzia Tornatore received a "Summa cum Laude" magister degree in chemical engineering in 2004 and PhD in 2007 from the University of Napoli Federico II. She is a tenured researcher at the Italian National Research Council (STEMS), where she has worked since 2004. Her work has been focused on the study of thermo-fluid dynamic processes that take place in internal combustion engines fueled with conventional and new-generation fuels. This research has been carried out mainly through the application of high-spatial- and -temporal-resolution optical techniques.

Luca Marchitto is a tenured researcher at the Italian National Research Council, where he has worked since 2007. His activity is mainly focused on the optimization of combustion processes in internal combustion engines and on the fluid dynamic characterization of sprays and air–fuel mixing in CI and SI engines by non-intrusive techniques.

Editorial

Advanced Technologies for the Optimization of Internal Combustion Engines

Cinzia Tornatore * and Luca Marchitto

Institute of Science and Technology for Sustainable Energy and Mobility (STEMS-CNR), Italian National Rsearch Council, Via Marconi 4, 80125 Napoli, Italy; luca.marchitto@stems.cnr.it
* Correspondence: cinzia.tornatore@stems.cnr.it

Citation: Tornatore, C.; Marchitto, L. Advanced Technologies for the Optimization of Internal Combustion Engines. *Appl. Sci.* **2021**, *11*, 10842. https://doi.org/10.3390/app112210842

Received: 10 November 2021
Accepted: 12 November 2021
Published: 17 November 2021

Publisher's Note: MDPI stays neutral with regard to jurisdictional claims in published maps and institutional affiliations.

Even in a scenario where electric vehicles gain market share and the sale of internal combustion engines is gradually reduced, at the present time, there are still no real options that can totally replace the internal combustion (IC) engine over the entire range of its applications. In the short-to-medium term, on-road and off-road transportation will most likely be characterized by a mix of solutions, including strongly electrified powertrain configurations, as well as conventional vehicles powered by IC engines.

The combustion engine will continue playing a central role; for this reason, there is a pressing need for its further optimization in terms of thermal efficiency and pollutant emissions. This goal can be achieved through more efficient and environment-friendly technologies.

The aim of this Special Issue is to put together the recent research in advanced technologies for the optimization of internal combustion engines in order to help the scientific community address the efforts toward the development of higher-power engines with lower fuel consumption and pollutant emissions. This Special Issue contains 11 technical articles that have been peer-reviewed under the journal's rigorous review criteria.

Papers [1,2] deal with water injection technology that represents a promising solution to make turbocharged spark-ignition engines operate with higher compression ratios, higher boost pressures, and stoichiometric combustions at high loads. This allows a reduction in brake-specific fuel consumption and, thus, in CO_2 emissions on the overall engine map. In [1], a detailed experimental analysis of a low-pressure water spray in the intake manifold (PWI technology) was presented, covering a lack of experimental data on automotive PWI systems. The obtained results evidenced how significant benefits in terms of atomization quality can be obtained by adopting injection pressure and water temperature levels compliant with standard low injection pressure technologies.

An alternative way of introducing water in an SI engine is in the form of an emulsion with gasoline (water-in-gasoline emulsions (WiGE)). This solution shows some advantages compared with the standard separate injections of water and gasoline. In [2], the authors performed an experimental investigation of WiGE effects on the performance and emissions of a turbocharged port fuel injection spark-ignition engine. WiGE was produced through a prototype micro-channel emulsifier that allows using a very small amount of surfactant to create stable emulsions. Water-in-gasoline emulsions demonstrated to be a suitable technique to reduce fuel consumption at medium/high loads of turbocharged spark-ignition engines, enabling stoichiometric combustions while preserving the turbine blades from severe thermal stresses.

Two contributions to this Special Issue deal with the use of alternative fuels for internal combustion engines as a promising solution for a more sustainable operation. The research work reported in [3] assessed the combustion and performance of a dual-fuel operation in a spark-ignition engine that simultaneously integrates acetone–butanol–ethanol (ABE) and hydroxy (HHO) doping in comparison with standalone gasoline operation. The implementation of the dual-fuel operation in spark-ignition engines proved to be a valuable tool

1

for controlling emissions and reducing fuel consumption while maintaining combustion performance and thermal efficiency.

In [4], an experimental study was carried out on the effects of diesel surrogates (hydrotreated vegetable oil (HVO) and oxymethylene ethers (OME_X)) on main spray and combustion characteristics. The different physical-chemical properties of HVO and OME_X make the study of their behavior in compression ignition engines essential. Quantitative parameters describing the evolution of diesel-like sprays such as liquid length, spray penetration, ignition delay, lift-off length, and flame penetration, as well as soot formation, were assessed in a constant high-pressure and high-temperature installation using optical diagnostics.

Combustion stability, engine efficiency, and emissions in a multi-cylinder spark-ignition internal combustion engine can be improved through the advanced control and optimization of individual cylinder operation. In [5], experimental and numerical analyses were carried out on a twin-cylinder turbocharged port fuel injection spark-ignition engine to evaluate the influence of cylinder-by-cylinder variation on performance and pollutant emissions. The proposed numerical methodology represents a valuable tool to support engine design and calibration, with the aim to improve both performance and emissions.

With respect to the optimization of compression ignition engines, two contributions to the Special Issue [6,7] deal with the optimization of the injection process since fuel mixing is critical in determining combustion efficiency and emissions levels.

In [6], simulation optimization methods and methods based on evolutionary algorithms were elaborated on for the optimization of the design and technology of injector nozzles in terms of minimizing energy losses on friction in compression ignition engines. The technical state of the injector apparatus significantly affected the engine performance, fuel consumption, toxicity, and smoke opacity of outlet gases. Alternatively, in [7], an analysis was carried out on the effects of the fuel injection rate shape on the diesel spray mixing process using a numerical simulation. The results show that a changing injection rate shape can enhance mixing during injection and achieve better mixing after the end of injection.

In [8], an investigation was carried out to improve energy conversion for a four-stroke free-piston engine. Unlike crankshaft-based engines, free-piston engines allow for a higher degree of freedom in shaping the piston trajectory, including adaptive compression ratios, which enables optimal operation with alternative fuels. The possibility of adapting the stroke course results in new degrees of freedom with which the combustion process can be optimized. In this work, four-stroke trajectories with different amplitudes and piston dynamics were proposed and analyzed regarding efficiency. The trajectories were described analytically so that they can be used for a prototype in future work.

Paper [9] deals with the use of a pre-turbine small-sized oxidation catalyst. The earlier activation of catalytic converters in internal combustion engines is becoming highly challenging due to the reduction in exhaust gas temperature caused by the application of CO_2 reduction technologies. In this context, the use of pre-turbine catalysts arises as a potential way to increase the conversion efficiency of the exhaust aftertreatment system. In this work, a small-sized oxidation catalyst consisting of a honeycomb thin-wall metallic substrate was placed upstream of the turbine to benefit from the higher temperature and pressure before the turbine expansion. The effects of this device on engine performance and emissions under driving conditions were shown.

The last two papers published in this Special Issue [10,11] refer to the use of numerical methods to aid the engine development process.

In [10], a computationally efficient tabulated chemistry solver for internal combustion engine optimization using stochastic reactor models was developed. The use of chemical kinetic mechanisms in computer-aided engineering tools for internal combustion engine simulations is of high importance for studying and predicting pollutant formation of conventional and alternative fuels. However, the usage of complex reaction schemes is accompanied by high computational costs in 0D, 1D, and 3D computational fluid dynamics frameworks. This work aimed to address this challenge and allow broader deployment of detailed chemistry-based simulations, such as in multi-objective engine optimization campaigns. A fast-running tabulated chemistry solver coupled to a 0D probability density function-based approach was proposed for the modeling of compression and spark-ignition engine combustion.

Finally, in [11], the authors introduced and validated a new transient simulation methodology of an ICE coupled to a hybrid architecture vehicle to predict internal combustion engine performance and emissions in real driving conditions. A one-dimensional computational fluid dynamic software was used and suitably coupled to a vehicle dynamics model. The results show that a crank angle resolution approach to the vehicle simulation is a viable and accurate strategy to predict engine emissions during any driving cycle with a computation effort compatible with the tight schedule of a design process.

Acknowledgments: Editors would like to acknowledge all the authors of the submitted papers for choosing this Special Issue and the peer reviewers for their dedication. Editors would also like to acknowledge the Section Managing Editor, Melon Zhang, for his precious support.

Conflicts of Interest: The authors declare no conflict of interest.

References

1. Postrioti, L.; Brizi, G.; Finori, G.M. Experimental Analysis of Water Pressure and Temperature Influence on Atomization and Evolution of a Port Water Injection Spray. *Appl. Sci.* **2021**, *11*, 5980. [CrossRef]
2. Tornatore, C.; Marchitto, L.; Teodosio, L.; Massoli, P.; Bellettre, J. Performance and Emissions of a Spark Ignition Engine Fueled with Water-in-Gasoline Emulsion Produced through Micro-Channels Emulsification. *Appl. Sci.* **2021**, *11*, 9453. [CrossRef]
3. Guillin-Estrada, W.; Maestre-Cambronel, D.; Bula-Silvera, A.; Gonzalez-Quiroga, A.; Duarte-Forero, J. Combustion and Performance Evaluation of a Spark Ignition Engine Operating with Acetone–Butanol–Ethanol and Hydroxy. *Appl. Sci.* **2021**, *11*, 5282. [CrossRef]
4. Pastor, J.V.; García-Oliver, J.M.; Micó, C.; García-Carrero, A.A.; Gómez, A. Experimental Study of the Effect of Hydrotreated Vegetable Oil and Oxymethylene Ethers on Main Spray and Combustion Characteristics under Engine Combustion Network Spray A Conditions. *Appl. Sci.* **2020**, *10*, 5460. [CrossRef]
5. Teodosio, L.; Marchitto, L.; Tornatore, C.; Bozza, F.; Valentino, G. Effect of Cylinder-by-Cylinder Variation on Performance and Gaseous Emissions of a PFI Spark Ignition Engine: Experimental and 1D Numerical Study. *Appl. Sci.* **2021**, *11*, 6035. [CrossRef]
6. Monieta, J.; Kasyk, L. Optimization of Design and Technology of Injector Nozzles in Terms of Minimizing Energy Losses on Friction in Compression Ignition Engines. *Appl. Sci.* **2021**, *11*, 7341. [CrossRef]
7. Naruemon, I.; Liu, L.; Liu, D.; Ma, X.; Nishida, K. An Analysis on the Effects of the Fuel Injection Rate Shape of the Diesel Spray Mixing Process Using a Numerical Simulation. *Appl. Sci.* **2020**, *10*, 4983. [CrossRef]
8. Tempelhagen, R.; Gerlach, A.; Benecke, S.; Klepatz, K.; Leidhold, R.; Rottengruber, H. Investigations for a Trajectory Variation to Improve the Energy Conversion for a Four-Stroke Free-Piston Engine. *Appl. Sci.* **2021**, *11*, 5981. [CrossRef]
9. Serrano, J.R.; Piqueras, P.; De la Morena, J.; Ruiz, M.J. Influence of Pre-Turbine Small-Sized Oxidation Catalyst on Engine Performance and Emissions under Driving Conditions. *Appl. Sci.* **2020**, *10*, 7714. [CrossRef]
10. Matrisciano, A.; Franken, T.; Gonzales Mestre, L.C.; Borg, A.; Mauss, F. Development of a Computationally Efficient Tabulated Chemistry Solver for Internal Combustion Engine Optimization Using Stochastic Reactor Models. *Appl. Sci.* **2020**, *10*, 8979. [CrossRef]
11. Marinoni, A.; Tamborski, M.; Cerri, T.; Montenegro, G.; D'Errico, G.; Onorati, A.; Piatti, E.; Pisoni, E.E. 0D/1D Thermo-Fluid Dynamic Modeling Tools for the Simulation of Driving Cycles and the Optimization of IC Engine Performances and Emissions. *Appl. Sci.* **2021**, *11*, 8125. [CrossRef]

Article

Experimental Analysis of Water Pressure and Temperature Influence on Atomization and Evolution of a Port Water Injection Spray

Lucio Postrioti [1,*], **Gabriele Brizi** [2] **and Gian Marco Finori** [1]

[1] Department of Engineering, University of Perugia, 06125 Perugia, Italy; gianmarco.finori@studenti.unipg.it
[2] STSe srl, 06125 Perugia, Italy; gabriele.brizi@stse.eu
* Correspondence: lucio.postrioti@unipg.it; Tel.: +39-075-5853733

Abstract: Port water injection (PWI) is considered one of the most promising technologies to actively control the increased knock tendency of modern gasoline direct injection (GDI) engines, which are rapidly evolving with the adoption of high compression ratios and increased brake mean effective pressure levels in the effort to improve their thermal efficiency. For PWI technology, appropriately matching the spray evolution and the intake system design along with obtaining a high spray atomization quality, are crucial tasks for promoting water evaporation so as to effectively cool down the air charge with moderate water consumption and lubricant dilution drawbacks. In the present paper, a detailed experimental analysis of a low-pressure water spray is presented, covering a lack of experimental data on automotive PWI systems. Phase doppler anemometry and fast-shutter spray imaging allowed us to investigate the influence exerted by the injection pressure level and by the water temperature on spray drop size and global shape, obtaining a complete database to be used for the optimization of PWI systems. The obtained results evidence how significant benefits in terms of atomization quality can be obtained by adopting injection pressure and water temperature levels compliant with standard low injection pressure technologies.

Keywords: water injection; phase doppler anemometry; knock control

Citation: Postrioti, L.; Brizi, G.; Finori, G.M. Experimental Analysis of Water Pressure and Temperature Influence on Atomization and Evolution of a Port Water Injection Spray. *Appl. Sci.* **2021**, *11*, 5980. https://doi.org/10.3390/app11135980

Academic Editor: Cinzia Tornatore

Received: 31 May 2021
Accepted: 21 June 2021
Published: 27 June 2021

Publisher's Note: MDPI stays neutral with regard to jurisdictional claims in published maps and institutional affiliations.

1. Introduction

The environmental impact, in terms of global greenhouse effect and pollutant emissions, of the automotive sector is considerable. Hence, severe limitations are in force and designed for, in particular, the adoption of challenging CO_2 emission targets for the next few years [1–3]. Consequently, the automotive industry is quickly adopting innovative technologies, globally aiming for a rapid increase in automotive powertrain efficiency. The leading method followed to improve powertrain efficiency is electrification, with the adoption of a progressively more significant energy storage capacity and electric propulsion system power. The final stage of this evolution is foreseen to be the battery electric vehicle, which ensures a zero CO_2 local emission operation. In this scenario, the internal combustion engine will still play a significant role for many years [4] due to the complexities and costs related to the actual implementation of the electrification path. Accordingly, along with electrification, a significant evolution of the gasoline engine to improve its efficiency is mandatory. The widespread adoption of gasoline direct injection (GDI) in spray-guided or pre-chamber configurations, turbocharging coupled with downsizing and downspeeding, higher compression ratios and application of the Miller cycle seem to be the most interesting innovation lines [5,6]. Unfortunately, many of the aforementioned technologies cause a drastic increase in the knocking tendency due to the increased charge temperature before and during combustion, thus restraining the potential benefits in terms of engine efficiency. As a matter of fact, particularly for high-performance engines, the knock tendency in high load conditions is currently controlled by reducing the spark advance and by enriching the fuel/air mixture, resulting in an efficiency penalty and in a restriction of the catalyst operating area with an increase of CO_2, CO and HC emissions.

In this frame, water injection technology can be used as an alternative way to control the charge temperature at the end of the intake process and during the combustion phase, thus reducing the risk of abnormal combustion in downsized and highly boosted GDI engines. Water injection can potentially enable $\lambda = 1$ operation in high-load and high-speed operation, even adopting high compression ratios [7–17], thereby gaining significant benefits in terms of engine efficiency while preserving the after-treatment system efficacy in controlling exhaust emissions.

Water injection can be implemented as low-pressure injection in the intake runners (PWI), as high-pressure injection in the combustion chamber (DWI) or as direct injection of a water/gasoline emulsion. Along with the potential benefits in terms of air charge temperature control, clearly water injection has different potential drawbacks, such as a possible lubricant dilution due to the liquid's impact on the cylinder liner. Potentially catastrophic lubricant dilution can be caused by an incorrect match among the spray's global shape with the inlet duct geometry (for PWI systems) or the combustion chamber (for DWI systems) or, in general, by a slow water evaporation rate. Furthermore, the water deposition on the duct and cylinder walls limits the air charge cooling effect, increasing the water flow rate required to effectively control the knock tendency, resulting in water-to-fuel rate ratios in the range of 0.2–0.5 in full load conditions. Among the possible different schemes, port water injection (PWI) is currently considered the most attractive as a compromise between efficacy and cost, being significantly higher for direct water and water/gasoline emulsion injection systems, including the eventual water recovery technologies from the exhaust stream [18].

Since the air charge temperature control potential under water injection is related to the high latent heat of vaporization of the water, the injected water evaporation rate is a crucial factor to be considered in PWI systems' design. In order to promote water evaporation, an adequate match of the water spray characteristics with the intake system design is required, and hence, a detailed knowledge of the spray characteristics is mandatory. In particular, given the moderate air charge temperature (typically in the range of 40–60 °C for intercooled engines) and the short spray residence time in the inlet runner, the water spray atomization quality is crucial for obtaining complete evaporation.

Unfortunately, in the technical literature, there is a substantial lack of detailed experimental data about low-pressure water sprays' global evolution and atomization level to support PWI system design and CFD simulation. In [9], Iacobacci et al. investigated the potential of a PWI system based on PFI injectors supplied with water at 25 °C ($P_{inj} = 4$ bar,g), changing the water/fuel ratio to mitigate the knock tendency at full load. Cordier et al. [10] tested different water injection technologies on a single-cylinder research engine, obtaining significant efficiency improvements. For the tested PWI configuration, P_{inj} was varied in the range of 5–20 bar,g, but no details about the resulting spray characteristics were reported. In [12], Paltrinieri et al. experimentally investigated the application of a PWI system operated at $P_{inj} = 7$ bar,g and $T_w = 55$ °C for a single-cylinder research engine with water-to-fuel ratios up to 60%. Different injector designs and positions along the inlet duct were used to explore the actuation timing effect. In this analysis, CFD simulations of the water spray evolution in the intake duct were carried out to investigate the spray–air interaction, but the effect of the spray characteristics with different operating conditions was not investigated. In other numerical analyses of water injection systems, the water spray characteristics, or even its evaporation rate, are assumed to be constant or similar to fuel sprays generated at the same injection pressure level [13,15,16].

In the present paper, a detailed experimental analysis of a low-pressure water spray is presented, discussing the effect of both the injection pressure and water temperature in the rail on the jet evolution and size characteristics. According to the current approach of the automotive industry for the definition of a PWI system's architecture, both these parameters were varied in ranges compliant with standard PFI technology to reduce the complexity and cost of key components such as injectors, rails and sensors. The injection pressure was varied from 5 bar,g to 11 bar,g, covering a pressure range explored by other Authors for PFI injectors [19]. Correspondingly, the water temperature in the rail was changed from 20 °C to 110 °C, approaching flash boiling conditions in order to promote the spray break-up and drop evaporation [20,21]. The analysis was carried out by a phase doppler anemometry (PDA) system and by a fast-shutter imaging apparatus in order to investigate both the drops' size quality and global spray characteristics obtained in a range of operating conditions. In the following sections, the experimental set-up and the test plan will be presented first, and the obtained results will be discussed.

2. Materials and Methods

The PWI injector (Bosch EV14) used for the present analysis was characterized in terms of the mean injected mass, global spray evolution, drop size and velocity. The main characteristics of the tested injector are reported in Table 1. The static flow rate was measured by the dINJ injection analyzer, a Zeuch method-type injection analyzer specifically designed to operate with low-pressure injection systems [8,22].

Table 1. PWI injector characteristics.

Injector	Bosch GS EV14
Holes	4×220 μm, symmetric to injector axis
Static Flow Rate	3.5 mm^3/ms @P$_{inj}$ 5 bar,g ; T$_w$ = 20 °C
Injector Driver	Darlington TIP121, V$_{supply}$ 14 V.

The injector under testing was fed with distilled water statically pressurized with nitrogen in the range from 5 bar,g to 11 bar,g (up to 15 bar,g only for the flow tests). Pressurized water was accumulated in a 100-cc reservoir, which was used as rail to directly feed the injector to have negligible pressure fluctuations during the injection event, according to the rules of JSAE2715. The rail structure was used as a fixture for an electric heater used to control water temperature. The feedback thermocouple was installed at the injector inlet in the same position as the Keller PAA M5 HB sensor (20 bar f.s., 50 kHz bandwidth, 1% accuracy) used to monitor the rail pressure. The injector current time history was acquired by a Pico TA189 probe (30 A, 100 kHz bandwidth) and averaged over 30 consecutive injection events. A schematic of the experimental setup is reported in Figure 1.

In each operating condition, the mean injected volume was measured by a precision balance (Radwag PS 1000/C/2, resolution of 1 mg, accuracy ±1.5 mg) during three repetitions of a 3000-shot sequence with a 10 Hz injection frequency.

In previous studies (e.g., [8,23]), the effect of the test vessel's pressure and temperature on low-pressure water sprays was investigated, concluding that the air temperature's variation from ambient to 50–60 °C (typical of boosted conditions with an intercooler) has negligible effects on the spray's evolution and size. On the other hand, boosted pressure levels have the similar effect of a corresponding injection pressure reduction, with the pressure differential across the injector being the main driving force affecting the spray's evolution. As a consequence, in this research, the test vessel's pressure and temperature were maintained at 1 bar,a and 25 °C for all the operating conditions.

In Table 2, the operating conditions used for the flow test are reported, evidencing the imposed water temperature at the injector's inlet.

The global spray evolution was investigated by a fast-shutter imaging technique, applied according to an ensemble averaging approach. The imaging apparatus was based on a pulsed Nd-Yag laser (Litron Nano L 200–20, 200 mJ/shot, shot duration 6 ns), which was used as a light source. The laser was synchronized with a fast-shutter, high-resolution CMOS camera (Dalsa Genie Nano M4020, resolution 3008 × 4112, 12-bit). According to the ensemble averaging approach, only one image per injection event was acquired at a given delay from the injection event's start (the TTL signal enabling the injector driver was used as a trigger). The statistical analysis of the spray's global development at an assigned delay was carried out by repeating the image acquisition over a series of consecutive injection events (30 in the present work). The repetition of the aforementioned acquisition sequence at different delays from the injection start allowed for characterization of the complete spray development throughout the entire injection event. The acquired images at the different timings were analyzed off-line by means of a proprietary digital analysis procedure developed in the LabVIEW™ environment, obtaining the spray tip penetration curves and global cone angle according to the JSAE2715 prescriptions. More details about the image analysis procedure are reported in [24,25].

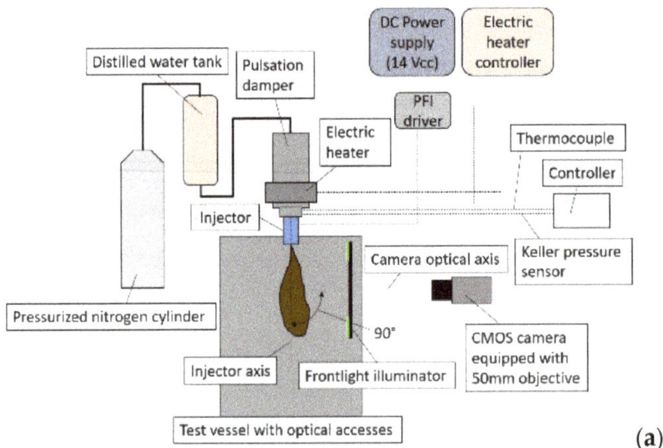

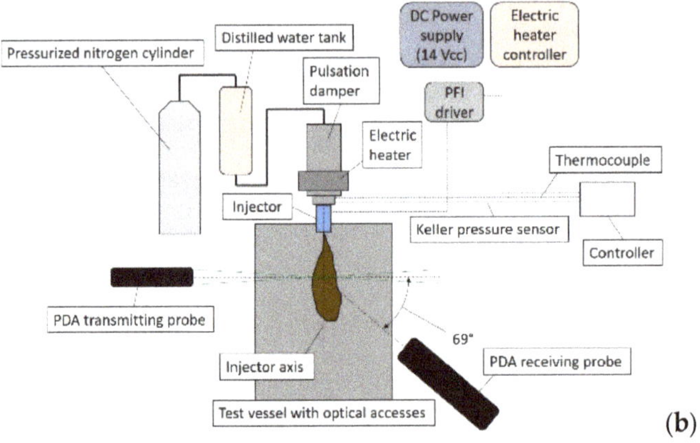

Figure 1. Schematic of the experimental setup for imaging (**a**) and PDA (**b**).

Table 2. Flow test plan showing the water temperature in °C at the injector's inlet.

Injection Pressure (bar,g)	Energizing Time (µs)							Water Temp. (°C)
	2000	3000	4000	5000	6000	8000		
5								
7								
9								
11							base	20
15							extended	20, 55, 90, 110

The test plan used for the global spray analysis by imaging was based on a full-factorial analysis of injection pressure P_{inj} levels of 5, 7 and 11 bar,g with water temperature levels of 20, 55, 90 and 110 °C. All spray imaging tests were carried out with an ET of 5 ms in order to ensure adequate steady flow operation for the injector and evidence the effects of both the water temperature and the injection pressure.

The effect of the water temperature and injection pressure on the drop size was evaluated using a phase doppler anemometer (Dantec Dynamics P80) along a measuring traverse composed of 18 stations. The traverse was positioned at Z = 50 mm downstream of the nozzle plate and aligned with the projection of two of the four nozzle holes on the examined plane to evidence the global spray symmetry. In Table 3, the main specifications for the PDA system used for the tests are reported.

Table 3. PDA system specifications.

Transmitter	60 mm
Receiver	HiDense 112 mm
Laser Source	Coherent Genesis MX514
Frequency Shift	40 MHz
Focal length (TX/RX)	310 mm/310 mm
Scattering Angle	110°
Drop Diameter Range	1–400 μm
Drop Velocity Range	−5–30 m/s
Ref. System Origin (X = Y = 0, Z = 0)	Nozzle tip center; Velocity positive: downward

The drop size and velocity characteristics were investigated in the operating conditions reported in Table 4. In all PDA tests, an ET of 5 ms was applied, acquiring data in a 30-ms time window after the start of the ET to capture the entire spray evolution, including its complete tail. Prescriptions from JSAE2715 were followed during the tests in terms of spray boundary detection and minimum number of samples per measuring position. The data were collected during 3000 consecutive shots, operating the injector at 8 Hz in each examined station.

Table 4. PDA test plan, with the water temperature measured at the injector inlet.

		Water Temperature (°C)			
		20	55	90	110
	5		X		
Injection Pressure (bar,g)	7	X	X	X	X
	9		X		
	11		X		

3. Results

3.1. Flow Test Results

In Figure 2, the results obtained in terms of the mean injected quantity, parametrically varying both the injection pressure level from 5 to 15 bar,g and the water temperature at the injector inlet from 20 °C to 110 °C, are reported. On the left in the same figure, the results are reported for when P_{inj} 7 = bar,g was assumed as the reference injection pressure level.

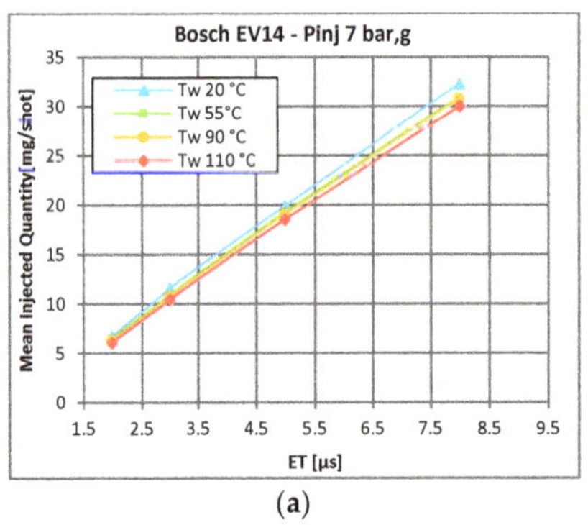

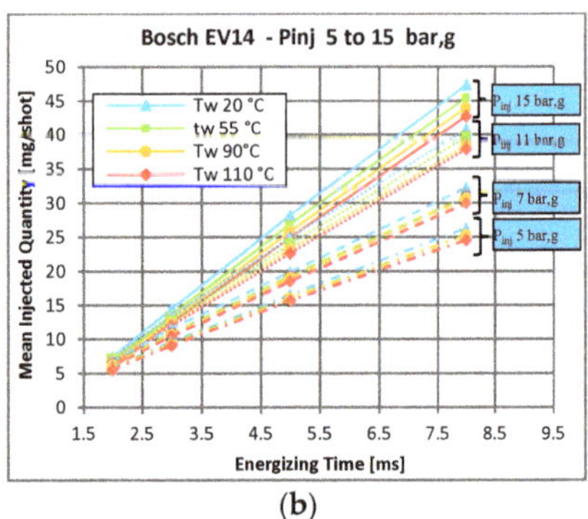

(a)   **(b)**

Figure 2. Mean injected quantity for T_w range 20–110 °C and P_{inj} = 7 bar,g (**a**) and from 5 to 15 bar,g (**b**).

From the obtained results, the effect of the water temperature increasing on the injected mass is evident. For a given injection pressure level, the higher the fluid temperature, the lower the injected quantity. With the ET at 5 ms and the water temperature changing from 20 °C to 110 °C, the mass percentage difference ranged from 5.7 % at P_{inj} = 7 bar,g to 11.5% at P_{inj} = 15 bar,g. This is likely to be ascribed mainly to the reduced magnetic force exerted by the solenoid on the injector needle, caused by the increased coil resistance. As is reported in Figure 3, for the reference injection pressure level, the injector current was significantly reduced in high water temperature conditions. Furthermore, the current time history slope—and presumably the needle rise—was slowed at high water temperatures, as is suggested by the delayed occurrence of the needle's fully raised position, evidenced by the sudden slope change in the current profile around 1.5 ms after the ET started.

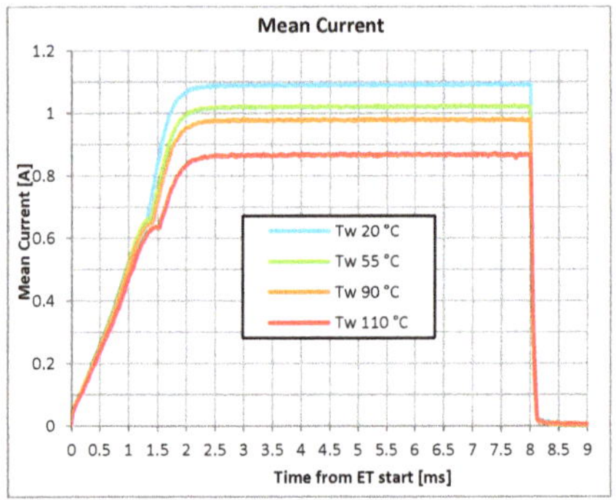

Figure 3. Injector current as a function of the water temperature. P_{inj} = 7 bar,g, and ET = 8.0 ms.

The effect of the fluid's temperature on the injector dynamics was particularly evident for short injection events, during which the injector operated in the so-called "ballistic regime". In ballistic operation, the needle did not attain the fully raised position, and consequently, the flow could not reach a static rate condition. As is reported in Figure 4a, where the flow results are plotted against the injection pressure level (with ET = 2 ms and P_{inj} = 15 bar,g), the injector's opening phase was drastically altered by high water temperature levels. In these conditions, the progressively reduced available magnetic force caused the mean injected mass to be lower at P_{inj} = 11 bar,g (and equal for T_w = 20 °C). Only for ET values longer than 3 ms was the expected rising dependence of the injected mass on P_{inj} obtained, given the presumable attained steady flow conditions and the reduced significance of the needle's opening phase with respect to the entire injection process.

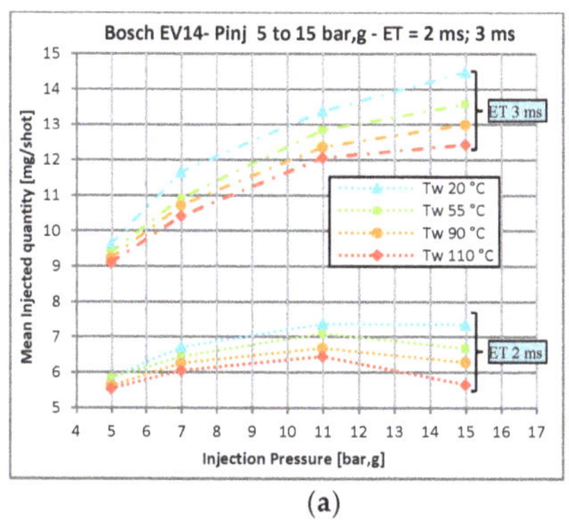
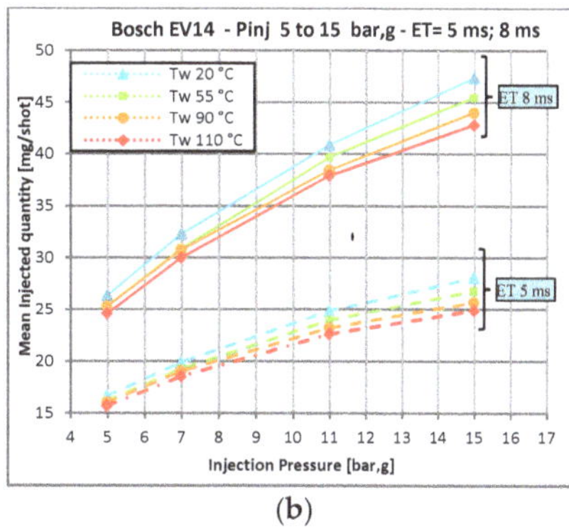

Figure 4. Mean injected quantity vs. injection pressure with water temperatures from 20 °C to 110 °C. (**a**) ET = 2 ms and 3 ms. (**b**) ET = 5 ms and 8 ms.

The obtained flow results seem to suggest that P_{inj} = 15 bar,g is not a feasible operating condition for this injector with a high water temperature and will no longer be investigated in terms of the spray evolution and drop size.

3.2. Spray Global Development

The imaging set-up described in Section 2 was used to investigate the effect of the injection pressure and water temperature on the global spray's evolution. For the sake of brevity, only a short sequence of the spray evolution in the reference condition (P_{inj} = 7 bar,g, 20 °C) is reported in Figure 5 along with pictures of the fully developed spray under high injection pressure and water temperature conditions (Figure 6).

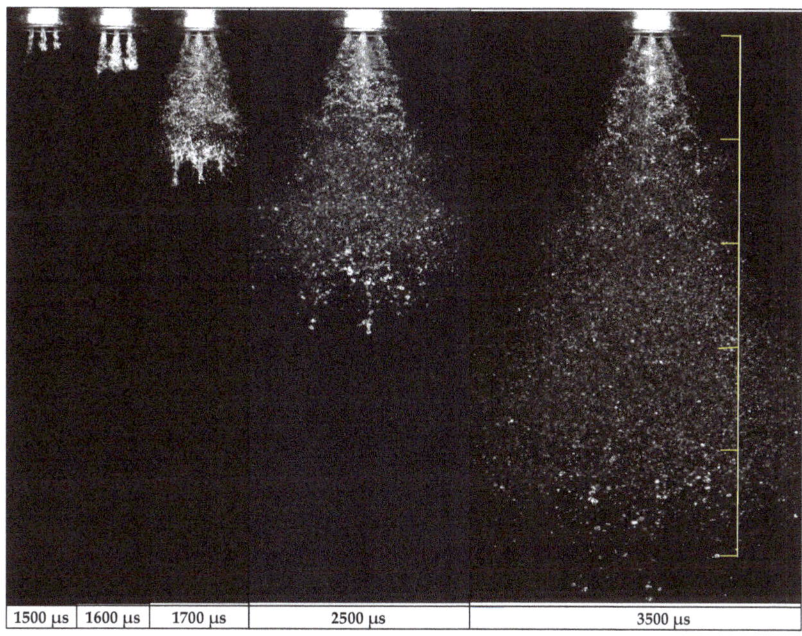

Figure 5. Spray evolution sequence at P_{inj} = 7 bar,g and 20 °C (timing from ET start, scale tick = 10 mm).

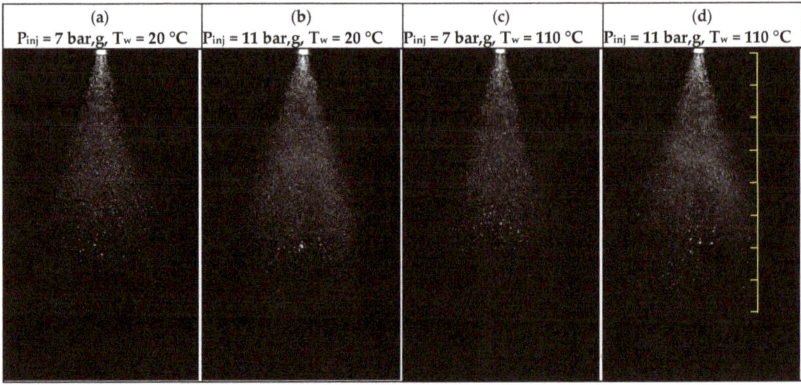

Figure 6. Fully developed spray at 4 ms after the ET's start. Scale tick = 10 mm.

As can be observed in Figure 5, four individual conical flow structures emerged from the nozzle at the beginning of the injection process, initially composed of relatively large ligaments. The four structures merged at a distance between 5 and 10 mm from the injector nozzle. After a short transition zone, at a 10-mm distance downstream, the primary break-up process seemed to be completed. The spray structure hereafter appears to be quite uniform and composed of relatively small drops; only in the most advanced and central part of the spray was the presence of significantly large ligaments clearly perceivable, probably originating from the initial injection transient when the flow velocity was restrained by the needle.

When the operating conditions were changed by increasing the injection pressure from 7 to 11 bar,g (Figure 6a,b), the spray structure was evidently altered, with an increased penetration and cone angle. Furthermore, the spray structure appeared to be composed of smaller droplets, with a reduced presence of large ligaments in the spray's bulk. Only in the spray tip were some large blobs originating from the injector's opening transient still present.

Conversely, the increase in water temperature (Figure 6c,d) did not seem to produce dramatic changes in the spray's overall structure with respect to the corresponding low-temperature operating condition. This evidence seems to suggest that in the tested operating conditions, flash boiling was not triggered despite the high temperature level at the injector inlet, possibly due to the progressive cooling of the water inside the injector body.

The complete set of results in terms of the spray tip penetration and spray cone angle are reported in Figures 7 and 8 for the 12 examined operating conditions, evidencing the effect on the spray's global development of the injection pressure and water temperature, respectively.

As reported in Figure 7, the injection pressure had a significant effect on both the spray tip penetration and global cone angle for all the examined water temperature levels. Globally, 5 ms after the ET's start, changing the injection pressure from P_{inj} = 5 bar,g to P_{inj} = 11 bar,g led the spray tip penetration to increase by 11% at T_w = 20 °C and by 12% at T_w = 110 °C. The spray tip penetration slightly increased from 5 bar,g to 7 bar,g, with this trend being more evident in the final part of the spray evolution and at a higher water temperature. When the injection pressure was raised to 11 bar,g, the final penetration increase was more evident, despite the relatively high injection pressure tending to slow the injector opening transient, consequently leading to the initial spray tip velocity being smaller. As a result, the spray penetration for P_{inj} = 11 bar,g was smaller than for P_{inj} = 7 bar,g up to 1.5 ms after the ET's start, in which time the initial penetration gap was recovered. The effect of the injection pressure on the spray cone angle was even more evident; for all the examined water temperature conditions, the spray diffusion angle progressively increased with higher injection pressure levels, obtaining almost parallel trends for this quantity. It is also interesting to observe how the monotonically decreasing trend for the spray cone angle changed its slope after the injector closure around 5.8 ms after the ET's start.

The effect of the water temperature at the injector inlet on the spray evolution is analyzed in Figure 8. As can be observed, only marginal effects were exerted by the water temperature on the spray's global structure for all the examined injection pressure levels. To be detailed, an unclear tendency was observed in terms of spray penetration for P_{inj} = 5 bar,g, while the effect was negligible for higher injection pressure levels. In terms of the spray cone angle, the increase in water temperature seemed to decrease the spray cone angle up to 90 °C, while a further increase to 110 °C seemed to attenuate or revert the trend. Globally, a moderate effect of the water temperature on the spray evolution was observed.

3.3. Spray Drop Size and Velocity

The PDA raw data for the measuring station corresponding to the injector axis projection on a plane at 50 mm from the nozzle (coordinates of X = Y = 0; Z = 50 mm) are reported in Figure 9a,b for the reference operating condition P_{inj} = 7 bar,g and Tw = 20 °C. In these plots, all the records relevant to 3000 consecutive injection events are reported (blue dots), along with the average values computed in 0.1 ms time bins (red dots). The ET's start was used as a time reference.

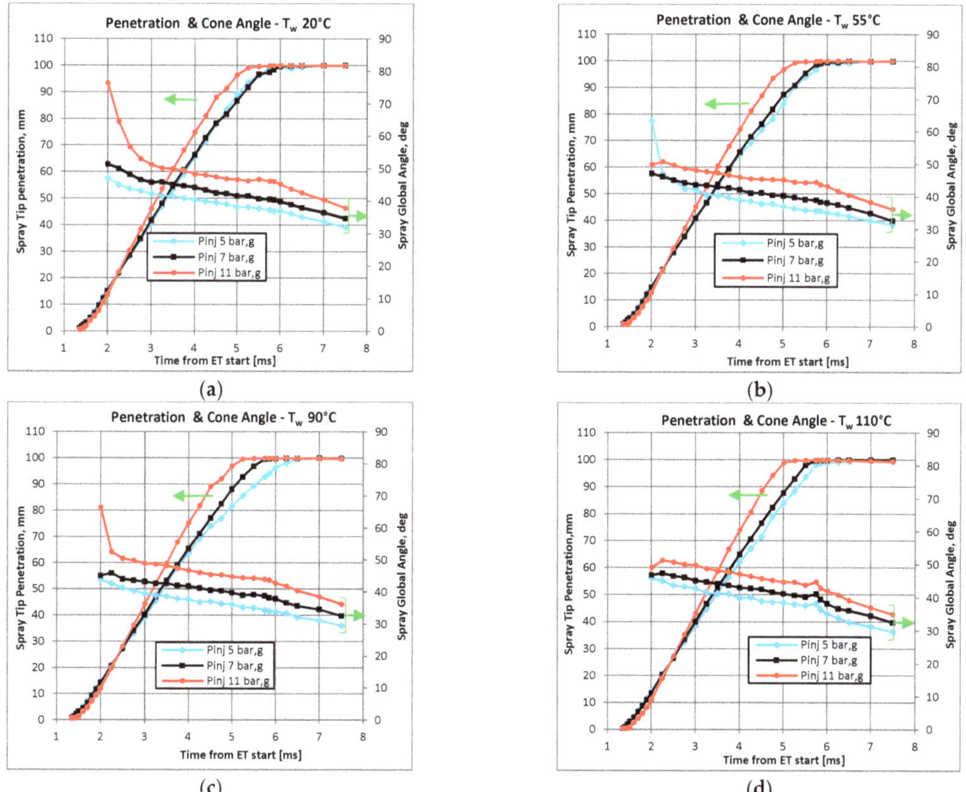

Figure 7. Effect of the injection pressure on the spray tip penetration and cone angle. (**a**) T_w = 20 °C; (**b**) T_w = 55 °C; (**c**) T_w = 90 °C; and (**d**) T_w = 20 °C.

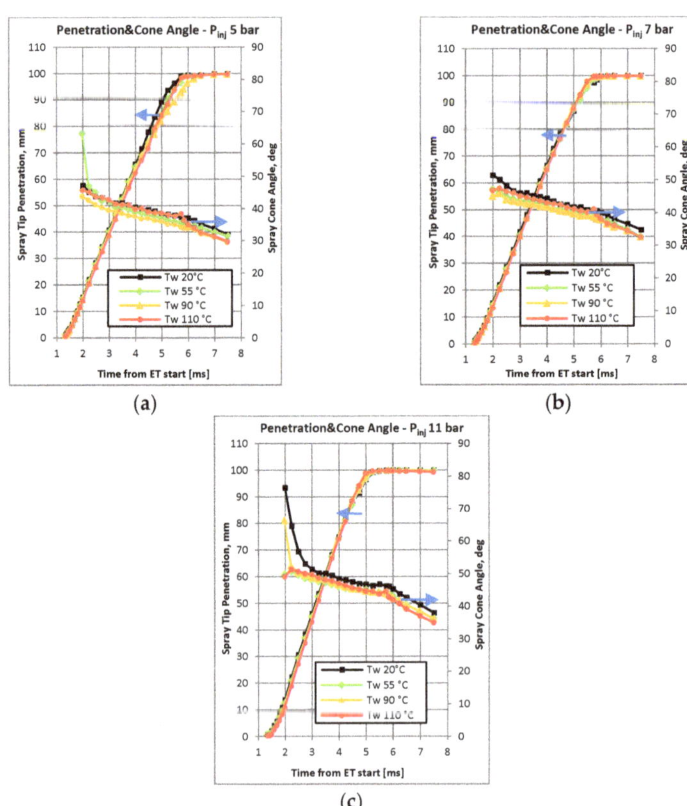

Figure 8. Effect of the water temperature on the spray tip penetration and cone angle. (**a**) P_{inj} = 5 bar,g; (**b**) P_{inj} = 7 bar,g; and (**c**) P_{inj} = 11 bar,g.

As can be seen, the drops' diameters and Z-velocity time histories evidence how the spray approached the considered measuring station about 4 ms after the ET's start. The main part of the spray structure flowed through the considered measuring station between 4 and 8 ms. During this time window, the drops' velocity ranged between approximately 5 and 22 m/s, while the observed drops' diameter range was predominantly between 15 and 150 µm. A significant number of drops with diameters up to 350 µm was also observed, possibly related to the defective primary break-up in the first part of the injection process, as was observed in the spray images. The presence of even such a restricted number of relatively large drops inherently had a non-marginal effect on the resulting Sauter mean diameter.

After 8 ms from the ET's start, the velocity data rapidly decreased to values around 2–3 m/s for the drops pertaining to the spray tail. This part of the spray's structure was composed of very small droplets (diameter values below 40 µm) featuring reduced momentum which continued flowing through the observed position for a long time.

When the injector's water temperature was raised to 110 °C, the drops' velocities and sizes time histories for the same measuring station, reported in Figure 9c,d, were obtained. Marginal effects due to the applied rise in temperature were observed in terms of both the drops' velocities and size ranges for the bulk spray evolution during the 4–8 ms time window, while a significant shift in the mean values was observed below 15 m/s and toward 50 µm.

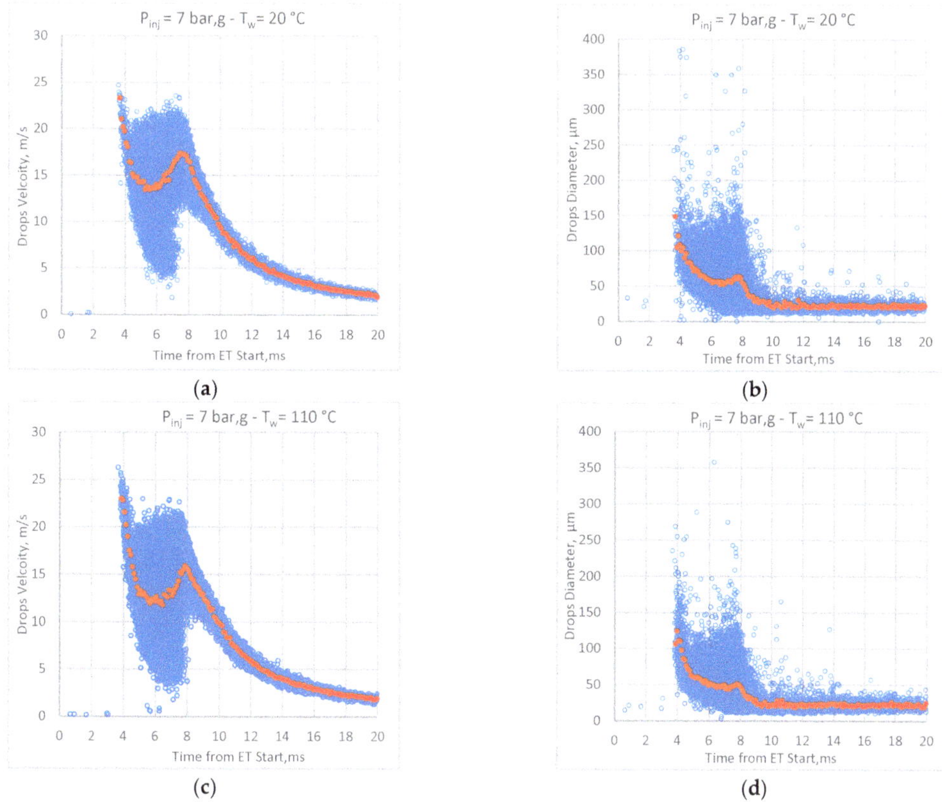

Figure 9. Effect of the water temperature on the drops' sizes and velocities. Raw data in X = Y = 0, Z = 50 mm, P_{inj} = 7 bar,g. (**a**) Drops' velocity for T_w = 20 °C. (**b**) Drops' diameter for T_w = 20 °C. (**c**) Drops' velocity for T_w = 110 °C. (**d**) Drops' diameter for T_w = 110 °C.

In Figure 10, the results obtained with four different water temperature levels at the injector inlet are reported for the complete measuring travers at Z = 50 mm and crossing the entire spray structure. In these plots, both the mean and Sauter mean diameter values are included, along with the mean velocity and relative drop count. All these quantities were computed with data relevant to the entire 30-ms time-window to have a full portrait of the spray quality in the examined positions.

The effect of the water temperature on the drops' sizes, which was already commented on for the X = Y = 0 and Z = 50 mm position, was substantially confirmed for the entire measuring traverse; a significant but not dramatic effect from 20 °C to 110 °C was observed, with size reductions ranging between 8 and 15 μm in terms of the SMD and between 4 and 9 μm in terms of the MD. Minor consequences of raising the water temperature were observed in terms of the drop count and mean velocity, which were proven to change only for the central positions of the spray structure. As was observed when commenting on the spray images in the same operating conditions in Section 3.2, the water temperature's rise at the injector inlet from 20 °C to 110 °C, assumed to be compatible with standard PFI technology (e.g., injector body, rail and pressure sensor), was not adequate to trigger a net flash boiling mechanism for the injection process. Consequently, the injection process resulted in a spray with a basically unaffected shape but appreciably improved atomization quality. Nevertheless, the drops' size improvement was not as significant as was presumably attainable in the case of completely developed flash boiling conditions [25].

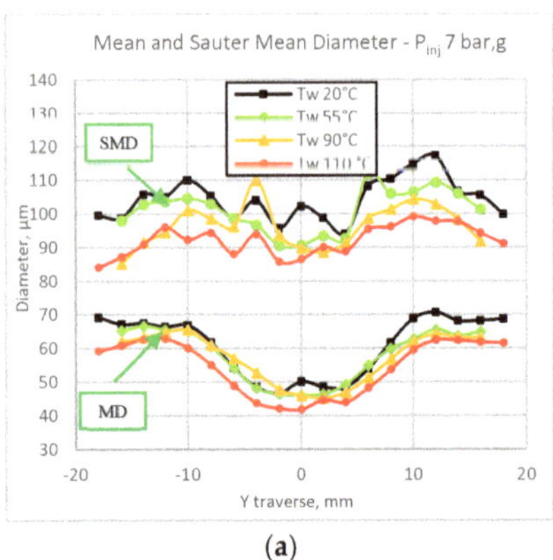

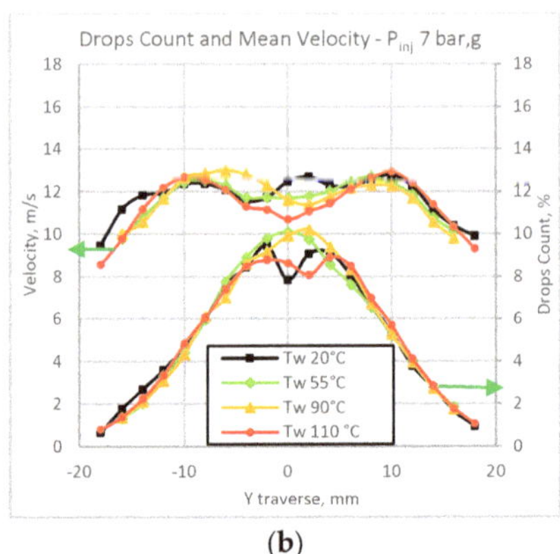

(**a**)   (**b**)

Figure 10. Effect of the water temperature on the drops' sizes and velocities over the measuring traverse at Z = 50 mm. P$_{inj}$ = 7 bar,g. (**a**) Drops' mean diameter. (**b**) Drop count and velocity.

A similar analysis is presented in order to separately analyze the potential effects exerted by an increase of the injection pressure in terms of the spray evolution and atomization quality. In Figures 11 and 12, the results obtained in terms of raw data for the measuring position X = Y = 0 and Z = 50 mm and in terms of the mean for the entire examined traverse are reported.

The injection pressure's effect on the drops' sizes is evident from the comparison of Figure 11b,d, with the P$_{inj}$ = 9 bar,g drop size consistently reduced for the spray bulk (from 4 to 8 ms) for time bin mean values close to 50 µm for a large part of the considered time window. Higher values were observed only for the very initial part of the injection process and around 7 ms when, presumably, the drops produced during the injector's closing phase approached the measuring station. Both the initial and final transients were characterized by a severe channel flow restriction, with consistent flow velocity reduction and drops breaking up the process penalization. For P$_{inj}$ = 5 bar,g, relatively large drops were attained at the Z = 50 mm measuring station with an almost constant velocity during the bulk spray time window, suggesting a moderate drag effect by the surrounding air. Conversely, with the higher injection pressure, the mean drops' velocities showed initially lower values, possibly due to a consistent drag exerted on the finely atomized drops, with the mean velocity rising only later when, presumably, the core of the completely developed spray attained the measuring station.

The analysis in Figure 12 confirmed the significant effect exerted by the injection pressure on drops' characteristics. The pressure increase from 5 to 11 bar,g caused a reduction of the SMD in excess of 20 µm in several stations, with minor effects in the spray's periphery. Smaller and more uniform effects were observed in terms of the mean diameter. The drops' mean velocities consistently increased for all the measuring stations as a direct consequence of the injection pressure's increase. As was observed in Figure 11, the velocity increase was observed mainly in the second part of the 4–8 ms time windows, when the fully developed spray crossed the measuring station. It is also interesting observing how, with P$_{inj}$ = 5 bar,g, the drop count profile was clearly asymmetrical, suggesting incipient off-design operation for the used injector. Correspondingly, the drops' atomization quality in the low-count region was particularly poor.

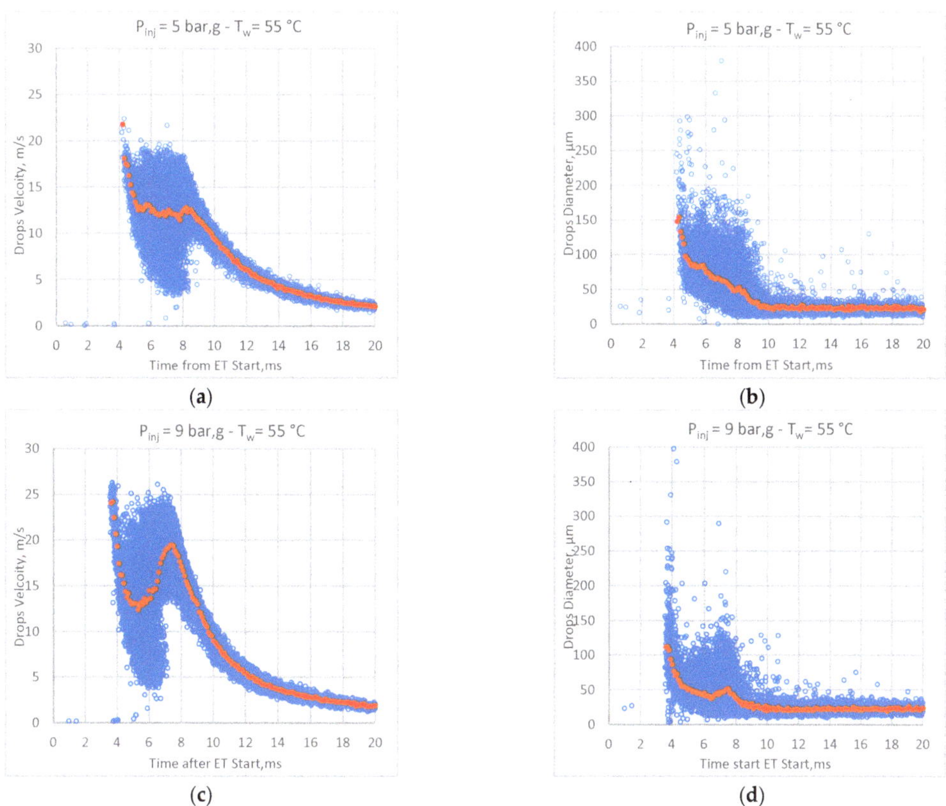

Figure 11. Effect of the injection pressure drops' sizes and velocities. Raw data in X = Y = 0, Z = 50 mm; T_w = 55 °C. (**a**) Drops' velocity for P_{inj} = 5 bar,g. (**b**) Drops' diameter for P_{inj} = 5 bar,g. (**c**) Drops' velocity for P_{inj} = 9 bar,g. (**d**) Drops' diameter for P_{inj} = 9 bar,g.

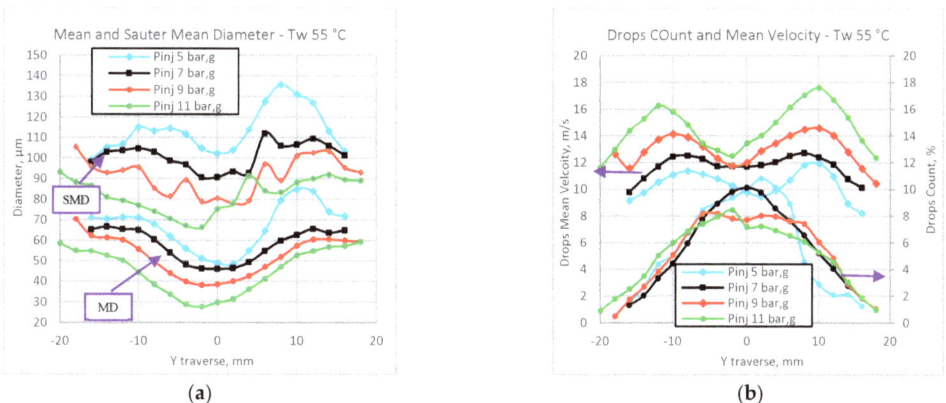

Figure 12. Effect of the injection pressure on drops' sizes and velocities over the measuring traverse at Z = 50 mm. T_w = 55 °C. (**a**) Drops' mean diameters. (**b**) Drops' counts and velocities.

Finally, in Figure 13, the diameter probability density function (PDF) for the droplet population is reported for the examined ranges of the injection pressure and water temperature. As can be observed, both the temperature and pressure variations caused a significant improvement of the atomization quality, with a consistent reduction in the PDF values for the intermediate diameter values (50–150 μm), while only marginal effects were obtained for the few very large drops produced during the initial part of the injection process. To be more detailed, it is interesting to point out how increasing the injection pressure (Figure 13b) from 5 to 11 bar,g also increased the PDF values in the range of 35–50 μm, confirming the modest atomization quality obtainable with low injection pressure values.

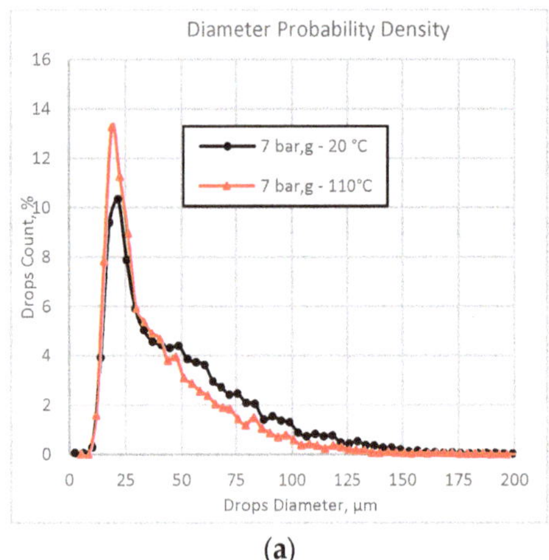

(a)

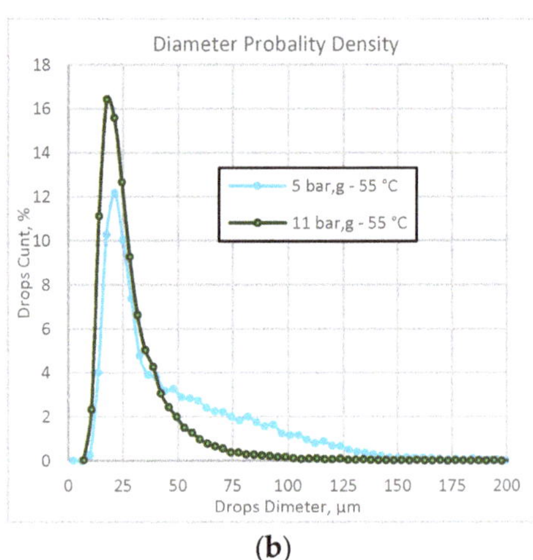

(b)

Figure 13. Effect of the injection pressure and water temperature on drops' diameter density probability functions. (a) P_{inj} = 7 bar,g. (b) T_w = 55 °C.

4. Conclusions

Water injection is promising technology for the mitigation of the knocking tendency of high-power density spark ignition engines, simplifying the adoption of high compression ratio levels and reducing the necessity for mixture enriching and spark retarding. The final goal is fully exploiting the efficiency potential of highly downsized, turbocharged spark ignition engines that have been rapidly spreading on the market in recent years.

The mixture temperature's control potential is related to the water's high latent heat of vaporization, and hence, despite the specific technology used to implement water injection, either injected directly in the combustion chamber or in the inlet runner, the water spray evaporation rate is crucial. To be more detailed, rapid and complete water spray evaporation enables controlling the mixture's temperature without significant drawbacks, such as excessive water consumption (water/fuel ratios in the range of 0.2–0.5 are quite common for full load operation) or potentially destructive lubricant oil dilution.

Port water injection (PWI) is currently considered the best compromise among system cost, complexity and efficacy. In the present paper, a deep analysis of the effects of both the water injection pressure and temperature on the resulting spray evolution and sizing characteristics was carried out separately. The ranges for both the pressure and temperature were chosen in order to be fully compliant with the low-pressure injection technology, with the water temperature and injection pressure below 110 °C and 11 bar,g, respectively. The obtained results can be summarized as follows:

- The injected mass per shot was significantly affected by the injected water's temperature. Increasing the water temperature from 20 °C to 110 °C caused a decrease of about 7% for the injected mass, depending on the injection pressure level and injection duration. This evidence seems to suggest the possible necessity of redesigning the injector coil in the case of high-temperature operation.

- In terms of the global spray evolution, no dramatic changes were observed for the spray tip penetration or cone angle in the examined injection pressure and temperature ranges which, in the present analysis, did not trigger flash boiling phenomena. Complete flash boiling development, with injection in sub-atmospheric conditions presumably not relevant for water injection operation, would require water temperature levels possibly not compatible with the standard injection components' integrity or durability.
- The spray atomization quality was sensibly affected by the water temperature's increase, with a benefit in terms of the Sauter mean diameter between 8 and 15 μm in the examined conditions. At the same time, the observed injection pressure effect was more significant, with the Sauter mean diameter reduction being approximately 20 μm and the injection pressure increasing from 5 bar,g to 11 bar,g.

The obtained results seem to confirm that there is still significant potential in the optimization of the water spray characteristics, mainly in terms of the atomization quality. To be more detailed, adequate water temperature control could assist the increase of the injection pressure in obtaining an adequate sizing level to speed up the evaporation process, contributing to water injection technology spreading.

Author Contributions: Conceptualization, L.P.; methodology, L.P. and G.B.; software, G.B.; validation, L.P., G.B. and G.M.F.; formal analysis, L.P., G.B. and G.M.F.; investigation, G.B. and G.M.F.; resources, L.P.; data curation, L.P., G.B. and G.M.F.; writing—original draft preparation, L.P.; writing—review and editing, L.P., G.B. and G.M.F.; visualization, L.P., G.B. and G.M.F.; supervision, L.P.; project administration, L.P.; funding acquisition, L.P. All authors have read and agreed to the published version of the manuscript.

Funding: This research received no external funding.

Institutional Review Board Statement: Not applicable.

Informed Consent Statement: Not applicable.

Data Availability Statement: Not applicable.

Conflicts of Interest: The authors declare no conflict of interest.

Abbreviations

GDI	Gasoline direct injection
ET	Energizing time (ms)
MD	Mean drop diameter (μm)
SMD	Sauter mean drop diameter (μm)
PDF	Probability density function
PFI	Port fuel injection
PWI	Port water injection
T_w	Water temperature (°C)
P_{inj}	Injection pressure, gauge (bar,g)

References

1. European Commission. Paris Agreement. December 2015. Available online: http://ec.europa.eu/clima/policies/international/negotiations/paris_en (accessed on 26 June 2021).
2. Commission Regulation (EU) No 459/2012 Amending Regulation (EC) No 715/2007 of the European Parliament and of the Council and Commission Regulation (EC) No 692/2008 as Regards Emissions from Light Passenger and Commercial Vehicles (Euro 6). 29 May 2012. Available online: https://eur-lex.europa.eu/legal-content/EN/TXT/?uri=celex%3A32012R0459 (accessed on 26 June 2021).
3. EEA. *Monitoring CO2 Emissions from New Passenger Cars and Vans in 2016*; EEA: Copenhagen, Denmark, 2017; ISBN 00100994.
4. Saliba, R.; Manns, J.; von Essen, C. *Groupes Motopropulseurs du Futur pour une Mobilité à faibles Émissions de Carbone*; CNAM: Paris, France, 2018.
5. Zhao, H. *Advanced Direct Injection Combustion Engine Technologies and Development: Gasoline Engines*; Elsevier: Amsterdam, The Netherlands, 2009; Volume 1.
6. Johnson, T.; Joshi, A. Review of Vehicle Engine Efficiency and Emissions. *SAE Int. J. Engines* **2018**, *11*, 1307–1330. [CrossRef]

7. Zhu, S.; Hu, B.; Akehurst, S.; Copeland, C.; Lewis, A.; Yuan, H.; Kennedy, I.; Bernards, J.; Branney, C. A review of water injection applied on the internal combustion engine. *Energy Convers. Manag.* **2019**, *184*, 139–158. [CrossRef]
8. Millo, F.; Mirzaeian, M.; Rolando, L.; Bianco, A.; Postrioti, L. A methodology for the assessment of the knock mitigation potential of a port water injection system. *Fuel* **2021**, *283*, 119251. [CrossRef]
9. Iacobacci, A.; Marchitto, L.; Valentino, G. Water Injection to Enhance Performance and Emissions of a Turbocharged Gasoline Engine under High Load Condition. *SAE Int. J. Engines* **2017**, *10*, 928–937. [CrossRef]
10. Cordier, M.; Lecompte, M.; Malbec, L.-M.; Reveille, B.; Servant, C.; Souidi, F.; Torcolini, N. Water Injection to Improve Direct Injection Spark Ignition Engine Efficiency. *SAE Tech. Pap.* **2019**. [CrossRef]
11. Gern, M.S.; Vacca, A.; Bargende, M. Experimental Analysis of the Influence of Water Injection Strategies on DISI Engine Particle Emissions. *SAE Int. J. Adv. Curr. Prac. Mobil.* **2020**, *2*, 598–606. [CrossRef]
12. Paltrinieri, S.; Mortellaro, F.; Silvestri, N.; Rolando, L.; Medda, M.; Corrigan, D. Water Injection Contribution to Enabling Stoichiometric Air-to-Fuel Ratio Operation at Rated Power Conditions of a High-Performance DISI Single Cylinder Engine. *SAE Tech. Pap.* **2019**. [CrossRef]
13. Cavina, N.; Rojo, N.; Businaro, A.; Brusa, A.; Corti, E.; De Cesare, M. Investigation of Water Injection Effects on Combustion Characteristics of a GDI TC Engine. *SAE Int. J. Engines* **2017**, *10*, 2209–2218. [CrossRef]
14. Khatri, J.; Denbratt, I.; Dahlander, P.; Koopmans, L. Water Injection Benefits in a 3-Cylinder Downsized SI-Engine. *SAE Int. J. Adv. Curr. Prac. Mobil.* **2019**, *1*, 236–248. [CrossRef]
15. Falfari, S.; Bianchi, G.M.; Cazzoli, G.; Ricci, M.; Forte, C. Water Injection Applicability to Gasoline Engines: Thermodynamic Analysis. *SAE Tech. Pap.* **2019**. [CrossRef]
16. Berni, F.; Breda, S.; Lugli, M.; Cantore, G. A Numerical Investigation on the Potentials of Water Injection to Increase Knock Resistance and Reduce Fuel Consumption in Highly Downsized GDI Engines. *Energy Proc.* **2015**, *81*, 826–835. [CrossRef]
17. Hoppe, F.; Thewes, M.; Seibel, J.; Balazs, A.; Scharf, J. Evaluation of the Potential of Water Injection for Gasoline Engines. *SAE Int. J. Engines* **2017**, *10*, 2500–2512. [CrossRef]
18. Sun, Y.; Fischer, M.; Bradford, M.; Kotrba, A.; Randolph, E. Water Recovery from Gasoline Engine Exhaust for Water Injection. *SAE Tech. Pap.* **2018**. [CrossRef]
19. Yasukawa, Y.; Okamoto, Y.; Kobayashi, N.; Saito, T.; Saruwatari, M. Multi-Swirl Type Injector for Port Fuel Injection Gasoline Engines. *SAE Tech. Pap.* **2014**. [CrossRef]
20. Van Vuuren, N.; Postrioti, L.; Brizi, G.; Picchiotti, F. Instantaneous Flow Rate Testing with Simultaneous Spray Visualization of an SCR Urea Injector at Elevated Fluid Temperatures. *SAE Int. J. Engines* **2017**, *10*, 2478–2485. [CrossRef]
21. Kapusta, Ł.J.; Rogoz, R.; Bachanek, J.; Boruc, Ł.; Teodorczyk, A. Low-Pressure Injection of Water and Urea-Water Solution in Flash-Boiling Conditions. *SAE Int. J. Adv. Curr. Prac. Mobil.* **2021**, *3*, 365–377. [CrossRef]
22. Postrioti, L.; Caponeri, G.; Buitoni, G.; van Vuuren, N. Experimental Assessment of a Novel Instrument for the Injection Rate Measurement of Port Fuel Injectors in Realistic Operating Conditions. *SAE Int. J. Fuels Lubr.* **2017**, *10*, 344–351. [CrossRef]
23. Brizi, G.; Postrioti, L.; van Vuuren, N. Experimental Analysis of SCR Spray Evolution and Sizing in High-Temperature and Flash Boiling Conditions. *SAE Int. J. Fuels Lubr.* **2019**, *12*, 87–108. [CrossRef]
24. Postrioti, L.; Malaguti, S.; Bosi, M.; Buitoni, G.; Piccinini, S.; Bagli, G. Experimental and numerical characterization of a direct solenoid actuation injector for Diesel engine applications. *Fuel* **2014**, *118*, 316–328. [CrossRef]
25. Postrioti, L.; Ubertini, S. An integrated experimental-numerical study of HSDI Diesel injection system and spray dynamics. *SAE Tech. Pap.* **2006**. [CrossRef]

*applied
sciences*

MDPI

Article

Performance and Emissions of a Spark Ignition Engine Fueled with Water-in-Gasoline Emulsion Produced through Micro-Channels Emulsification

Cinzia Tornatore [1], Luca Marchitto [1,*], Luigi Teodosio [1], Patrizio Massoli [1] and Jérôme Bellettre [2]

[1] Institute of Science and Technology for Sustainable Energy and Mobility (STEMS-CNR), Via Marconi 4, 80125 Naples, Italy; cinzia.tornatore@stems.cnr.it (C.T.); luigi.teodosio@stems.cnr.it (L.T.); patrizio.massoli@stems.cnr.it (P.M.)

[2] Laboratoire de Thermique et Énergie de Nantes, Université de Nantes, CNRS, LTeN, UMR 6607, F-44000 Nantes, France; jerome.bellettre@univ-nantes.fr

*   Correspondence: luca.marchitto@stems.cnr.it

Abstract: This paper presents an experimental study investigating the effects of water-in-gasoline emulsion (WiGE) on the performance and emissions of a turbocharged PFI spark-ignition engine. The emulsions were produced through a micro-channels emulsifier, potentially capable to work inline, without addition of surfactants. Measurements were performed at a 3000 rpm speed and net Indicated Mean Effective Pressure (IMEP) of 16 bar: the engine point representative of commercial ECU map was chosen as reference. In this condition, the engine, fueled with gasoline, runs overfueled ($\lambda = 0.9$) to preserve the integrity of the turbocharger from excessive temperature, and the spark timing corresponds to the knock limit. Starting from the reference point, two different water contents in emulsion were tested, 10% and 20% by volume, respectively. For each selected emulsion, at $\lambda = 0.9$, the spark timing was advanced from the reference point value to the new knock limit, controlling the IMEP at a constant level. Further, the cooling effect of water evaporation in WiGE allowed it to work at stoichiometric condition, with evident benefits on the fuel economy. Main outcomes highlight fuel consumption improvements of about 7% under stoichiometric mixture and optimized spark timing, while avoiding an excessive increase in turbine thermal stress. Emulsions induce a slight worsening in the HC emissions, arising from the relative impact on combustion development. On the other hand, at stoichiometric condition, HC and CO emissions drop with a corresponding increase in NO.

Keywords: water-in-gasoline emulsion; micro-channel emulsifier; experiments; spark-ignition engine; performance; fuel consumption; emissions

Citation: Tornatore, C.; Marchitto, L.; Teodosio, L.; Massoli, P.; Bellettre, J. Performance and Emissions of a Spark Ignition Engine Fueled with Water-in-Gasoline Emulsion Produced through Micro-Channels Emulsification. *Appl. Sci.* **2021**, *11*, 9453. https://doi.org/10.3390/app11209453

Academic Editors: Georgios Karavalakis and Ricardo Novella Rosa

Received: 31 May 2021
Accepted: 6 October 2021
Published: 12 October 2021

Publisher's Note: MDPI stays neutral with regard to jurisdictional claims in published maps and institutional affiliations.

1. Introduction

Water injection (WI) technology is seen as a promising solution to make turbocharged spark ignition (SI) engines operate with higher compression ratios, higher boost pressures and stoichiometric combustions at high loads. This allows a reduction in brake-specific fuel consumption (BSFC) and thus in CO_2 emission on the overall engine map [1,2].

In downsized turbocharged SI engines, at stoichiometric conditions and full load, it is necessary to control knock tendency and to fulfill the maximum allowable temperature at the turbine intake. In commercial SI engines, this is commonly performed by introducing an excess fuel in the combustion chamber and retarding the ignition timing, leading to lower thermodynamic efficiency. Water injection can replace the cooling effect of the air/fuel mixture enrichment with benefits in the BSFC. The introduction of water in the combustion chamber of an SI engine reduces the gas temperature level before the combustion takes place due to the water's high latent heat of vaporization that cools down the air/fuel mixture. Hoppe et al. [2] numerically investigated the relative weight of water vaporization and the sole dilution effect in the charge-cooling process. They compared the effect of liquid- and vapor-water injection on the air–fuel mixture temperature, at the spark timing. They found a reduction in air–fuel mixture temperature with liquid-water injection five times higher than in the case of vapor injection, indicating the evaporation as the main driver of charge cooling. Moreover, the evaporated water acts similar to an inert gas during the combustion process and it reduces the combustion temperature by increasing the global heat capacity [3].

Starting in the 1920s, research has been progressing on the use of water in internal combustion engines [4], and many researchers presented works on water injection in the intake manifold (port injection) or directly in the cylinder (direct injection) of spark-ignition engines [5].

In Reference [6], Zhuang et al. experimentally analyzed the influence of port water injection on the consumption and the thermal efficiency of a turbocharged direct injection engine. The tests were conducted at low speed and medium–high load, at stoichiometric condition. A reduction of about 3.5% in specific fuel consumption and an increase in thermal efficiency of about 1.5% were obtained when the maximum water/gasoline percentage of 50% was injected, thanks to the knock mitigation. The influence of advanced spark timing was found to be dominant compared to the effect of charge dilution on the combustion duration. Therefore, the combustion center was advanced of about 5 CAD. Further advantages can be obtained at high engine speed and full load. As aforementioned, in these conditions turbocharged engines run overfueled to keep the turbine temperature below the mechanical failure threshold. Advancing the knock-limited spark advance allow us to reduce the exhaust temperature, avoiding charge enrichment. Sun et al. investigated the influence of port water injection on the performance of turbocharged direct-injection spark-ignition engine at full-load (18 bar BMEP) conditions. The port fuel injection allowed us to reduce the fuel enrichment of about 13% with a thermal efficiency increase of about 4.5% [7].

Similar results were found by Tornatore et al., who experimentally tested the behavior of a small displacement turbocharged spark-ignition engine equipped with a port-water-injection system. They observed a reduction in fuel consumption between 6 and 12% at full load conditions, at varying engine speeds [1].

In Reference [8], Hunger et al. investigated the influence of direct water injection on knock mitigation and thermal efficiency. They found a relationship between the injection timing and charge cooling effect. When water injection is too advanced, the spray impinges against the cylinder walls, generating a liquid film. The vaporization process starts during the compression stroke; therefore, the water deposited against the liner evaporates extracting heat in part from the end gas and in part from cylinder walls with a drop in knock-suppression efficacy. Furthermore, a proper combustion chamber design is necessary to locate the nozzle, considering the limited available space left by inlet and exhaust four valves, the spark plug and the gasoline injector. On the other hand, port water injection does not allow us to fully take advantage of vaporization heat extraction from the end gas, and, at the same time, it has a limited influence on the volumetric efficiency, as the intake temperature is too low to induce water evaporation in the manifold.

An alternative way of introducing water in a SI engine is under the form of an emulsion with gasoline. This solution shows some advantages compared to the standard separate injections of water and gasoline. Indeed, the water-in-gasoline emulsion (WiGE) technique requires a single injector per cylinder, a lower number of injection control variables and, consequently, a simpler control of fuel injection. Referring to the WiGE very few papers are available in the technical literature discussing its production, employment in SI engines and relative influence on combustion, performance and pollutant emissions. On the other hand, the use of water-in-diesel emulsion (WiDE) in diesel engines has been found to be an economical technique for improved combustion and fuel economy; the presence of water in diesel engines also leads to a drop in NOx formation and in the rate of soot particles [9]. In the case of WiGE, old research studies [10,11] focused on the enhanced knock mitigation capabilities of emulsion fuel and its influence on performance of spark-ignition engines.

As a starting point of Reference [10], the authors stated that when WiGEs are correctly used and the engine is properly adjusted to give optimum performance, water addition lowers gasoline octane number requirement and reduces the thermal stress on the parts of the cylinder–piston group without any loss of maximum engine power or torque and without increasing the specific gasoline consumption. The study reported in Reference [11] also confirmed the improved gasoline octane number with the employment of water addition in the form of emulsion. In Reference [12], an analysis on the effects of supplementing gasoline with water on spark-ignition engine performance and emissions is reported. Experiments on a single-cylinder engine, engine cycle simulations, and vehicle tests were performed. This research showed that the concept of adding water to gasoline presents some negative aspects consisting in the increased hydrocarbon emissions and decreased vehicle drivability. The resulting main benefits of water–gasoline fuels were the higher octane ratings and the decreased nitric oxide emissions. A recent paper reports the outcomes of a basic research on direct injection of gasoline/water emulsion. It shows the influence of the amount of water in a WiGE on the spray evolution in a high-pressure chamber [13]. The main evidence of this study is represented by the opportunity to optimize the direct injection of WiGE to improve fuel consumption and emissions in DISI engines.

On the other hand, the scientific literature is very poor in the use of water–gasoline emulsions in modern turbocharged SI engines, and WIGE potential in fuel enrichment suppression has not yet been explored. Moreover, there is still a lot of progress to be made in the selection of the emulsifiers and in surfactant with the aim to guarantee the stability of the emulsion fuels for a regular on-board vehicle storage before its use [14]. Surfactants have an additional cost, and, according to the literature papers, they can lead to hazardous emissions during combustion, based on their chemical nature [15]; thus, the use of a stable emulsion with a very low level of surfactant is desired.

Starting from the above discussed state-of-the-art, this paper investigates the effects of water-in-gasoline emulsions on the performance and emissions of a turbocharged port fuel injection (PFI) spark-ignition engine. WiGE is produced through a prototype micro-channel emulsifier [16] that allows for the use of a very small amount of surfactant to create stable emulsions.

Further development of the prototype emulsifier will be aimed at using this device inline by the integration with the engine fuel injection system; this will allow for us to have a very flexible fueling system and to completely quit the use of surfactants, since it will not be necessary to face the issues associated with a long-term WiGE storage. In this way, using WiGEs will allow us to introduce water into the combustion chamber, without equipping the engine with a secondary injection system.

2. Experimental Setup and Procedures

The experiments for this work were carried out on a downsized and turbocharged SI engine equipped with port fuel injectors. The engine has 2 cylinders and 4 valves for each cylinder; its main characteristics are shown in Table 1.

Table 1. PFI twin-cylinder engine's main characteristics.

Characteristic	Dimension
Displacement (cm^3)	875.4
Compression ratio	10:1
Bore (mm)	80.5
Stroke (mm)	86.0
Connecting Rod (mm)	136.8
Rated Power (kW)	63.7 @ 5500 rpm
Max Torque (Nm)	146.7 @ 2000 rpm

The Mitsubishi turbocharger allows us to reach boost pressures up to 2.4 bar. The turbine volute presents a waste-gate valve that allows us to control the maximum admissible turbocharger speed and, thus, turbine inlet and compressor outlet pressures. The compressor is equipped with a dump-valve to prevent surge occurrence.

A sketch of the engine test bench is shown in Figure 1. An air-handling unit constantly supplies intake air to the engine compressor at a temperature of 293 K.

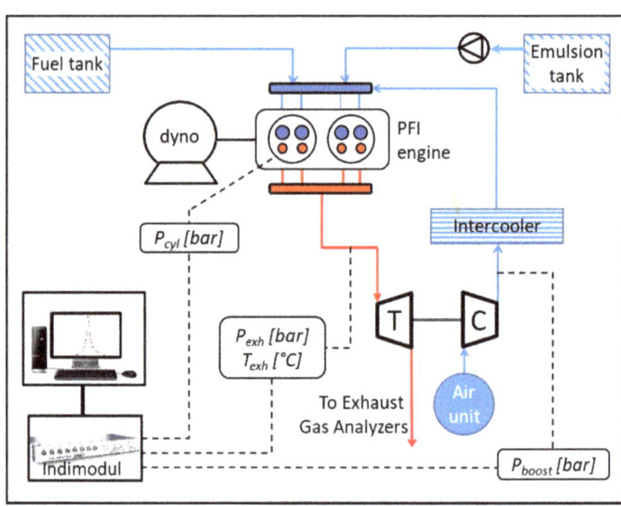

Figure 1. General layout of the experimental setup.

In-cylinder pressure signals are detected by using one piezo-quartz pressure transducer (accuracy of ±0.1%) for each cylinder. The boost pressure and the pressure upstream of the turbine are measured through piezo-resistive low-pressure indicating sensors installed at the compressor outlet and in the exhaust manifold. Exhaust flow temperature is monitored by using a thermocouple to check that the value does not exceed the allowable temperature limits for the turbine inlet.

The emulsion injection system is mainly composed of a commercial liquid pump to pressurize the fluid at 4 bar and a rail to accumulate a proper amount of emulsion upstream of the injectors.

The engine operation can be controlled by a commercial engine control unit (ECU) or alternatively by a prototype driver managed by a LabView software tool that is able to switch from the reference commercial ECU to the external mode [17]. Using the second option, it is possible to control the main engine parameters, such as spark advance (SA), start and duration of gasoline and emulsion injection. The fuel injection quantity is fine monitored to get the selected air/fuel (A/F) ratio according to the lambda sensor placed at the engine exhaust.

Crank-angle-degree resolved data are acquired by using an indicating system (AVL IndiModul) coupled to a shaft encoder. A combustion analysis software, AVL INDICOM, allows for signal processing and data saving. In-cylinder pressure data are collected with a high sampling resolution (0.1 CAD), from −90 to 90 CAD ATDC; outside of this angular interval, the sampling resolution is set at 1 CAD.

It is worth highlighting that the engine speed/load condition was actually controlled through the dynamometric brake, while a refined monitoring of IMEP level was performed through the indicating software. Engine IMEP was measured as the mean of single cylinders IMEPs. Therefore, controlling IMEP instead of BMEP allowed us to monitor the effects of cycle-resolved, cycle-to-cycle and cylinder-to-cylinder variations.

Regulated exhaust gaseous emissions and CO_2 were measured: hydrocarbon (THC), CO and CO_2 by means of a nondispersive infrared (NDIR) sensor, while an electrochemical sensor measured NOx.

During experiments, a reference engine operating condition was chosen among the points stored in the commercial ECU (full gasoline conditions). The selected point represents a knock-limited engine operation and is characterized by a speed of 3000 rpm, a high net IMEP of 16.0 bar, a rich relative A/F mixture ($\lambda = 0.9$) and the spark timing (ST) at -10 CAD ATDC. This condition was set as representative of the overfueling strategies adopted in downsized engines at high loads: knock occurrence is avoided by delaying the combustion phasing towards the expansion stroke with an increase in the temperature at the turbine inlet. For this reason, in order to preserve exhaust engine hardware, the ECU adopts fuel-enrichment strategies which reduce the exhaust gas temperature, with penalties over fuel consumptions. As an alternative knocking suppression technique to overfueling strategies, water was injected in the engine through emulsions with gasoline. Two different WiGEs were tested: WiGE 10 and WiGE 20. The first one consists of an emulsion containing 10% of water and 90% of gasoline by volume, while the second one contains 20% water and 80% gasoline.

As is well-known, an emulsion is a fine dispersion of two liquid phases. Details concerning the adopted emulsification technique can be found in Reference [18]. Briefly, two controlled fluxes of water and oil impact in a cross-section through properly dimensioned micro-channels. As a result of the impact, the water is atomized in microdroplets inside the oil matrix. Depending on the physical properties of the oil (viscosity, surface tension, etc.), the emerging emulsion is stable for a variable time, which is long enough to be used in inline feeding systems (some hours before water separation). In the present case, the emulsion is produced offline, and a small amount of nonionic surfactant (SPAN80 0.2%v) is added to preserve the emulsion stability.

First, the engine was run with gasoline in the reference engine operating condition. Then, at the same speed, gasoline was replaced with WiGE 10. The IMEP and the lambda values were kept the same as the reference condition by fine-tuning the injection duration (DOI) and plenum pressure, for a peer-to-peer comparison with reference gasoline case. Then, the spark timing was advanced until reaching the new knocking limited spark advance. Aim of the present work is to investigate the potential of WiGE injection as a knock mitigation strategy alternative to fuel enrichment. After the tests under a rich relative A/F ratio ($\lambda = 0.9$), WiGE injection duration was changed in order to reach a stoichiometric A/F mixture ($\lambda = 1.0$), while the IMEP was kept always the same of reference level. A spark timing sweep was performed under stoichiometric condition starting from the reference spark timing until reaching the new knock limit. This procedure was repeated for WiGE 20, and the overall test matrix is shown in the following Table 2. Gasoline and WiGE presented the same injection timings: -165 CAD ATDC.

During the experiments, the turbine inlet temperature (TIT) was always kept below 950 °C, and the maximum in-cylinder pressure was below 85 bar (± 5 bar), as that is the maximum allowable peak pressure to preserve the engine from mechanical failure of components. To this aim, the maximum boost level was automatically controlled in the range 1.8–2.0 bar, acting on the waste-gate valve opening. The coolant temperature was set at 85 ± 1 °C, using a water heat exchanger.

Table 2. Test conditions.

Fuel	λ [± 0.01]	ST [cad atdc]	DOI [cad]	P_{INT} [bar]
gasoline	0.90	-10	195	1.72
WiGE 10	0.90	-10	230	1.74
WiGE 10	0.90	-11	226	1.70
WiGE 10	0.90	-12	223	1.70
WiGE 10	0.90	-13	220	1.69
WiGE 10	1.00	-10	215	1.78

Table 2. *Cont.*

Fuel	λ [$\pm$0.01]	ST [cad atdc]	DOI [cad]	P_{INT} [bar]
WiGE 10	1.00	−11	213	1.77
WiGE 10	1.00	−12	210	1.75
WiGE 10	1.00	−13	208	1.73
WiGE 10	1.00	−14	206	1.76
WiGE 20	0.90	−10	255	1.70
WiGE 20	0.90	−12	255	1.70
WiGE 20	0.90	−14	255	1.70
WiGE 20	0.90	−16	255	1.70
WiGE 20	1.00	−10	250	1.84
WiGE 20	1.00	−12	243	1.81
WiGE 20	1.00	−14	235	1.78
WiGE 20	1.00	−16	230	1.74
WiGE20	1.00	−19	225	1.72

3. Results

The effect of WiGE on the engine performance was investigated through the in-cylinder pressure analysis. As aforementioned and further discussed in the following, with reference to the TIT results, the gasoline case at stoichiometric A/F mixture, and SA = −10 CAD ATDC is missing because it is not feasible with the maximum TIT target. Figure 2 shows a comparison between the gasoline and WIGE in-cylinder pressure traces at the reference engine operating condition (λ = 0.9, SA = −10 CAD ATDC). As aforementioned, this condition is representative of commercial ECU map and the spark timing under rich air/fuel mixture corresponds to the gasoline knock limit. When switching to WiGE, the injection duration and the plenum pressure were adjusted to keep the A/F ratio and IMEP load constant. The first effect of WiGE is the charge cooling, due to the evaporation of water droplets; moreover, the water in the combustion chamber acts as an inert during the combustion process; this causes a slowdown of the rate of energy release with a reduction in the pressure peak proportional to the water content.

Figure 3 shows the relationship between water content and combustion duration and phasing, at different spark timings and relative A/F ratios and WiGEs, considering a representative engine cylinder (Cyl #2). For each spark timing sweep, a clear indication of knock-limit (KL) point is depicted in Figure 3a. This indication of KL points extends to the other figures proposed below. In agreement with the pressure traces shown in Figure 2, at SA = −10 CAD ATDC and λ = 0.9, the use of WiGE prolongs the combustion duration (Figure 3a) and delays the combustion phasing (Figure 3b). On the other hand, proportionally to the water content, the cooling and dilution effects of WiGE on the in-cylinder charge mitigate the knock tendency and allow us to advance the spark timing up to −13 and −16 CAD ATDC (for WiGE 10 and WiGE 20, respectively); consequently, also the combustion center is advanced with respect to gasoline reference case. At stoichiometric condition, the knock-limited spark timing can be further advanced with the result of a better combustion phasing (MFB50 = 12.4 CAD ATDC @ SA = −19 for WiGE20).

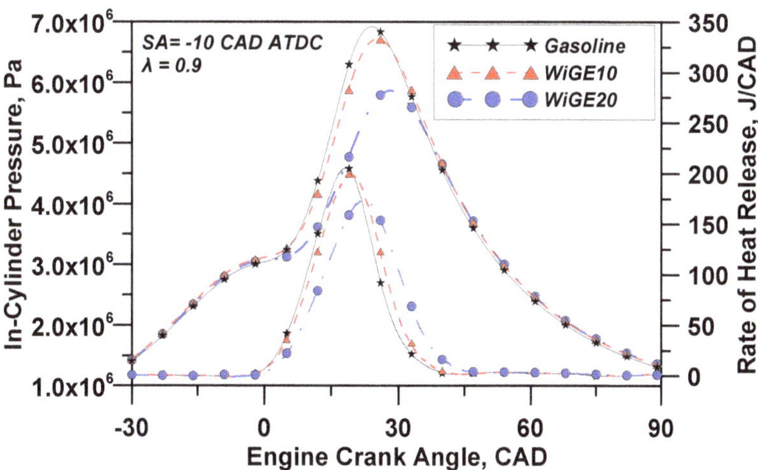

Figure 2. In-cylinder pressure trace and rate of heat release at SA = −10 CAD ATDC, λ = 0.9.

Figure 4 shows the trend of TIT against spark timing and relative air-to-fuel ratio. As shown, at a reference spark timing of −10 CAD ATDC, the turbine inlet temperature is almost the same for gasoline and WiGE in rich mixture condition. Of course, the rich A/F mixture at the reference point was selected in the manufacturer calibration to avoid, with a certain safety margin, the knock onset and the excessive thermal stress to the turbine. On the other hand, in the case of the engine mounted on the test bench, a lower heat transfer can be realized for the turbine compared to the case of an engine on the real vehicle. Based on this consideration, even if the allowable maximum TIT is 950 °C, a target of 819 °C, which is representative of the gasoline reference condition, was set for WiGE. In light of the above discussion, the criterion followed for the engine tests is the identification of knock-limited operations with WiGE and stoichiometric mixture realizing a TIT level equal or lower to the one achieved at the reference ECU condition and a lower ISFC. By increasing the spark advance, the earlier combustion phasing allowed by WiGE results in a significant reduction in the turbine inlet temperature, because the temperature of the in-cylinder gases is lower at the opening of the exhaust valves if the combustion takes place early in the engine cycle. The use of WiGE10 does not allow us to reach the TIT target at stoichiometric condition (TIT = 827 °C at knock limit; SA = −14 CAD ATDC), while WiGE20 resulted in a lower TIT than gasoline reference case at the two most advanced spark timing: 818 °C at SA = −16 CAD ATDC, and 804 °C at SA = −19 CAD ATDC.

The Indicated Specific Fuel Consumption (ISFC) for all the investigated WiGE fuels, spark timings and relative air-to-fuel ratio is shown in Figure 5. It is worth pointing out that fuel consumption is estimated by considering only the gasoline content of WiGEs, as the water is inert.

As expected, introducing water in combustion chamber prolongs the combustion duration with a worsening in efficiency. Hence, at fixed spark timing (−10 CAD ATDC) and at the same air/fuel ratio (λ = 0.9), fuel consumption associated to WiGE is higher than the one measured in the reference condition. When leaning WIGE/air mixture the fuel efficiency is improved. On the other hand, this improvement is not large enough to eliminate the gap with gasoline reference case. Therefore, at fixed spark timing (−10 CAD ATDC), the WiGE fuel consumption at stoichiometric condition is still higher than the one measured for gasoline at rich condition (λ = 0.9).

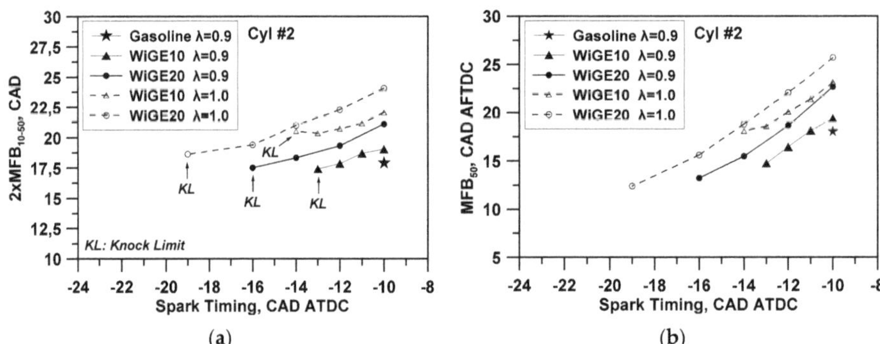

Figure 3. Combustion duration (2xMFB10-50) (a) and combustion center (MFB50) (b) against spark timing and relative air-to-fuel ratio for WiGE 10 and WiGE 20.

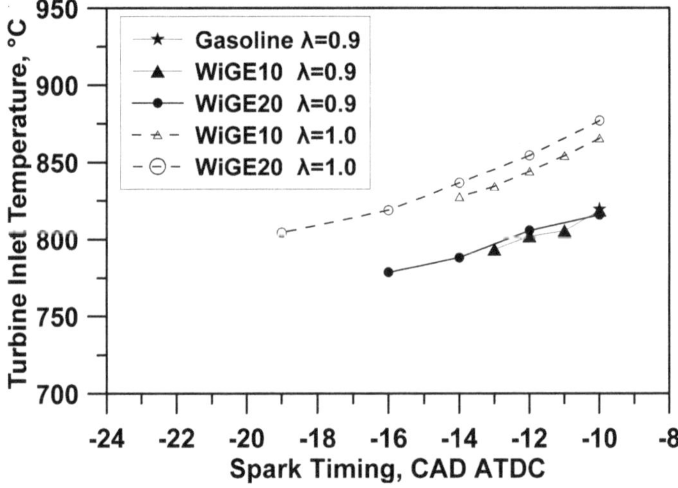

Figure 4. Turbine inlet temperature against spark timing and relative air-to-fuel ratio.

Thus, to achieve the same IMEP as the reference case, it is necessary to inject a higher amount of gasoline. Higher consumption is measured for WiGE20 compared to WiGE10. This behavior is probably ascribed to the water-induced lengthening of combustion process.

A decreasing trend of ISFC with advancing the spark timing is observed for both air-to-fuel ratios and both emulsions. At $\lambda = 0.9$ and reference spark timing (SA = -10 CAD ATDC), WiGE employment induces a worsening in the ISFC values compared to gasoline reference condition, due to combustion duration extension induced by water. On the other hand, the improved knock resistance related to water presence allows us to optimize the combustion phasing. The discussed water effects are mutually conflicting for the engine efficiency, and ultimately the ISFC level of KL points (for WiGE 10 and 20) under the rich A/F mixture never reaches the one attained by the ECU-reference rich gasoline case. At engine operation under stoichiometric A/F mixture, the combined effect of WiGE knock mitigation and the leaner mixture allows for an ISFC reduction of about the 3.7% and 7.1% compared to ECU-reference gasoline point at rich mixture condition, for WiGE 10 and WiGE 20, respectively.

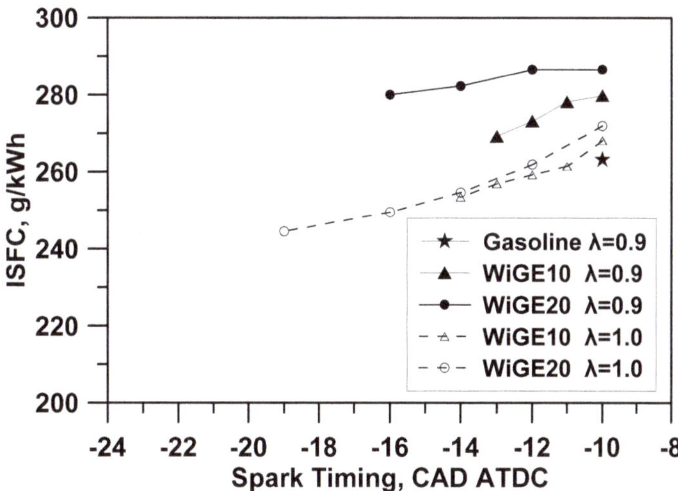

Figure 5. Indicated Specific Fuel Consumption against spark timing and relative air-to-fuel ratio.

Figure 6 shows the correlation between the CO and CO_2 exhaust emissions and emulsion fueling. For the rich A/F mixture condition, coherently with fuel consumption results, an increase in CO_2 is measured for emulsions (both with WiGE 10 and 20) if compared with the ECU reference gasoline case, while the CO is almost similar. In stoichiometric A/F mixture condition, CO_2 emissions for emulsions are higher than the ones measured under rich condition, due to a more complete combustion process.

Figure 6a also shows that the main CO variations have to be attributed to the modification of the A/F mixture quality. As expected, a strong reduction in CO of one order of magnitude is measured when switching from the rich to stoichiometric A/F mixture.

Figure 7 shows the variation of HC and NO emissions against the spark advance at the selected relative air-to-fuel ratios and for all the investigated fuels. At the rich mixture condition, the WiGE combustion produces larger HC, proportional to the water content, almost independent of the spark timing. Similarly, at stoichiometric condition, the increase in water content in the emulsion induces slightly higher unburned HC emissions. On the other hand, for a fixed water-in-gasoline emulsion, a reduction in the unburned HC is observed passing from the rich to stoichiometric A/F mixture.

Referring to NO emissions, at rich condition, the switch from full gasoline to WiGE involves a reduction in NO. This reduction is more marked for WiGE 20 than WiGE 10, and it is due to the combined effect of charge cooling and thermal dilution.

It is the case to highlight that, even with WiGE 20, the water quantity contained in the in-cylinder charge is low and the corresponding vapour conversion (evaporation process) is constrained by thermodynamic conditions, available time and full humidity conditions. However, the high latent heat of vaporization for water has a relevant role on the reduction of in-cylinder mixture temperature. As reported in Reference [19], at the spark timing, a reduction in charge temperature of 13 °C can be achieved at 3500 rpm and 17 bar BMEP when injecting 17%w of water in the intake manifold.

In stoichiometric conditions, the increase in oxygen content and the higher combustion temperature due to the lack in any diluent (excess fuel or water) promote NO formation, with a maximum NO level at KLSA for both WiGE 10 and WiGE 20, due to the increase of in-cylinder peak temperature by advancing the spark timing.

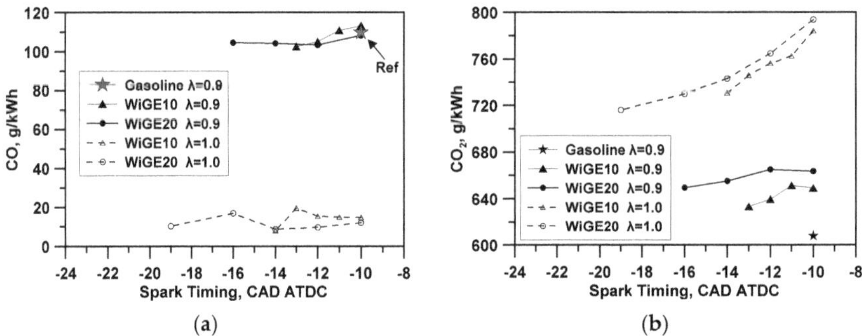

Figure 6. CO (**a**) and CO$_2$ (**b**) emissions against spark timing and relative air-to-fuel ratio.

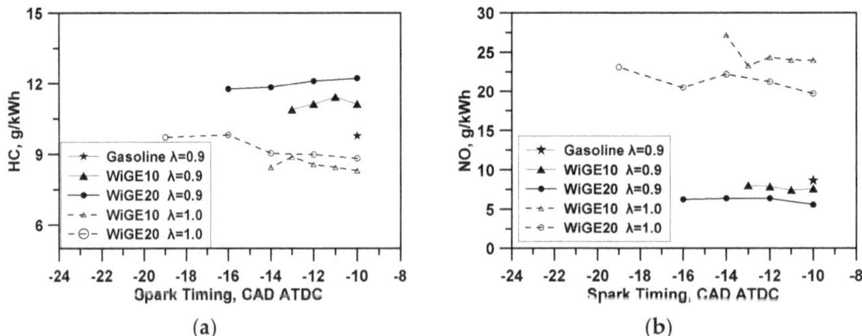

Figure 7. HC (**a**) and NO (**b**) emissions against spark timing and relative air-to-fuel ratio.

4. Conclusions

In this work, an experimental study was carried out to quantify the influence of WiGEs on combustion, performance and exhaust emissions at a high-load knock-limited operation of a small turbocharged PFI spark-ignition engine. The emulsions are produced through a micro-channels emulsifier, potentially capable to work inline, without the addition of surfactants. Two different water contents in emulsion were tested, 10% and 20% by volume, respectively. As in the present work, the emulsion is produced offline, and a small amount of nonionic surfactant (SPAN80 0.2%v) is added to preserve emulsion stability. Engine tests are performed with full gasoline and emulsions injected in the intake runners and considering a reference engine point at a speed of 3000 rpm and 16 bar IMEP. For the selected operating condition, the standard ECU calibration is applied in gasoline mode, running the engine under rich air/fuel mixture (λ = 0.9) and a spark advance (SA = −10 CAD AFTDC) at knock limit. Starting from the above reference point, a spark timing sweep is realized for each WiGE up to the new knock-limited condition, keeping the IMEP constant. Further, the cooling and dilution effects of water evaporation in WiGE has allowed us to work at stoichiometric condition. Engine overall performance, in-cylinder pressure traces and pollutant emissions are measured in each tested condition.

The analysis of experimental in-cylinder pressure cycles and burn-rate profiles show that the water presence in the combustion chamber produces a cooling and dilution effect of charge, inducing an increasing slowdown of combustion velocity and a lowering of pressure peak with the water content. On the other hand, the cooling and dilution effects of WiGE allow us to mitigate the knock occurrence and, consequently, to advance the spark timing, reaching an optimized combustion phasing.

A decreasing TIT trend with spark timing is observed, and the measured TIT level under stoichiometric mixture at the most advanced spark timing reaches a quite similar value to the one attained in the reference gasoline condition.

Relevant ISFC benefits are realized with WiGE 10 (3.7%) and WiGE 20 (7.1%) in stoichiometric mixture and optimized combustions.

Concerning the exhaust emissions, a comparison with the reference gasoline mode highlights a slight increase in HC emissions with a corresponding reduction in NO when using WiGEs. When switching at stoichiometric A/F ratio, the more complete combustion results in a certain reduction in HC, while major penalties for NO are found. One order of magnitude reduction in CO levels is obtained.

Summarizing, water-in-gasoline emulsions demonstrated to be a technique to improve fuel consumption at medium/high loads of turbocharged spark-ignition engines, enabling the stoichiometric combustions, while preserving the turbine blades from severe thermal stresses.

Author Contributions: Conceptualization, methodology and formal analysis, L.M., C.T., L.T., P.M. and J.B.; experimental setup, investigation, data acquisition and data curation, L.M., C.T. and L.T.; writing—original draft preparation, L.M., C.T., L.T., P.M. and J.B.; writing—review and editing, visualization and supervision, L.M., C.T., L.T., P.M. and J.B. All authors have read and agreed to the published version of the manuscript.

Funding: This research was partially funded by the Region Pays de la Loire, Chaire Connect Talent ODE program for outstanding research in the field of energetics.

Institutional Review Board Statement: Not applicable.

Informed Consent Statement: Not applicable.

Data Availability Statement: Not applicable.

Acknowledgments: The authors gratefully acknowledge the support of the Region Pays de la Loire, Chaire Connect Talent ODE program, for outstanding research in the field of energetics. The authors thank Alfredo Mazzei and Bruno Sgammato for the technical support in the experimental campaign.

Conflicts of Interest: The authors declare no conflict of interest.

Abbreviations

A/F	air-to-fuel
ATDC	After Top Dead Center
BSFC	brake-specific fuel consumption
CAD	crank angle degree
DISI	direct-injection spark ignition
DOI	duration of injection
ECU	engine control unit
IMEP	indicated mean effective pressure
ISFC	Indicated Specific Fuel Consumption
KL	knock limit
KLSA	knock-limited spark advance
MFB10	10% of mass fraction burned
MFB50	50% of mass fraction burned
MFB10-50	combustion core duration
NDIR	nondispersive infrared
PFI	port fuel injection
ROHR	rate of heat release
SA	spark advance
SI	spark ignition
ST	spark timing
TDC	Top Dead Center
THC	Total Hydrocarbon Content

TIT	turbine inlet temperature
WI	water injection
WiDE	water-in-diesel emulsion
WiGE	water-in-gasoline emulsion

References

1. Tornatore, C.; Siano, D.; Marchitto, L.; Iacobacci, A.; Valentino, G.; Bozza, F. Water Injection: A Technology to Improve Performance and Emissions of Downsized Turbocharged Spark Ignited Engines. *SAE Int. J. Engines* **2017**, *10*, 2319–2329. [CrossRef]
2. Hoppe, F.; Thewes, M.; Baumgarten, H.; Dohmen, J. Water injection for gasoline engines: Potentials, challenges, and solutions. *Int. J. Engine Res.* **2015**, *17*, 86–96. [CrossRef]
3. da Rocha, D.D.; Radicchi, F.D.C.; Lopes, G.S.; Brunocilla, M.F.; Gomes, P.C.D.F.; Santos, N.D.S.A.; Malaquias, A.C.T.; Filho, F.A.R.; Baêta, J.G.C. Study of the water injection control parameters on combustion performance of a spark-ignition engine. *Energy* **2020**, *217*, 119346. [CrossRef]
4. Fu, Z.; Liu, M.; Xu, J.; Wang, Q.; Fan, Z. Stabilization of water-in-octane nano-emulsion. Part I: Stabilized by mixed surfactant systems. *Fuel* **2010**, *89*, 2838–2843. [CrossRef]
5. Falfari, S.; Bianchi, G.M.; Cazzoli, G.; Forte, C.; Negro, S. Basics on Water Injection Process for Gasoline Engines. *Energy Procedia* **2018**, *148*, 50–57. [CrossRef]
6. Huang, Y.; Sun, Y.; Teng, Q.; He, B.; Chen, W.; Qian, Y. Investigation of water injection benefits on downsized boosted direct injection spark ignition engine. *Fuel* **2019**, *264*, 116765. [CrossRef]
7. Sun, Y.; Fischer, M.; Bradford, M.; Kotrba, A.; Randolph, E. *Water Recovery from Gasoline Engine Exhaust for Water Injection*; SAE Technical Paper; SAE International: Warrendale, PA, USA, 2018. [CrossRef]
8. Hunger, M.; Böcking, T.; Walther, U.; Günther, M.; Freisinger, N.; Karl, G. Potential of Direct Water Injection to Reduce Knocking and Increase the Efficiency of Gasoline Engines. In Proceedings of the International Conference on Knocking in Gasoline Engines, Berlin, Germany, 12–13 December 2017; Springer: Berlin/Heidelberg, Germany, 2017; pp. 338–359.
9. Vellaiyan, S.; Amirthagadeswaran, K. The role of water-in-diesel emulsion and its additives on diesel engine performance and emission levels: A retrospective review. *Alex. Eng. J.* **2016**, *55*, 2463–2472. [CrossRef]
10. Nikitin, I.M.; Zavodov, V.S.; Puzdyrev, M.K.; Yakovlev, A.V. Water/gasoline emulsion production technology. *Chem. Technol. Fuels Oils* **1981**, *17*, 34 36. [CrossRef]
11. Zavodov, V.S.; Luksho, V.A.; Neudakhin, V.I.; Novitskii, G.G.; Shatrov, E.V. Operation of carburetor engines on water/gasoline emulsions. *Chem. Technol. Fuels Oils* **1982**, *18*, 25–29. [CrossRef]
12. Peters, B.D.; Stebar, R.F. Water-Gasoline Fuels-Their Effect on Spark Ignition Engine Emissions and Performance. *SAE Trans.* **1976**, *85*, 1832–1853. [CrossRef]
13. Wagner, T.; Rottengruber, H.; Beyrau, F.; Dragomirov, P.; Schaub, M. Measurement of a Direct Water-Gasoline-Emulsion-Injection. In Proceedings of the Ilass Europe 28th European Conference on Liquid Atomization and Spray Systems, Valencia, Spain, 6–8 September 2017; pp. 330–338.
14. Li, A.; Zheng, Z.; Peng, T. Effect of water injection on the knock, combustion, and emissions of a direct injection gasoline engine. *Fuel* **2020**, *268*, 117376. [CrossRef]
15. Dittmann, P.; Dörksen, H.; Steiding, D.; Kremer, F. Influence of Micro Emulsions on Diesel Engine Combustion. *MTZ Worldw.* **2015**, *76*, 38–44. [CrossRef]
16. Marchitto, L.; Calabria, R.; Tornatore, C.; Bellettre, J.; Massoli, P.; Montillet, A.; Valentino, G. Optical investigations in a CI engine fueled with water in diesel emulsion produced through microchannels. *Exp. Therm. Fluid Sci.* **2018**, *95*, 96–103. [CrossRef]
17. Tornatore, C.; Bozza, F.; De Bellis, V.; Teodosio, L.; Valentino, G.; Marchitto, L. Experimental and numerical study on the influence of cooled EGR on knock tendency, performance and emissions of a downsized spark-ignition engine. *Energy* **2019**, *172*, 968–976. [CrossRef]
18. Belkadi, A.; Tarlet, D.; Montillet, A.; Bellettre, J.; Massoli, P. Water-in-oil emulsification in a microfluidic impinging flow at high capillary numbers. *Int. J. Multiph. Flow* **2015**, *72*, 11–23. [CrossRef]
19. De Bellis, V.; Bozza, F.; Teodosio, L.; Valentino, G. Experimental and Numerical Study of the Water Injection to Improve the Fuel Economy of a Small Size Turbocharged SI Engine. *SAE Int. J. Engines* **2017**, *10*, 550–561. [CrossRef]

applied sciences — MDPI

Article

Combustion and Performance Evaluation of a Spark Ignition Engine Operating with Acetone–Butanol–Ethanol and Hydroxy

Wilson Guillin-Estrada [1], Daniel Maestre-Cambronel [2], Antonio Bula-Silvera [1], Arturo Gonzalez-Quiroga [1] and Jorge Duarte-Forero [2,*]

[1] UREMA Research Unit, Universidad del Norte, Km. 5 Vía Puerto Colombia, Puerto Colombia, Barranquilla 080007, Colombia; wguillin@uninorte.edu.co (W.G.-E.); abula@uninorte.edu.co (A.B.-S.); arturoq@uninorte.edu.co (A.G.-Q.)

[2] KAÍ Research Unit, Universidad del Atlántico, Carrera 30 Número 8-49, Puerto Colombia, Barranquilla 080007, Colombia; dmaestre@est.uniatlantico.edu.co

* Correspondence: jorgeduarte@mail.uniatlantico.edu.co; Tel.: +57-5385-3002

Citation: Guillin-Estrada, W.; Maestre-Cambronel, D.; Bula-Silvera, A.; Gonzalez-Quiroga, A.; Duarte-Forero, J. Combustion and Performance Evaluation of a Spark Ignition Engine Operating with Acetone–Butanol–Ethanol and Hydroxy. *Appl. Sci.* **2021**, *11*, 5282. https://doi.org/10.3390/app11115282

Academic Editors: Cinzia Tornatore and Luca Marchitto

Received: 31 March 2021
Accepted: 7 May 2021
Published: 7 June 2021

Abstract: Alternative fuels for internal combustion engines (ICE) emerge as a promising solution for a more sustainable operation. This work assesses combustion and performance of the dual-fuel operation in the spark ignition (SI) engine that simultaneously integrates acetone–butanol–ethanol (ABE) and hydroxy (HHO) doping. The study evaluates four fuel blends that combine ABE 5, ABE 10, and an HHO volumetric flow rate of 0.4 LPM. The standalone gasoline operation served as the baseline for comparison. We constructed an experimental test bench to assess operation conditions, fuel mode, and emissions characteristics of a 3.5 kW-YAMAHA engine coupled to an alkaline electrolyzer. The study proposes thermodynamic and combustion models to evaluate the performance of the dual-fuel operation based on in-cylinder pressure, heat release rate, combustion temperature, fuel properties, energy distribution, and emissions levels. Results indicate that ABE in the fuel blends reduces in-cylinder pressure by 10–15% compared to the baseline fuel. In contrast, HHO boosted in-cylinder pressure up to 20%. The heat release rate and combustion temperature follow the same trend, corroborating that oxygen enrichment enhances gasoline combustion. The standalone ABE operation raises fuel consumption by around 10–25 g • kWh^{-1} compared to gasoline depending on the load, whereas HHO decreases fuel consumption by around 25%. The dual-fuel operation shows potential for mitigating CO, HC, and smoke emissions, although NOx emissions increased. The implementation of dual-fuel operation in SI engines represents a valuable tool for controlling emissions and reducing fuel consumption while maintaining combustion performance and thermal efficiency.

Keywords: acetone–butanol–ethanol; dual-fuel operation; electrolyzer; emissions levels; hydroxy gas; spark ignition engine

1. Introduction

The escalation of the world population and the unprecedented trend of energy consumption represent significant challenges of the current century. The massive utilization of internal combustion engines (ICE) has led to worldwide modernization while supporting the current living standards; however, such high-scale utilization has also resulted in uncontrolled fossil fuel consumption and alarming environmental pollution [1–3]. The latter represents a complex problem since ICEs play a central role in different sectors such as transportation, agriculture, power generation, and industry, thus setting intensified pressure on pollution deceleration [4–6]. Governmental and international organizations have made a tremendous effort to potentialize global energy transition to renewables while simultaneously setting restrictions in various sectors to mitigate the rate of greenhouse emissions [7,8].

Appl. Sci. **2021**, *11*, 5282. https://doi.org/10.3390/app11115282 — https://www.mdpi.com/journal/applsci

The transition to alternative fuels in both compression ignited (CI) and spark-ignited (SI) engines is a pressing need, which emerges as a feasible solution to promote a more reliable operation, minimize global emissions and reduce fossil-fuel dependence. The investigation around CI engines is extensive and diverse as more than 144 biodiesel blends have been reported in the literature with promising results for a reliable and cleaner operation [9].

In the same way, the implementation of alternative fuels in SI engines has been broadly discussed. In this scenario, there is an inclined trend towards implementing oxygenated compounds such as bioethanol and biobutanol. Biobutanol offers different advantages as it can be blended in gasoline at relatively high mixing ratios without modifying the engine functionality. However, the main drawback of biobutanol is the high energy consumption associated with its production, which is based on the acetone–butanol–ethanol (ABE) fermentation process. Therefore, the direct implementation of ABE is a more reliable option from a techno-economic viewpoint. The utilization of ABE and other alcohol additives in SI engines has already been documented in the literature.

For instance, Masum et al. [10] investigated the influence on overall emissions and combustion performance of partial fuel substitution in SI engines with oxygenated fuels such as methanol, butanol, and pentanol in a volumetric replacement of 20%. The engine torque was maximized by using all the fuel blends, whereas emissions levels were minimized moderately. Similarly, Yacoub et al. [11] experimentally evaluated the overall performance of mixing gasoline with different straight-chain alcohol chains from methanol to pentanol (C1–C5) in a SI engine. This study unravels the importance of achieving optimal operating conditions within the engine to guarantee CO and HC emissions minimization. In contrast, the NOx emissions presented both upward or downward trends depending on the engine operating conditions. Nithyanandan et al. [12] examined the overall performance of ABE solution in different volumetric ratios, namely 3:6:1, 6:3:1, and 5:14:1. The study outlines the predominant role of increasing acetone content (6:3:1) as the brake thermal efficiency is improved since the combustion phasing resembles that of pure gasoline.

In the same vein, different perspectives in ethanol implementation have been derived from promoting sustainable operation in ICEs. Di Blasio et al. [13] implemented advanced optical methodologies to analyze the main structural and chemical characteristics of soot particles emitted by ethanol-fueled engines. The results demonstrated the low incidence of ethanol incorporation on the quality and nanostructure of soot emissions. Likewise, Gargiulo et al. [14] outlined that ethanol fumigation positively impacts the greenhouse emissions levels while reducing the concentration of emitted particles. Beatrice et al. [15] revealed the central role of engine calibration, pilot injection, and rail pressure to optimize the benefits of ethanol towards emissions minimization and higher thermal efficiency in CI engines. Similarly, Vassallo et al. [16] allocated the pressing need for research on advanced injection systems as a concrete driver of CO_2 emissions targets in the future state of ICEs while maintaining high power density. Belgiorno et al. [17] elaborated on recent advances that integrate gasoline partially premixed combustion in Euro 5 diesel engines. The study focused on describing the effects of appropriate calibration parameters, pilot quantity, and exhaust gas recirculation to maximize thermal efficiency and reduce global pollutants.

On the other hand, hydrogen technology gradually becomes a prominent candidate as an energy carrier that can promote an enhanced operation in both CI and SI engines based on environmental and operational perspectives. Hydrogen production is primarily led by gas reforming technologies representing nearly 60% of the global production [18]. Nonetheless, the carbon footprint of reforming-based production schemes features several challenges to contribute to greenhouse emissions minimization. Therefore, the role of hydrogen production via water-splitting and biomass technologies will be of increased interest in the mid-long term of the hydrogen market [19]. The continuous research on electrolyzers has facilitated the construction of sophisticated and feasible components that maintain proper operation, high-purity reactant agents, and reasonable production rates [18–20].

Therefore, since water electrolysis produces hydrogen and oxygen, the gaseous fuel enrichment in ICEs can be performed either with standalone hydrogen operation and hydroxy gas (HHO).

Shivaprasad et al. [21] experimentally evaluate the influence of hydrogen doping from 5% to 25% in a single-cylinder. Increasing hydrogen replacement increases the in-cylinder pressure while minimizing both HC and CO emissions; however, the adverse effect of such implementation was the intensification of NOx formation. Ismail et al. [22] envisioned HHO enrichment as a secondary fuel in SI engines encountering promising results towards enhancing thermal efficiency and power output and decreasing emissions. Yilmaz et al. [23] revealed that a constant volumetric HHO enrichment in the engine triggers adverse effects in power output, fuel consumption, and emissions levels. Therefore, the authors implemented a hydroxy control unit to control the volumetric rates of gaseous fuel replacement via voltage and current variations to guarantee the optimal rate based on the engine operation. In this way, they managed to reduce fuel metrics, overall emissions and enhance thermal performance.

The main contribution of this investigation is to evaluate the performance of the dual-fuel operation in spark ignition (SI) engines while simultaneously implementing hydroxy (HHO) gas enrichment and acetone–butanol–ethanol (ABE) as additive. The study incorporates evaluation metrics based on combustion performance, thermal efficiency, fuel consumption, and emissions levels. The novelty of this paper relies on the incorporation of a complete methodology to predict combustion performance and energy/exergy distributions. This study examines a combined fuel operation mode in SI, which has not drawn sufficient attention in published studies. In the development of the experimental assessment, ABE is used in different volumetric ratios, namely 5% (ABE 5) and 10% (ABE 10), whereas hydroxy gas is implemented as gaseous fuel in volumetric flow additions of 0.4 LPM. Moreover, the study includes a complete characterization of the experimental test bench, hydroxy generation system, instrumentation, and measuring uncertainty. Therefore, this work represents a further effort on closing the knowledge gap in the implementation of alternative fuels in SI engines while pinpointing experimental and numerical guidelines to evaluate the performance of dual-fuel technologies.

This paper is structured as follows: Section 2 outlines the main features of the experimental test bench, tested fuels, instrumentation characteristics, and describes the constitutive formulation of the combustion and thermodynamic modeling. Section 3 provides the core findings while critically discussing the outcomes. Finally, Section 4 states the concluding remarks while describing the limitations and future avenues in the field.

2. Materials and Methods

2.1. Experimental Test Bench

The experiments were performed in a 4T, naturally aspirated spark-ignition engine (model MZ175, YAMAHA®). The engine has a volumetric capacity of 171 cm^3 and a compression ratio of 8.5:1. It is worth discussing the relevance of the volumetric capacity in ICEs since it provides a clear perspective of the context of the present application. In essence, this matter is essential, considering that international and governmental regulations concerning greenhouse emissions in the transportation sector are based on the volumetric capacity. The typical taxation margin is classified as low (<1000 cc), middle (1200–1500 cc), and high (>1500 cc) capacity [24]. Note that the engine used in this study is intended for power generation applications. However, its operational characteristics resemble those in commercial vehicles, extending the impact spectrum of the proposed dual-fuel technology. Table 1 lists the main features of the SI engine.

Table 1. Specifications of the gasoline engine.

Specification	Value
Engine type	4T OHV
Max. Power	3.5 kW
Bore × Stroke	66 × 50 mm
Max. Torque	10.5 Nm/2400 rpm
Compression ratio	8.5:1
Fuel capacity	4.5 L
Ignition system	T.C.I
Displacement	171 cc

A picture of the test bench is shown in Figure 1. The experimental setup consists of the SI engine, hydroxy generation system, and DAQ system, as shown in Figure 2. Firstly, the SI engine is integrated into a dynamometer to control the load condition. A crankshaft angle sensor (Beck Arnley 180–042) allows measuring engine speed. The in-cylinder pressure is measured with a piezoelectric transducer (KISTLER type 7063-A) placed in the cylinder head. The engine fuel consumption rate is obtained via a scale (OHAUS PA313) and a chronometer. On the other hand, intake airflow is measured employing a hot-wire type mass sensor (BOSCH 22,680 7J600). Additionally, the temperatures of the exhaust gases were measured using K-type thermocouples. Lastly, the measurement of CO, NOx, and HC emissions was carried out using two different gas analyzers, namely BrainBee AGS-688 and PCA® 400. An additional gas analyzer (BrainBee OPA-100) measured the opacity levels of the exhaust gases. The measuring instruments were integrated into a data acquisition system that processes the output data. Table 2 lists the main features of the measuring instruments of the test bench.

Figure 1. Experimental test bench.

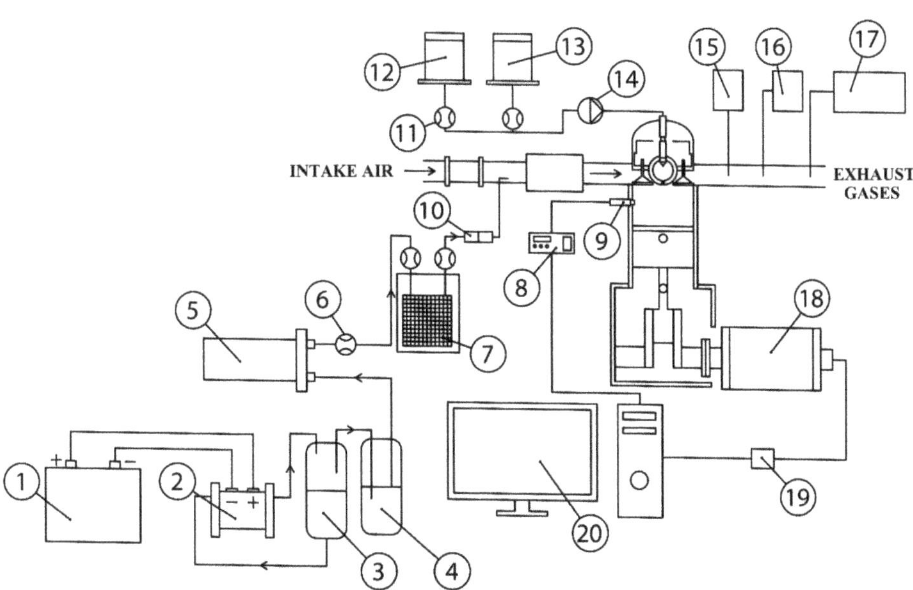

Figure 2. Schematic representation of the experimental test bench. (1) AC-DC converter; (2) electrolyzer cell; (3) electrolytic tank; (4) bubbler; (5) HHO storage tank; (6) HHO flowmeter; (7) flame arrester; (8) charge amplifier; (9) pressure sensor; (10) silica gel filter; (11) flowmeter; (12) gasoline tank; (13) ABE tank; (14) fuel pump; (15) BrainBee AGS-688 emission gas; (16) PCA 400 emission gas analyzer; (17) opacimeter BrainBee OPA-100; (18) alternator; (19) encoder; (20) data acquisition (DAQ) system.

Table 2. Specifications of measuring instruments.

Parameter	Instrument	Manufacturer	Range
Cylinder pressure	Piezoelectric transducer	KISTLER type 7063-A	0–250 bar
Airflow	Air mass sensor	BOSCH OE-22680 7J600	0–125 g/s
Angle	Crankshaft angle	Beck Arnley 180–0420	5–9999 RPM
Fuel measuring	Gravimetric meter	OHAUS-PA313	0–310 g
HHO gas flow	HHO flow rate	GT-556-MTR-ICV	0–3 LPM
Temperature	Temperature sensor	Type K	-200–1370 °C
CO	Exhaust gas analyzer	BrainBee AGS-688	0–9.99%
HC	Exhaust gas analyzer	BrainBee AGS-688	0–9999 ppm
NOx	Exhaust gas analyzer	PCA-400	0–3000 ppm
Smoke opacity	Exhaust gas analyzer	BrainBee OPA-100	0–99.9%

The hydroxy gas addition was carried out by means of an HHO gas generator installed in the engine intake system (see Figure 2). Hydroxy gas is obtained through a dry cell made of stainless steel, with the ability to withstand high temperatures and currents. Additionally, this type of material does not cause a chemical reaction with the electrolytic substance. To improve cell performance, KOH was used as a catalyst at a concentration of 20% (gram of solute/volume of solution). This allows improving the conductivity of the dry cell.

An electrolytic tank constantly supplies a flow of water to the dry cell to maintain constant hydroxy production. Additionally, a bubbler tank was installed to retain the water content in the hydroxy gas. Two flame arresters and a silica gel filter were installed to prevent flashback.

The measurement uncertainty results from various factors such as measuring instrumentation, calibration, and external environmental conditions [25]. The type A evaluation method, which comprises the statistical evaluation of a series of measurements, was employed. The type A method calculates the best estimate (b_i) of a set of measurements (x_1, x_2, x_3, ... x_n) using Equation (1):

$$b_i = \overline{x} = \frac{1}{n} \bullet \sum_{i=1}^{n} x_i \tag{1}$$

In the uncertainty model, the standard deviation (S) assists in calculating the error dispersion of a set of measurements (x_1, x_2, x_3, ... x_n) as expressed in Equation (2).

$$S = \sqrt{\frac{1}{n-1} \bullet \sum_{i=1}^{n} (x_i - \overline{x})^2} \tag{2}$$

Lastly, the model uses the standard uncertainty to measure the mean experimental standard deviation $u(x_i)$ from the measurements.

$$(x_i) = \frac{1}{\sqrt{n}} \bullet \sqrt{\frac{1}{n-1} \bullet \sum_{i=1}^{n} (x_i - \overline{x})^2} \tag{3}$$

where n refers to the number of repetitions in the measurements.

Our study set a total of five repetitions ($n = 5$) in the experiments for each operational variable following previous studies [26]. Table 3 shows the uncertainty associated with each operational variable.

Table 3. Measurement uncertainty of measured variables.

Variable	Uncertainty (%)
Pressure chamber	±0.4
Air mass	±1.2
Crankshaft angle	±1.1
Gravimetric meter	±1.2
HHO flow rate	±1.0
CO_2	±1.1
HC	±1.5
Smoke opacity	±2.0
NOx	±1.5
Total uncertainty	±3.8

2.2. Tested Conditions and Fuel Characteristics

Table 4 lists the properties of the gasoline and the ABE blend oxygenates. Here we implemented ABE additive in a mixing ratio of 3:6:1, which has been implemented in SI engines [27].

The study also used two different alcohol blends, namely ABE 5 and ABE 10, representing the replacement rate in gasoline fuel. Table 5 shows the main properties of these blends.

Table 4. Properties of the fuels used in this study [28].

Parameter	Units	Gasoline	Acetone	Butanol	Ethanol
Chemical formula	-	$C_4 - C_{12}$	C_3H_6O	C_4H_9OH	C_2H_5OH
LHV	$(MJ \bullet kg^{-1})$	43.4	29.6	33.1	26.8
Density	$(kg \bullet m^{-3})$	737	788	810	789
Vaporization latent heat	$(kJ \bullet kg^{-1})$	440	518	716	904
Autoignition temperature	$(°C)$	300	465	343	420
Laminar flame speed	$(cm \bullet s^{-1})$	33	34	48	39

Table 5. Properties of the ABE blends.

Parameter	Units	ABE5	ABE10
LHV	$(MJ \bullet kg^{-1})$	42.79	42.20
Density	$(kg \bullet m^{-3})$	740.38	743.76
Latent Vaporization heat	$(kJ \bullet kg^{-1})$	451.77	463.54
Autoignition temperature	$(°C)$	304.36	308.73
Laminar flame velocity	$(cm \bullet s^{-1})$	33.49	33.99

On the other hand, hydroxy doping has been implemented directly in the intake air system, and the volumetric flow replacement follows the methodology proposed by Ismail et al. [29]. The latter states that a suitable gas substitution rate is 0.25 LPM for an engine capacity of 1000 cm^3. Therefore, this study used a 0.04 LPM replacement rate. The main properties of the hydroxy gas that serves within the modeling are: density $(0.49 \text{ kg} \bullet m^{-3})$ and the LHV $(21.99 \text{ MJ} \bullet kg^{-1})$. The experimental assessment alternates the dual-fuel operation in the engine to relate the influence of hydroxy and ABE compounds. A series of 15 runs were established, as shown in Table 6.

Table 6. Nomenclature and composition of fuels.

Test	RPM	Load (%)	Fuel Mixture Composition	Symbology
1			100% Gasoline	G
2			95% Gasoline + 5% ABE	ABE5
3		50	90% Gasoline + 10% ABE	ABE10
4			95% Gasoline + 5% ABE + 0.04 LPM Hydroxy	ABE5 + HHO
5			90% Gasoline + 10% ABE + 0.04 LPM Hydroxy	ABE10 + HHO
6			100% Gasoline	G
7			95% Gasoline + 5% ABE	ABE5
8	2400	75	90% Gasoline + 10% ABE	ABE10
9			95% Gasoline + 5% ABE + 0.04 LPM Hydroxy	ABE5 + HHO
10			90% Gasoline + 10% ABE + 0.04 LPM Hydroxy	ABE10 + HHO
11			100% Gasoline	G
12			95% Gasoline + 5% ABE	ABE5
13		100	90% Gasoline + 10% ABE	ABE10
14			95% Gasoline + 5% ABE + 0.04 LPM Hydroxy	ABE5 + HHO
15			90% Gasoline + 10% ABE + 0.04 LPM Hydroxy	ABE10 + HHO

2.3. Fundamentals of the Combustion and Thermodynamic Models

The fundamental formulation implemented in this paper represents a simplified model of the physical phenomena due to required modeling assumptions. First, all combustion gases in all stages follow ideal gas behavior [30]. Moreover, the flame propagation speed is considered to operate below the supersonic condition, assuming a uniform pressure in the combustion chamber [31]. The combustion reactants are assumed in stoichiometric amounts, considering that a significant part of the combustion results from diffusion interactions. Similarly, reactant's properties are calculated using an average temperature in the combustion chamber, which entails thermal stabilization as a result of the diffusion process. Finally, the model accounts for heat exchange interactions in the cylinder liner to reinforce the model prediction capabilities.

As a first insight, the model establishes the first law of thermodynamics for an open system, which constitutes the combustion chamber. In essence, this model enables the characterization of the heat release curves as a function of the engine operating conditions. Accordingly, Equation (4) gives an energy balance for the control volume neglecting macroscopic effects [32]:

$$\frac{dU}{d\theta} = \frac{dQ}{d\theta} - \frac{dW}{d\theta} + \sum_i \frac{dH_i}{d\theta} \tag{4}$$

where U refers to internal energy, Q and W represent heat and mechanical work, respectively. H refers to the system enthalpy, and θ relates to the crank angle.

The heat release rate (H_{RR}) can be expressed as defined in Equation (5):

$$H_{RR} = \frac{m_{comb} \bullet C_v \bullet \frac{dT}{d\theta} + P \bullet \frac{dV}{d\theta} + R \bullet T \bullet \frac{dm_{bb}}{d\theta} + \frac{dQ_r}{d\theta} - \frac{dm_{fuel}}{d\theta} \bullet (h - u)}{m_{comb} \bullet LHV} \tag{5}$$

where m_{comb}, m_{fuel}, and m_{bb} refer to the mass of combustion gases, fuel, and blow-by gas, respectively. Additionally, R, h, u, V, and T represent the ideal gas constant, specific enthalpy, specific internal energy, volume, and temperature, respectively, which are parameters that assist in determining the thermodynamic state of the fuel mixture. Lastly, Q_r, C_v, and LHV refer to rejected heat, specific heat at constant volume, and lower heating value, respectively.

2.3.1. Calculation of Combustion Gases Properties

Firstly, the average temperature inside the combustion chamber is calculated via Equation (6), which includes the universal gas constant (R). The latter is calculated via Equation (7).

$$T = \frac{P \bullet V}{m_{comb} \bullet R} \tag{6}$$

$$R = X_{air} \bullet R_{air} + X_{st} \bullet R_{st} + X_g \bullet R_g \tag{7}$$

where R_{air}, R_{st} and R_g refer to the air gas constants of air, stoichiometric combustion, and gaseous fuel, respectively. Similarly, X_{air}, X_{st} and X_g correspond to the mass fraction of the gases mentioned above.

On the other hand, the specific heat ratio of combustion gases uses the Zucrow and Hoffman correlation [33], as described in Equation (8).

$$\gamma(T) = 1.46 - 1.63 \bullet 10^{-4} \bullet T + 4.14 \bullet 10^{-8} \bullet T^2 \tag{8}$$

Subsequently, the specific heat at constant volume of the combustion gases is given by Equation (9):

$$C_v(T) = \frac{R}{\gamma(T) - 1} \tag{9}$$

Lastly, both the specific enthalpy and internal energy are calculated as a function of the specific heat ratio, which is temperature-dependent, as shown in Equations (10) and (11), respectively.

$$h(T) = R \int \frac{\gamma(T)}{\gamma(T) - 1} \, dT \tag{10}$$

$$u(T) - R \int \frac{1}{\gamma(T) - 1} \, dT \tag{11}$$

2.3.2. Blow-by Gas Losses

As previously discussed, the blow-by gas losses phenomena account for a significant share of energy losses. Therefore, incorporating such effects in the thermodynamic model becomes a determinant factor in predicting the operating conditions [34]. Hence, the study implements the formulation introduced by Irimescu [35] to predict the energy losses derived from exhaust gas leakage inside the combustion chamber as described in Equations (12) and (13).

$$\frac{dm_{bb}}{d\theta} = \frac{P_{ext} \bullet A_v \bullet C_D}{N \bullet (R \bullet T_{ext})^{1/2}} \bullet \left(\frac{P_{int}}{P_{ext}}\right)^{1/\gamma(T)} \bullet \left[2 \bullet \frac{\gamma(T)}{\gamma(T)-1} \bullet \left(1 - \left(\frac{P_{int}}{P_{ext}}\right)^{\frac{\gamma(T)-1}{\gamma(T)}}\right)\right]^{1/2} \tag{12}$$

$$\frac{dm_{bb}}{d\theta} = \frac{P_s \bullet A_v \bullet C_D}{\gamma(T)^{1/2} \bullet N \bullet (R \bullet T_{ext})^{1/2} \bullet \left[1 - \left(\frac{2}{\gamma(T)+1}\right)^{\frac{\gamma(T)+1}{2\bullet(\gamma(T)-1)}}\right]}$$

$$if, \frac{P_{int}}{P_{ext}} \leq \left(\frac{2}{\gamma(T)-1}\right)^{\frac{\gamma(T)-1}{\gamma(T)}} \tag{13}$$

where P_{ext} and P_{int} represent the exhaust and intake flow pressures, respectively. N refers to the engine speed. C_D and A_v account for the discharge coefficient and engine valve area calculated via Equations (14) and (15), respectively.

$$C_D = \frac{\dot{m}_{fuel}}{\dot{m}_t} \tag{14}$$

$$A_v = \frac{\pi \bullet D_v{}^2}{4} \tag{15}$$

where $\dot{m}_{fuel}$ is the experimental mass flow of the inlet valve and $\dot{m}_t$ is the theoretical mass flow rate, calculated considering a constant compressible flow through a valve orifice. Besides, D_v is the diameter of the valve and amounts to 28 mm.

2.3.3. Rejected Heat

The thermodynamic model used Equation (16) to predict the heat transfer rate from the combustion gases to the combustion chamber walls.

$$\frac{dQ_r}{d\theta} = \frac{h_{wall} \bullet A_{wall} \bullet (T - T_{wall})}{2 \bullet \pi \bullet N} \tag{16}$$

where T_{wall} and A_{wall} are the temperature and the wall surface area. Besides, h_{wall} is the heat transfer coefficient, calculated via the correlation proposed by Woschni [36] as given by Equation (17):

$$h_{wall} = 3.26 \bullet (w \bullet P)^{0.8} \bullet T^{-0.55} \bullet b^{-0.2} \tag{17}$$

where b accounts for the internal diameter of the combustion chamber and w represents the average velocity of the combustion gases. The latter is calculated via Equation (18).

$$w = k_1 \bullet (2 \bullet N \bullet S_t) + k_2 \bullet \frac{T_o \bullet V_d \bullet (P - P_m)}{P_o \bullet V_o} \tag{18}$$

Subscript "o" relates the initial state for volume (V_o), pressure (P_o) and temperature (T_o). P and P_m are the actual and mean pressure in the combustion chamber. k_1 and k_2 are model constants, whose values are defined as 2.30 and 3.25×10^{-3}, respectively. Finally, S_t represents the engine stroke.

2.3.4. Combustion Chamber Volume

Here we define the combustion chamber volumetric characteristics, which involve different volume contributions that incorporate operational and geometrical patterns to define the instantaneous volumetric displacement within each cycle [37]. Equation (19) gives the instantaneous volume of the combustion chamber.

$$V = V_{mv} + V_{disp} + \bullet V_{dp} + \bullet V_{inf} + \bullet V_c \tag{19}$$

The term V_{mv}, calculated via Equation (20), represents the free space encountered once the piston reaches the top-dead center (TDC).

$$V_{mv} = \frac{\pi \bullet D^2}{4} \bullet \left[\frac{2 \bullet L_{cr}}{r_c - 1} \right] \tag{20}$$

where r_c and D are the compression ratio and the internal diameter of the piston, respectively. L_{cr} represents crankshaft length and V_{disp} the volume displaced by the connecting rod-crank mechanism. The latter is predicted using Equation (21).

$$V_{disp} = \frac{\pi \bullet D^2}{4} \bullet \left[L_{cr} + L_{rod} - R_y \bullet \theta \right] \tag{21}$$

L_{rod} is the longitude of the crankshaft and R_y represents the vertical position of the piston. Consequently, the volumes denoted as ΔV_{dp} and ΔV_{inf} are calculated using Equations (22) and (23), respectively. The former represents the variation of the instantaneous volume induced by pressure-deformation effects imposed by the combustion gases. The latter accounts for the volume related to the inertial forces of the connecting rod-crank shaft mechanism.

$$\Delta V_{dp} = \frac{\pi \bullet D^2 \bullet L_{rod}}{4 \bullet A_c} \bullet \left(\frac{k_{def}}{E_s} \right) \bullet (P \bullet A_p) \tag{22}$$

$$\Delta V_{inf} = \frac{\pi \bullet D^2 \bullet L_{rod}}{4 \bullet A_c} \bullet \left(\frac{k_{def}}{E_s} \right) \bullet (m \bullet a_p) \tag{23}$$

where k_{def}, E_s, A_c, A_p and a_p are the deformation constant, the elastic modulus of steel, connecting rod critical area, the piston cross-sectional area, and piston acceleration, respectively.

Finally, ΔV_c is calculated according to Equation (24) and represents the variation of the instantaneous volume produced by the clearances in the combustion chamber [37].

$$\Delta V_c = -\frac{\pi \bullet D^2}{4} \bullet \sum_{i=1}^{2} (e_i \bullet \sin \varphi_i \bullet \cos \alpha_i) \tag{24}$$

where e represents the eccentricity between journal and bearing, measured along their centerline. Similarly, φ relates to the angle of rotation, and α represents the angle between the connecting rod and the piston.

2.3.5. Energy Distribution and Emissions Processing

The engine inlet heat energy ($\dot{Q}_{int}$) is defined based on the fuel mode, which can be entirely liquid using pure gasoline or oxygenated fuel blends with ABE as indicated in Equation (25), or alternately liquid-gaseous fuel, which constitutes the dual-fuel operation mode given by Equation (26).

$$\dot{Q}_{int} = \dot{m}_{fuel} \bullet LHV_f \tag{25}$$

$$\dot{Q}_{int} = \dot{m}_{fuel} \bullet LHV_{fuel} + \dot{m}_{HHO} \bullet LHV_{HHO} \tag{26}$$

where the subscripts fuel refers to the fuel mixture and HHO represents the hydroxy gas. Similarly, the model defines the exhaust gas energy as follows:

$$\dot{Q}_{exh} = \dot{m}_{exh} \bullet C_{p,exh} \bullet T_{exh} \tag{27}$$

where $C_{p,exh}$ represents the specific heat capacity of the combustion gases at constant pressure. On the other hand, the mechanical efficiency of the engine according to the fuel mode is defined as follows.

Liquid fuel standalone operation:

$$\eta_{mech} = \frac{PW}{\dot{m}_{fuel} \bullet LHV_{fuel}} \bullet 100 \tag{28}$$

Dual-fuel mode:

$$\eta_{mech} = \frac{PW}{\dot{m}_{fuel} \bullet LHV_{fuel} + \dot{m}_{HHO} \bullet LHV_{HHO}} \bullet 100 \tag{29}$$

where PW is the power output of the engine calculated according to Equation (30).

$$PW = \frac{2 \bullet \pi \bullet \omega \bullet T_r}{60 \bullet 1000} \tag{30}$$

where T_r is the torque condition, and ω refers to the engine angular speed. Lastly, this section concludes with the calculation of the unit conversion of the overall emissions.

On the other hand, the measuring instrumentation for emissions levels is commonly reported in ppm and %vol, which are the default characteristics of gas analyzers. However, it is necessary to apply further processing to display these results according to the international standards (i.g. European legislation) that describe pollutants in terms of $g \bullet km^{-1}$ for light-dutty and passenger vehicles and $g \bullet kWh^{-1}$ for heavy-duty vehicles [38]. Therefore, the study converts the output data from the gas analyzers into $g \bullet kWh^{-1}$ according to the empirical correlations proposed by Heseding and Daskalopoulos [39]. The relevance of the emissions above processing is that they assist in relating emissions and fuel metrics, facilitating comparison. The correlation for emissions processing is based on the following formulation:

$$EP_i = EV_{i,dry} \bullet \left(\frac{M_i}{M_{exh,dry}} \bullet k_d \right) = EV_{i,wet} \bullet \left(\frac{M_i}{M_{exh,wet}} \bullet k_w \right) \tag{31}$$

where EP_i, $EV_{i,dry}$ and $EV_{i,wet}$ represents the pollutant mass in the power unit ($g \bullet kWh^{-1}$), exhaust emissions on a dry basis and wet basis, respectively. The term M_i accounts for the molecular mass, while $M_{exh,d}$ and $M_{exh,w}$ relates to the molecular mass of exhaust emissions on a dry and wet basis, respectively. Finally, the terms k_{dry} and k_{wet} are empirical constants with a value of 3.873 $g \bullet kWh^{-1}$ and 4.160 $g \bullet kWh^{-1}$, respectively, and relate the power unit and the exhaust emissions on a dry basis and wet basis, respectively. The conversion of the three primary pollutants treated in experimental evaluation is described in Equation (32) to (34) [39].

$$CO \left[\frac{g}{kWh} \right] = 3.591 \bullet 10^{-3} \bullet CO(\% \, vol) \tag{32}$$

$$NOx \left[\frac{g}{kWh} \right] = 6.636 \bullet 10^{-3} \bullet NOx(ppm) \tag{33}$$

$$HC \left[\frac{g}{kWh} \right] = 2.002 \bullet 10^{-3} \bullet HC(ppm) \tag{34}$$

3. Results and Discussion

3.1. Cylinder Pressure

The experimental evaluation of the dual-fuel operation in the SI engine begins with the characterization of the pressure gradients developed within the combustion chamber. This parameter provides a clear perspective of the fuel mode performance while relating the appropriate mixing interaction between the base fuel (gasoline/ABE), hydroxy gas (gaseous fuel), and air. Figure 3 shows the overall behavior of the in-cylinder pressure at different load conditions. Notice that the standalone gasoline operation has been set as the baseline fuel for comparison purposes.

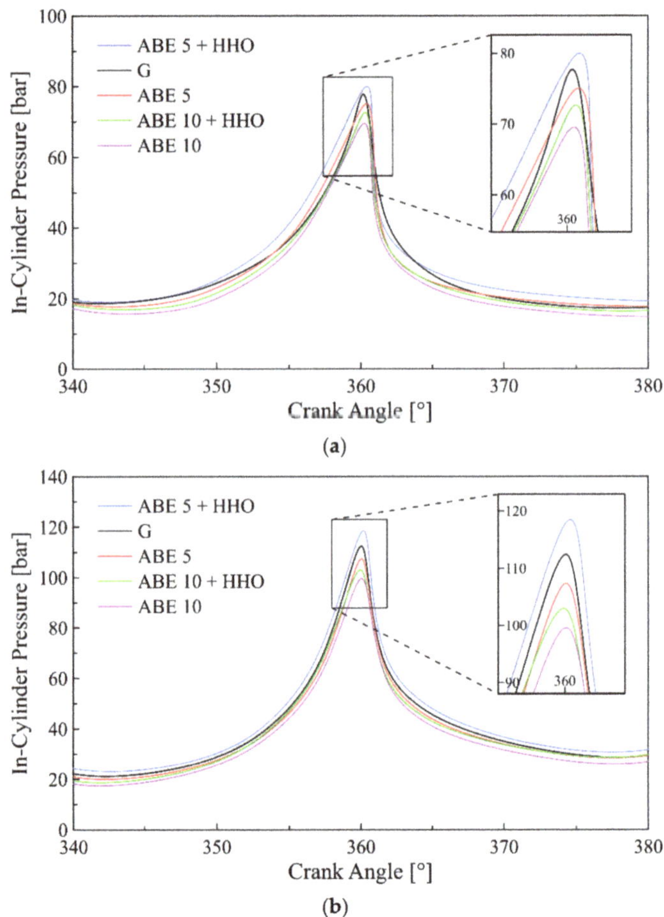

(a)

(b)

Figure 3. *Cont.*

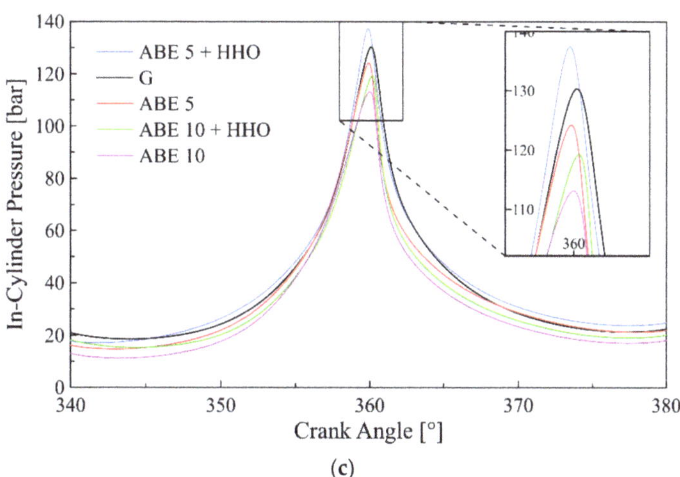

Figure 3. Influence of fuel on cylinder pressure for an engine load of (**a**) 50%, (**b**) 75%, and (**c**) 100%.

According to the results, the maximum in-cylinder pressure within the engine operating range is achieved by the ABE 5 + HHO with values between 60–135 bar. This result is consistent with the magnification of the laminar flame speed of this blend. Results in Figure 3 assist in the evaluation of the combustion phasing from the different fuel blends. Based on Figure 3a, all the blends present a retarded phasing between 0.2°–0.6° compared to the baseline fuel, which implies that a low engine load inhibits the fast combustion for the alternative fuels. The overall pattern encountered in the phasing is consistent with related investigations of ABE at different blend ratios [12]. Next, in Figure 3b, all the tested fuels feature a relatively similar phasing condition. However, at high engine loads (Figure 3c), the enhancement on the laminar flame speed from the oxygenated fuels becomes evident as the combustion phasing is advanced, ranging from 0.8° to 1.6° compared to the baseline, which facilitates the completion of combustion and faster peak pressures achievement even before reaching 360°. The latter is a direct indication of a significant improvement in combustion efficiency and thermal performance.

The pressure curve features an increasing trend as the engine load increases since more fuel is mixed to meet the power demand. Notably, increasing the fuel substitution with ABE (ABE 10) limited the pressure produced during combustion between 30–70 bar, which can be associated with the lower calorific value of this blend. This result agrees with similar investigations related to diesel engine combustion phenomena while operating with biodiesel blends [40]. Another contributor to the reduced pressure of ABE-based blends could be associated with a higher-octane rating that extends the initiation delay and reduces the laminar flame speed. In contrast, hydroxy doping enables a significant improvement in the combustion pressure in both blends (ABE 5/10) up to 75%. This pattern can be explained considering that incorporating hydroxy in the engine facilitates a homogeneous air-fuel mixture. The positive effects on the gasoline octane rating by incorporating HHO in the intake air system can be mentioned as another contributor to the enhanced combustion performance since the compression ratio is maximized. It should be noted that the overall trend of the pressure curves after combustion features a sharp decrement which guarantees that the knocking condition is not reached. On average, the combustion pressure developed in the standalone gasoline operation is higher than that of ABE 5, ABE 10, and ABE 10 + HHO by 12%, 18%, and 24%, respectively. Contrarily, the ABE 5 + HHO shows an enhanced pressure range compared to commercial gasoline between 10–15% for the engine load ranges analyzed.

3.2. Heat Release Rate (H_{RR})

The H_{RR} relates to the fuel conversion efficiency since it shows how much chemical energy is transformed into thermal energy. Figure 4 depicts the H_{RR} at different engine load conditions as a function of the crankshaft angle.

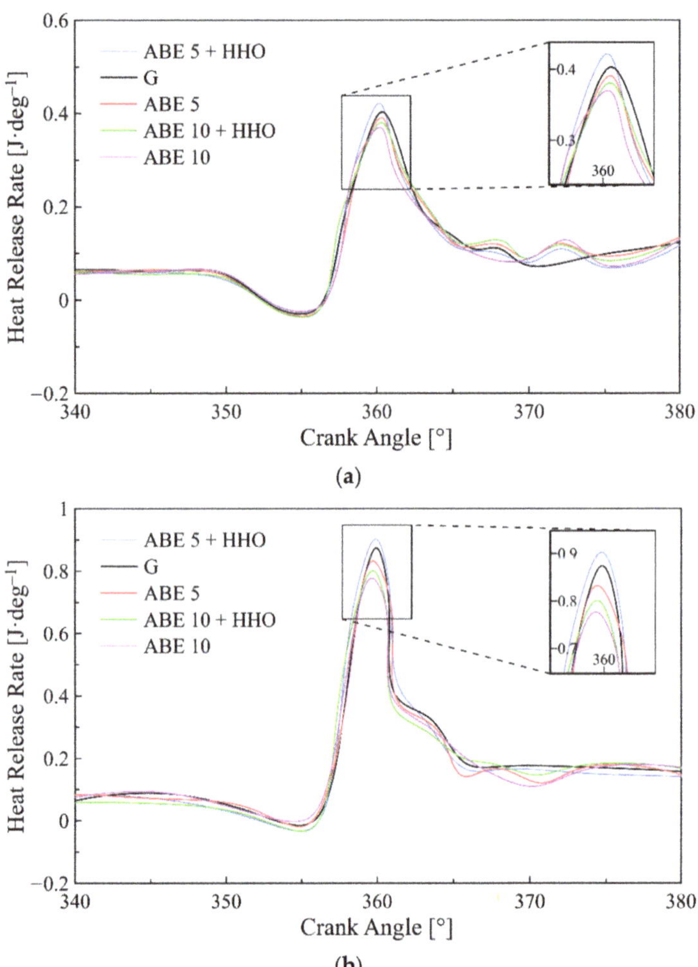

Figure 4. *Cont.*

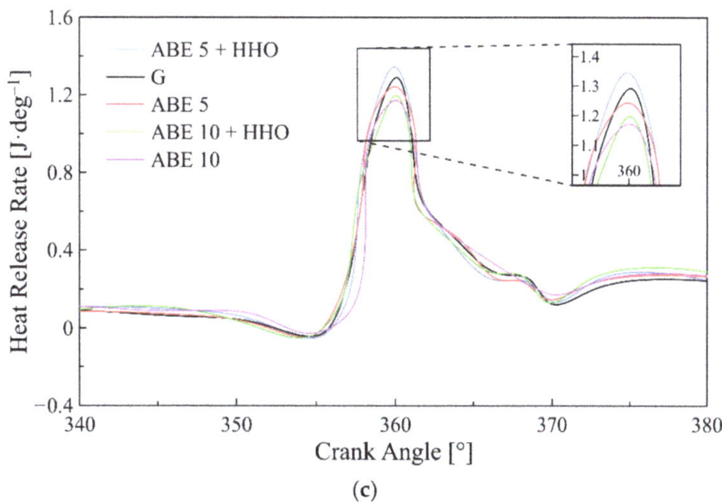

(c)

Figure 4. Influence of fuel on heat release rate for (**a**) 50%, (**b**) 75%, (**c**) 100% of the engine load.

According to the results, the general pattern in all the load conditions states that ABE 5 + HHO features the highest heat release from the fuel blends, followed by gasoline, ABE 5, ABE 10 + HHO, and ABE 10. It can be observed that a higher engine load promotes the increase of heat release in the combustion chamber, which implies extreme conditions and maximum chemical energy conversion. The average reduction of the H_{RR} for the ABE 5 and ABE 10 was 3.5% and 6.78%, respectively, compared to gasoline as the baseline fuel. The latter can be explained considering the higher viscosity of these blends that promote a slower combustion process, thus limiting heat release.

HHO enrichment increases heat release in all the blends, which implies integrating gaseous fuel mitigates the cooling effect associated with utilizing oxygenates (ABE). The higher calorific value and subsequent enhancement in the laminar flame speed resulting from HHO doping can be mentioned as contributors to the enhanced behavior in the heat release that offset the decrement experienced by ABE blends when compared to pure gasoline. It is worth mentioning that the fuel chemical structure supports the enhanced behavior from HHO doping since both hydrogen and oxygen coexist in the air/fuel mixture, whereas gasoline consists of hydrocarbon molecules [41]. Therefore, the gaseous fuel incorporation promotes an improved combustion performance due to the direct interaction of the diatomic molecules that suppress the ignition delay. In this sense, HHO doping also fosters the massive bond-breaking trend of the gasoline molecules, thus facilitating the heat release rate, the laminar flame speed, and subsequently improving combustion efficiency.

3.3. Combustion Chamber Temperature

Figure 5 shows the average temperature of the combustion chamber for the tested fuels. This parameter indicates the ability of a blend for combustion phasing. Notice that the study only presents the temperature distribution for a full load rate, representing the critical condition from the analyzed cases since it holds the highest heat release rate and maximum pressure peaks.

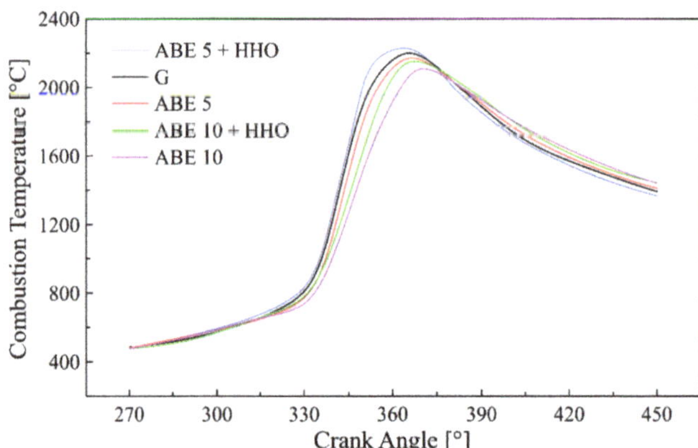

Figure 5. In-cylinder temperature at a full load rate.

According to the results, the maximum temperature is achieved by ABE 5 + HHO, followed by the gasoline, ABE 5, ABE 10 + HHO, and ABE 10. Interestingly, the temperature after combustion presents a reverse trend concerning the blends, which can be associated with the higher heat release rate during combustion, minimizing the temperature in this stage. The maximum temperature reaches a value of 2229 °C for ABE 5 + HHO, supporting the energetic contribution derived from the dual-fuel operation. Moreover, as the ABE content escalates, the combustion temperature drops between 20–45%. This pattern can be explained considering the lower heating value and intensified latent heat of the ABE blends. Notably, increasing the latent heat promotes a temperature drop in the intake stage, resulting in a lower temperature at the end of the compression stage. Moreover, incorporating hydroxy gas in the blends intensifies the temperature due to the higher hydrogen and oxygen content that stimulates the chemical energy conversion of the air/fuel mixture [42]. On average, the combustion chamber temperature of ABE 5 and ABE 10 decreased by 6.21% and 12.23% compared to the gasoline fuel. In contrast, HHO enrichment increases the temperature of ABE 5 and ABE 10 by up to 394 °C and 341 °C, respectively.

3.4. Engine Performance

The study points out the brake specific fuel consumption (BSFC) overall trend in Figure 6 that is driven to examine the fuel mass consumption per power unit while providing a global perspective of this parameter.

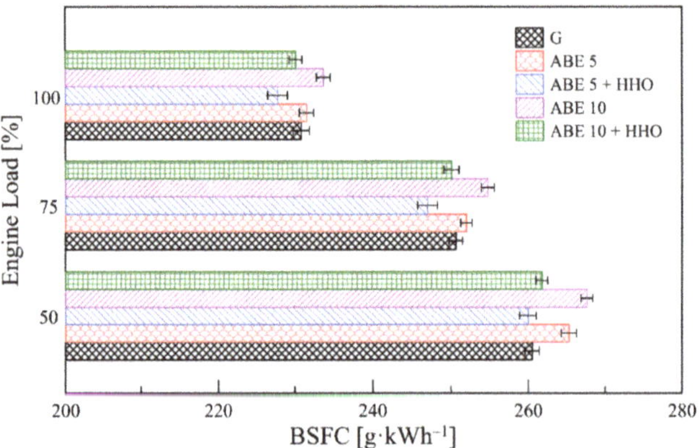

Figure 6. Diagram of Brake Specific Fuel Consumption (BSFC).

According to the results, the implementation of oxygenated blends (ABE 5 and ABE 10) promotes higher fuel consumption when compared to the standalone gasoline operation. The latter is a direct consequence of the lower energy density and lower calorific value relative to gasoline. Therefore, a higher fuel amount is required to obtain the same unit power, which escalates the BSFC. For comparison, increasing the ABE ratio fosters the BSFC up to 15%. This pattern is in line with other investigations [12,43]. Note that the ABE-based blends might influence the octane rating rise, supported by longer ignition delays (Figure 4c) that critically limits the power output, thus magnifying BSFC. The fuel conversion efficiency might be a determinant factor altering the fuel metrics since it was corroborated in the in-cylinder pressure curves that at low engine loads, the combustion center retarded up to 0.6°. Therefore, the altered combustion phasing further deteriorates fuel conversion efficiency, thus increasing fuel consumption. The investigation of Nithyanandan [12] leads to similar findings when implementing blends of ABE 20–40. In this sense, the integration of control strategies that facilitates spark timing could be a feasible solution to avoid the magnification of BSFC in dual-fuel operation [44,45].

The highest fuel consumption was achieved by ABE 10, followed by the ABE 5, Gasoline, ABE 10 + HHO, and ABE 5 + HHO. This result demonstrates that HHO enrichment surpasses the increment in the fuel metrics derived from ABE replacement implementation. The latter implies that gaseous fuel promotes chemical conversion efficiency while acting as a heat intensifier, as verified in Figure 5.

3.5. Emission Characteristics

This section aims to examine the influence of dual-fuel operation on the overall emissions of the SI engine. Reducing greenhouse emissions facilitates an eco-friendly operation in the ICEs. The study mainly examines the pollutant levels of CO, HC, NOx, and smoke opacity while variating fuel operation mode and the engine load.

3.5.1. CO Emissions

First, the overall carbon monoxide emissions (CO) for all the tested fuels are displayed in Figure 7 at different load conditions.

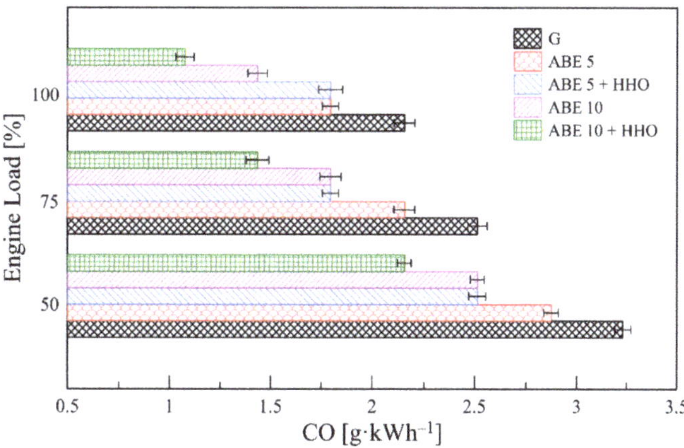

Figure 7. CO emissions for fuels tested.

The load condition features an inverse relation with the CO emission levels. The engine load plays a central role in the emissions behavior since, at high engine loads, it can be reached a non-oxygenated condition that promotes CO formation. However, the high CO formation remains high at low engine loads due to a highly lean mixture that hinders fuel burning. Thus, the flame cannot be maintained due to the limited propagation speed.

Specifically, the ABE implementation reduces CO emissions between 11% to 33% compared to gasoline, depending on the engine load. The HHO implementation enables a further reduction of up to 22%. The enhanced behavior from the fuel blends results from promoting complete combustion, meaning that more CO is converted into CO_2. Additionally, the ABE incorporation facilitates oxidation due to the increased laminar flame speed and the leaning effect of its oxygenated nature. The higher volatility derived from the acetone in the ABE blends further promotes CO minimization. Li et al. [46] encountered that the ABE features a post-flame oxidation trend that reduces CO levels.

The implementation of HHO doping in the intake air minimizes CO levels, which can also be explained considering the direct oxygen enrichment in the air/fuel mixture. CO emission is directly linked to the air/fuel ratio within the engine and the fuel consumption. Hence, since hydroxy enrichment reduces the BSFC (Figure 6), fewer CO levels are evidenced in the exhaust gases.

It is worth underlying the trade-off of the proposed fuel methodology since the CO minimization resulted in intensified CO_2 formation. Considering that international regulations claimed for integral solutions to meet CO_2 standards, it triggers a collateral cost impact in the proposed dual-fuel technology due to the imminent necessity to engage additional methods to promote a sustainable operation in future scenarios. Di Blasio et al. [13] have pointed out the predominant role of advanced fuel injection systems in this goal. Moreover, the latter technology is accompanied by additional improvements in fuel economy and combustion noise that reinforce its implementation. Undoubtedly, implementing such sophisticated fuel injection systems increases the investment cost that can be only supported by substantial fuel savings while maintaining high-efficiency operation. Moreover, considering the effect of combustion phasing, the necessity to control spark timing to promote higher combustion efficiencies requires in-depth exploration of engine design characteristics and fuel injection systems that foster a techno-economic operation.

3.5.2. HC Emissions

Figure 8 shows the overall emissions of hydrocarbons (HC) as a function of the engine load.

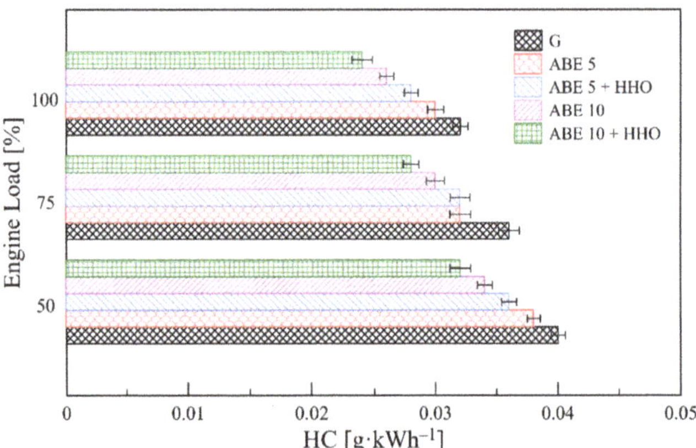

Figure 8. HC emissions for fuels tested.

In terms of unburned hydrocarbons, pure gasoline features the highest emissions levels with nearly $0.042 \text{ g} \bullet \text{kWh}^{-1}$. Dual-fuel operation facilitates HC minimization since ABE reduces emissions levels while hydroxy doping upgrades this share. Noticeably, HC emissions are significantly lower than CO levels but still represents an undesired pollutant that affects human health and air quality. The overall trend of HC emissions can be explained based on the same fundamentals of CO emissions. It is important to note that alcohols additives (ABE) feature improved oxygenated characteristics that further improve combustion efficiency and promote a homogeneous air/fuel mixture, which reduces HC formation. These concluding remarks are consistent with that of Masum et al. [10], which unravels the effect of engine speed on ABE blend overall performance. The enhanced laminar flame speed of ABE and HHO is another contributor supporting HC minimization.

3.5.3. NOx Emissions

Figure 9 shows NOx formation for all the tested fuels as a function of engine load. The increment in combustion temperature sets the appropriate conditions to promote nitrogenates oxidation, thus producing NOx. Therefore, it can be stated that NOx formation depends significantly on the in-cylinder temperature, the concentration of oxygen in the fuel, and the residence time of the reaction.

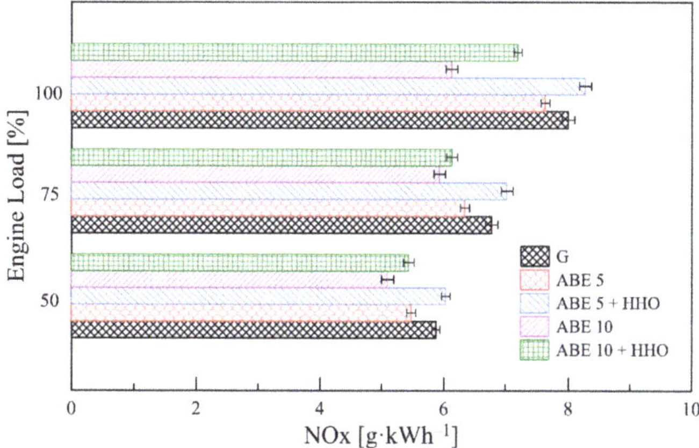

Figure 9. NOx emissions for fuels tested.

Based on the results, the maximum NOx emissions are reached at a full engine load. Specifically, ABE implementation facilitates reducing the overall emissions of NOx, but HHO upgrades this share. The latter implies that the magnification of the in-cylinder temperature provided by the hydroxy doping suppresses the positive effect of the ABE. The enhanced behavior of ABE standalone blends can be attributed to the higher vaporization heat that limits the air-fuel mixture temperature in the intake stage, which reflects on reduced combustion temperature, thus reducing the potential of NOx formation. This pattern is consistent with the experimental outcome of a similar investigation that directly measures the temperature at the intake valve closing [12,46].

The highest NOx emission levels were achieved by the ABE 5 + HHO, followed by pure gasoline, ABE 5, ABE 10 + HHO, and ABE 10. Incorporating the oxygenated compounds in the air/fuel mixture reduces NOx emissions between 0.36 to 1.87 g • kWh^{-1} compared to gasoline. In contrast, the integration of hydroxy doping in the intake air for the ABE 5 case maximizes the emissions levels between 0.16 to 0.27 g • kWh^{-1} but ABE 10 + HHO remains behind the baseline fuel during all the engine loads. The higher oxygen content in the air/fuel mixture can directly contribute to the rise in NOx emissions by incorporating HHO in the fuel blends.

It is essential to mention the trade-off between the effect of HHO doping based on fuel economy and emission levels. Based on the fuel metrics (Figure 6), the integration of hydroxy gas minimizes fuel consumption, which ratifies prospective benefits in terms of fossil-fuel depletion and economic viewpoints. Moreover, since ABE is replacing up to 10% of fossil fuel, the net fuel saving escalates based on a global perspective. In the counterpart, hydroxy doping stimulates NOx formation, as corroborated in this section. A potential solution to mitigate the impact of NOx emissions magnification could be integrating waste heat recovery (WHR) technologies that further enable a higher fuel utilization ratio. The integration of exhaust gas recirculation systems can be another feasible opportunity to promote the sustainable operation of ICEs.

3.5.4. Smoke Emissions

The emissions section concludes with the smoke opacity emission depicted in Figure 10 for the different tested fuels.

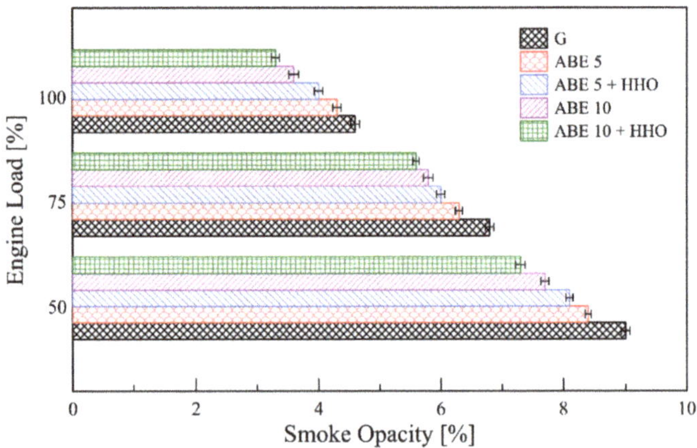

Figure 10. Smoke emission for fuels tested.

According to the results, the emissions levels of smoke opacity are higher at low engine loads. In this sense, as the engine load increases from 50% to 100%, the emissions are reduced up to 54%, demonstrating that smoke emissions depend primarily on the engine operating conditions. Interestingly, the standalone gasoline operation features the highest smoke formation in the exhaust stream. On the contrary, implementing alcohol compounds and HHO doping promotes smoke opacity minimization. This result implies that the extension of the oxygenated conditions limits the emission levels. In other words, it is observed that by only using ABE additive in the blends, a decent reduction of smoke is achieved when compared to the baseline fuel, and the HHO replacement furthers enlarge the smoke drop. Specifically, gasoline presents up to 9% of smoke emissions, followed by ABE 5 between 4.2 to 8.3% and ABE 10 with less than 7.6%. The implementation of hydroxy enrichment in ABE 5 and ABE 10 provides a further reduction of up to 0.3% and 0.4%, respectively.

Exhaust after-treatment technologies have reported salient results towards emissions reduction. However, the main advantage of the proposed dual-fuel methodology relies on the simplified implementation that does not intervene in engine structure and requires negligible modification on its functionality. The latter elucidate clear advantages from a techno-economic viewpoint and the importance of promoting dual-fuel operation in future platforms of ICEs.

3.5.5. Fuel Energy Distribution

This section concludes by examining the fuel energy distribution within the engine for the tested fuels. Accordingly, Figure 11 summarizes the fuel energy allocation based on the power output and exhaust gases directly measured on the experimental assessment. Notice that the energy loss comprises lubrication oil, refrigerant, among other energy losses such as blow-by gas that limit energy conversion [34].

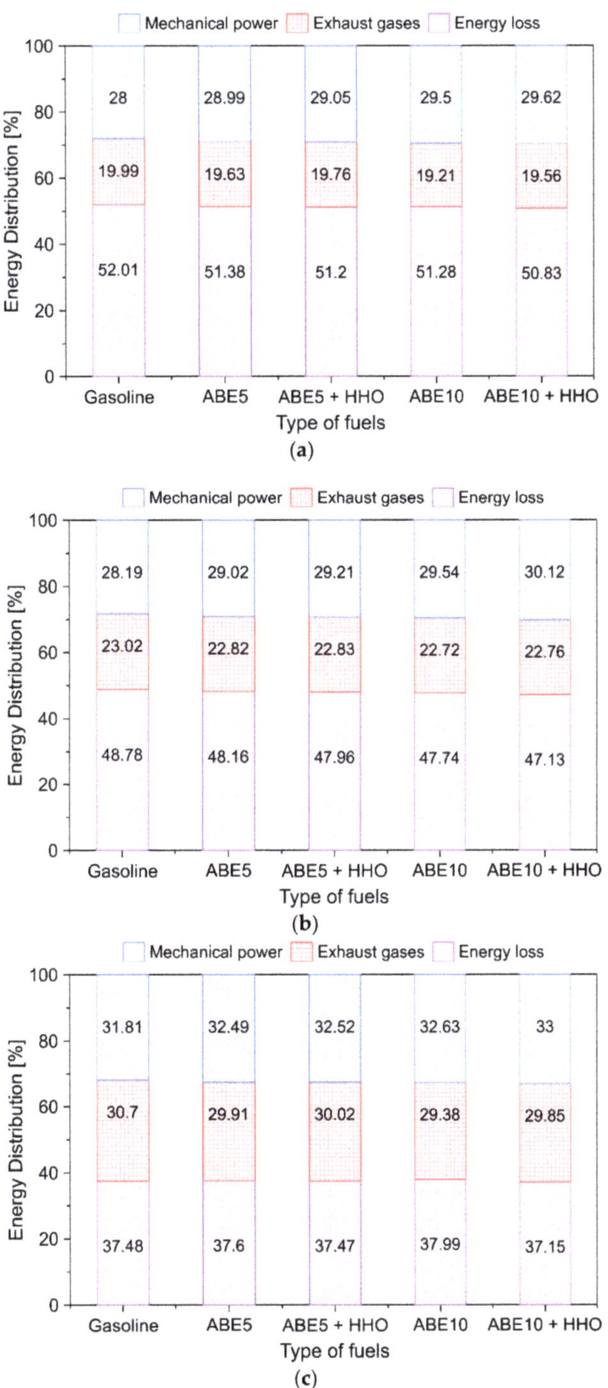

Figure 11. Energy distribution in the SI engine, (**a**) load 50%, (**b**) load 75% and (**c**) load 100%.

According to the results, increasing the engine load intensifies the energy contribution of both the power output and exhaust gases. This result is consistent with WHR applications where higher loads facilitate the fuel utilization ratio. The exhaust streams feature higher temperatures which corroborate the imminent escalation of NOx emission as depicted in Figure 9 [3,5]. However, the negative pattern of increasing the engine load is the significant rise in fuel consumption, as ratified in Figure 6. The results demonstrated that almost half of the fuel energy results in energy losses and other sources. Partial fuel substitution with ABE reduces the heat recovery potential from exhaust gases compared to conventional gasoline, which can be associated with the lower calorific value of this fuel blend.

Overall, ABE 10 + HHO features the best performance on the power output distribution from the tested fuels ranges in all the engine load conditions. An average increment of nearly 2.4% can be achieved compared to the baseline fuel. Contrarily, the maximum heat potential in the exhaust gases is higher in the gasoline fuel, with an improvement between 0.6–0.9% compared to the rest tested fuels. Subsequently, Figure 12 describes the behavior of thermal efficiency for the different fuel blends.

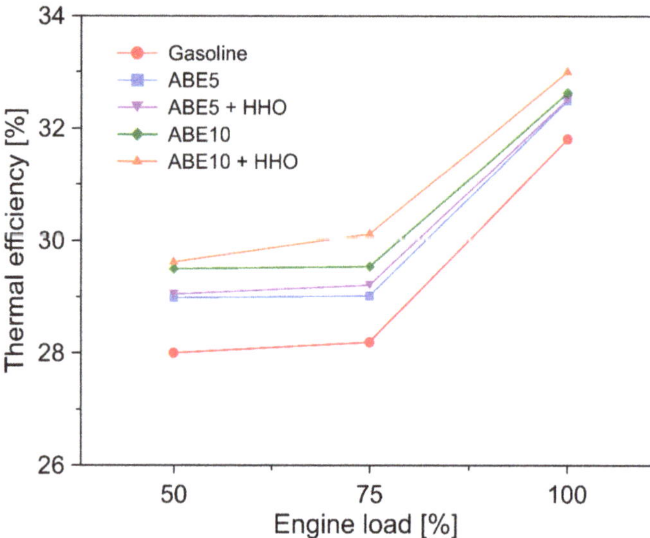

Figure 12. Thermal efficiency for different types of fuels.

According to the results, the thermodynamic efficiency directly relates to the engine load while representing a sensitivity between 4–6%. The latter can be attributed to non-optimized combustion derived from lower H_{RR} and pressure ranges. This pattern is consistent with similar applications in diesel engines operating in dual-fuel mode [15,16]. Moreover, the comparative assessment between fuel blends demonstrates that the addition of ABE in gasoline endorses engine efficiency, while hydroxy doping favors efficiency escalation. For the tested conditions, the maximum thermal efficiency of 31.81%, 32.49%, 32.52%, 32.63%, and 33.01% was obtained when operating with gasoline, ABE5, ABE5 + HHO, ABE10, and ABE10 + HHO, respectively. The above behavior is mainly attributed to a complete combustion process due to the presence of ABE and hydroxy, which is reflected in the reduction of CO emissions (see Figure 7).

Figure 13 shows the exergy distribution of the engine for the different operating conditions and fuel blends.

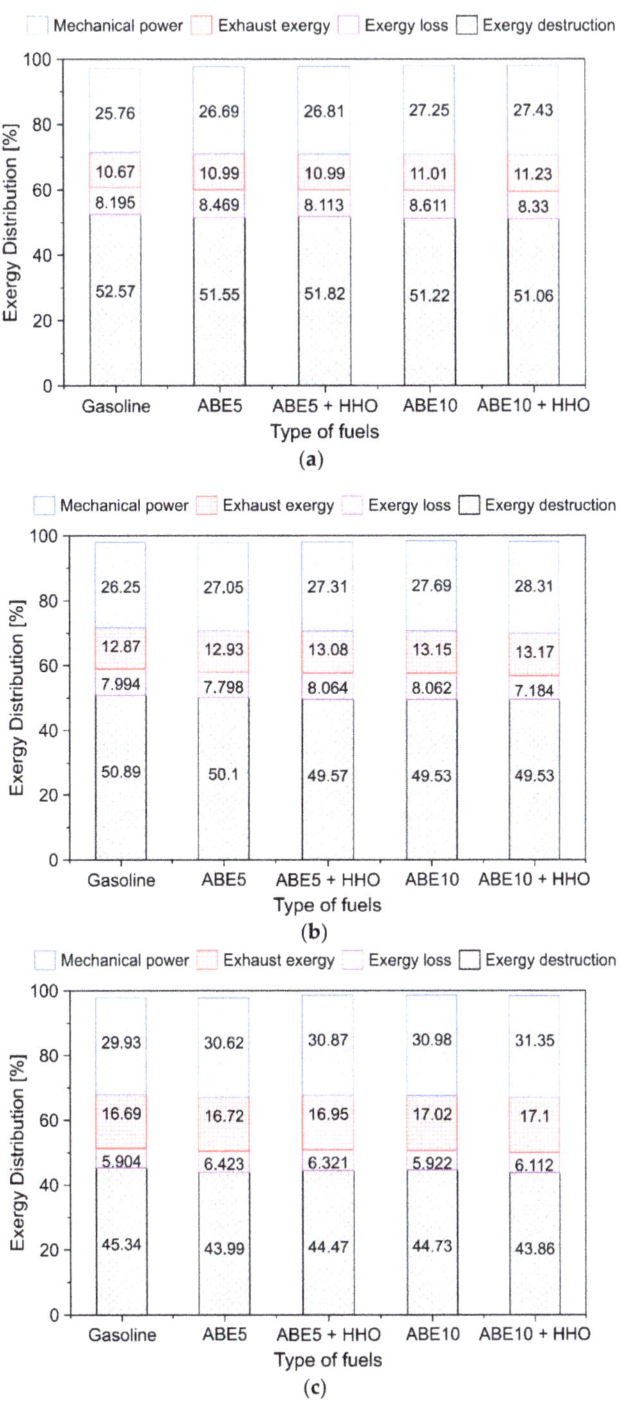

Figure 13. Exergy distribution in the SI engine, (**a**) load 50%, (**b**) load 75% and (**c**) load 100%.

Results show that exergy destruction represents the greatest share from the sources analyzed. The latter can be attributed to several factors such as residual fuel mixing, turbulence flow instabilities during combustion, among other irreversibilities that limit the chemical energy conversion [47]. Besides, as the engine load increases, both the useful exergy (mechanical power) and exhaust gas exergy rise, which implies improved combustion efficiency as both internal and external irreversibilities decrease [47,48]. The engine reaches a maximum of 29.9%, 30.6%, 30.9%, 31.0%, and 31.4% for the useful exergy when implementing gasoline fuels ABE5, ABE5 + HHO, ABE10, and ABE10 + HHO, respectively.

4. Conclusions

This investigation reports the potential application of dual-fuel operation in SI engines using acetone–butanol–ethanol (ABE) and hydroxy gas (HHO) enrichment. We characterized the main operation and design constraints of an experimental setup that combines a 3.5 kW YAMAHA engine and HHO generation. The evaluated fuel blends were ABE 5, ABE 5 + HHO, ABE 10, and ABE 10 + HHO, and gasoline was the baseline fuel. The core findings of the investigation can be summarized as follows:

- ABE standalone blends reduced both in-cylinder pressure and heat release rate compared to pure gasoline. Contrarily, hydroxy enrichment intensified the former and the latter while promoting a homogeneous fuel mixture.
- Engine load directly affected the combustion phasing leading to advanced or retarded combustion in the range of $0.2°$–$1.2°$.
- ABE-based blends increase BSFC between 10–25 $g \bullet kWh^{-1}$ compared to pure gasoline due to lower calorific value and lower energy density. The partial fuel substitution with hydroxy gas counterbalanced this rise while obtaining a net BSFC reduction compared to the baseline fuel.
- The implementation of dual-fuel operation promoted a significant minimization of CO, HC, and smoke levels. However, CO_2 and NOx emissions escalated due to enhanced combustion oxidation and higher combustion temperatures, which opens a new path for incorporating advanced fuel injection systems and after-exhaust treatment technologies.
- Energy losses represented a predominant share (37–52%) from the chemical energy input depending on the load. Increasing ABE and HHO content in the dual-fuel operation maximizes the power output by up to 2.2%. In contrast, high-load conditions promoted the minimization of energy losses, which implies higher combustion efficiency.
- ABE 10 + HHO featured the highest thermal efficiency (28–33%) from the fuel blends. Moreover, hydroxy doping increased efficiency up to 1.8%.
- Exergy destruction represents up to half of the exergy distribution, demonstrating the predominant share of internal irreversibilities in the combustion phenomena. Dual-fuel mode and higher engine loads result in enhanced useful exergy and power output.

The implementation of dual-fuel operation in SI engines demonstrated promising results towards emissions reduction and enhanced combustion performance. Further studies should assess the effect of engine speed on engine performance and control of the spark timing. Emissions maps are a robust tool to predict emissions levels and thermal performance. The incorporation of advanced technologies such as control strategies, waste heat recovery, and exhaust gas recirculation stand as promising avenues in the long-term perspectives of ICEs to further enhance the fuel utilization ratio.

Author Contributions: Conceptualization, W.G.-E., D.M.-C. and J.D.-F.; Methodology, W.G.-E., D.M.-C., A.B.-S. and J.D.-F.; Software, W.G.-E., D.M.-C., A.G.-Q. and J.D.-F.; Validation, A.B.-S., A.G.-Q. and J.D.-F.; Formal Analysis, W.G.-E. and D.M.-C.; Investigation, W.G.-E. and D.M.-C.; Resources, A.B.-S., A.G.-Q. and J.D.-F.; Writing—Original Draft Preparation, W.G.-E., D.M.-C. and A.G.-Q.; Writing—Review and Editing, A.B.-S. and J.D.-F.; Funding Acquisition, W.G.-E. and A.B.-S. All authors have read and agreed to the published version of the manuscript.

Funding: This research received no external funding.

Appl. Sci. **2021**, 11, 5282

Institutional Review Board Statement: Not applicable.

Informed Consent Statement: Not applicable.

Data Availability Statement: Not applicable.

Acknowledgments: Acknowledgments to Universidad del Atlántico through the project ING81-CII2019 "Estudio experimental de la sustitución parcial de combustible con HIDROXY (HHO) en motores térmicos de encendido por compresión y la influencia sobre sus prestaciones", Universidad del Norte, Sphere Energy Company for the support provided for the investigative internship of Daniel Maestre, and Colombian Institute for Scientific and Technological Development (COLCIENCIAS) through the "Convocatoria 809 Minciencias, Formación de capital humano de alto nivel para las regiones- Atlántico" for the support provided.

Conflicts of Interest: The authors declare no conflict of interest.

Abbreviations

The following abbreviations are used in this manuscript:

ABE	Acetone–Butanol–Ethanol
BSFC	Break specific fuel consumption
CI	Compression ignition
HHO	Hydroxy gas
LHV	Lower heating value
ICE	Internal combustion engine
H_{RR}	Heat release rate
SI	Spark ignition
BSFC	Brake specific fuel consumption
CO	Carbon monoxide
CO2	Carbon dioxide
NOx	Nitrogen oxides
HC	Hydrocarbons
WHR	Waste heat recovery
Nomenclature	
A	Area
b	Internal diameter of the combustion chamber
b_i	Best estimate of measurement
P	Mean combustion chamber pressure
V	Combustion chamber volume
m	Gas mass
k_{dry}/k_{wet}	Empirical emission gas constants
k_1/k_2	Model constants
k_{def}	Deformation constant
C_v/C_p	Specific heat at constant volume/pressure
T	Combustion chamber gas temperature
Q	Heat release
Q_r	Heat rejected by convection
H	Enthalpy
h	Specific enthalpy
h_{wall}	Heat transfer coefficient of the wall
R	Ideal gas constant
N	Engine speed
n	Number of repetitions
PW	Power output
R_y	Vertical position of the piston
S_t	Engine stroke
S	Standard deviation
T_r	Engine torque

U	Internal energy
u	Specific Internal Energy
X	Gas Mass Fraction
A_{wall}	Heat transfer surface area of the combustion chamber
A_c	Connecting rod's critical area
D	Diameter
L	Length
EP	Pollutant emissions in power unit
EV	Exhaust emissions in ppm/%vol.
r_c	Compression ratio
E_s	Elastic modulus of steel
a_p	Piston acceleration
e	Eccentricity between the stump and the bearing, located in its centerline
M	Gas molecular weight
W	Mechanical work
w	Average velocity of the combustion chamber
x_i	Measurement
Greek Letters	
θ	Crankshaft angle
Δ	Differential variation
ρ	Fluid density
α	Angle between the connecting rod and piston
φ	Rotational angle
γ	Specific heat ratio
ω	Angular speed
η	Efficiency
Subscripts	
0	Initial conditions
$comb$	Combustion chamber gas
bb	Blow-by gas
cr	Crankshaft
D	Discharge
dp	Pressure deformation
$disp$	Displaced
ext	Exhaust
st	Stoichiometric combustion
m	Mean
$mech$	Mechanical
mv	Top-dead center volume
t	Theoretical
g	Gaseous fuel
v	Valve
int	Intake/inlet
inf	Inertial forces

References

1. Matrisciano, A.; Franken, T.; Mestre, L.C.G.; Borg, A.; Mauss, F. Development of a Computationally Efficient Tabulated Chemistry Solver for Internal Combustion Engine Optimization Using Stochastic Reactor Models. *Appl. Sci.* **2020**, *10*, 8979. [CrossRef]
2. Ochoa, G.V.; Rojas, J.P.; Forero, J.D. Advance Exergo-Economic Analysis of a Waste Heat Recovery System Using ORC for a Bottoming Natural Gas Engine. *Energies* **2020**, *13*, 267. [CrossRef]
3. Ochoa, G.V.; Gutierrez, J.C.; Forero, J.D. Exergy, Economic, and Life-Cycle Assessment of ORC System for Waste Heat Recovery in a Natural Gas Internal Combustion Engine. *Resources* **2020**, *9*, 2. [CrossRef]
4. Forero, J.D.; Taborda, L.L.; Silvera, A.B. Characterization of the performance of centrifugal pumps powered by a diesel engine in dredging applications. *Int. Rev. Mech. Eng. (IREME)* **2019**, *13*, 11–20. [CrossRef]
5. Valencia, G.; Duarte, J.; Isaza-Roldan, C. Thermoeconomic Analysis of Different Exhaust Waste-Heat Recovery Systems for Natural Gas Engine Based on ORC. *Appl. Sci.* **2019**, *9*, 4017. [CrossRef]
6. Diaz, G.A.; Forero, J.D.; Garcia, J.; Rincon, A.; Fontalvo, A.; Bula, A.J.; Padilla, R.V. Maximum Power From Fluid Flow by Applying the First and Second Laws of Thermodynamics. *J. Energy Resour. Technol.* **2017**, *139*, 032903. [CrossRef]

7. Ochoa, G.V.; Isaza-Roldan, C.; Forero, J.D. Economic and Exergo-Advance Analysis of a Waste Heat Recovery System Based on Regenerative Organic Rankine Cycle under Organic Fluids with Low Global Warming Potential. *Energies* **2020**, *13*, 1317. [CrossRef]

8. Orozco, W.; Acuña, N.; Duarte, J. Characterization of Emissions in Low Displacement Diesel Engines Using Biodiesel and Energy Recovery System. *Int. Rev. Mech. Eng. (IREME)* **2019**, *13*, 420–426. [CrossRef]

9. Tamilselvan, P.; Nallusamy, N.; Rajkumar, S. A comprehensive review on performance, combustion and emission characteristics of biodiesel fuelled diesel engines. *Renew. Sustain. Energy Rev.* **2017**, *79*, 1134–1159. [CrossRef]

10. Masum, B.M.; Kalam, M.A.; Masjuki, H.H.; Palash, S.M.; Fattah, I.M.R. Performance and emission analysis of a multi cylinder gasoline engine operating at different alcohol–gasoline blends. *RSC Adv.* **2014**, *4*, 27898–27904. [CrossRef]

11. Yacoub, Y.M.; Bata, R.M.; Gautam, M. The performance and emission characteristics of C1-C5 alcohol-gasoline blends with matched oxygen content in a single-cylinder spark ignition engine. *Proc. Inst. Mech. Eng. Part A J. Power Energy* **1998**, *212*, 363–379. [CrossRef]

12. Nithyanandan, K.; Zhang, J.; Li, Y.; Wu, H.; Lee, T.H.; Lin, Y.; Lee, C.-F.F. Improved SI engine efficiency using Acetone–Butanol–Ethanol (ABE). *Fuel* **2016**, *174*, 333–343. [CrossRef]

13. Di Blasio, G.; Viscardi, M.; Alfè, M.; Gargiulo, V.; Ciajolo, A.; Beatrice, C. Analysis of the Impact of the Dual-Fuel Ethanol-Diesel System on the Size, Morphology, and Chemical Characteristics of the Soot Particles Emitted from a LD Diesel Engine. *SAE Tech. Pap. Ser.* **2014**. [CrossRef]

14. Gargiulo, V.; Alfe, M.; Di Blasio, G.; Beatrice, C. Chemico-physical features of soot emitted from a dual-fuel ethanol–diesel system. *Fuel* **2015**, *150*, 154–161. [CrossRef]

15. Beatrice, C.; Denbratt, I.; Di Blasio, G.; Di Luca, G.; Ianniello, R.; Saccullo, M. Experimental Assessment on Exploiting Low Carbon Ethanol Fuel in a Light-Duty Dual-Fuel Compression Ignition Engine. *Appl. Sci.* **2020**, *10*, 7182. [CrossRef]

16. Vassallo, A.; Beatrice, C.; Di Blasio, G.; Belgiorno, G.; Avolio, G.; Pesce, F.C. The Key Role of Advanced, Flexible Fuel Injection Systems to Match the Future CO_2 Targets in an Ultra-Light Mid-Size Diesel Engine. *SAE Tech. Pap. Ser.* **2018**. [CrossRef]

17. Belgiorno, G.; Dimitrakopoulos, N.; Di Blasio, G.; Beatrice, C.; Tunestål, P.; Tunér, M. Effect of the engine calibration parameters on gasoline partially premixed combustion performance and emissions compared to conventional diesel combustion in a light-duty Euro 6 engine. *Appl. Energy* **2018**, *228*, 2221–2234. [CrossRef]

18. Milani, D.; Kiani, A.; McNaughton, R. Renewable-powered hydrogen economy from Australia's perspective. *Int. J. Hydrogen Energy* **2020**, *45*, 24125–24145. [CrossRef]

19. Escobar-Yonoff, R.; Maestre-Cambronel, D.; Charry, S.; Rincón-Montenegro, A.; Portnoy, I. Performance assessment and economic perspectives of integrated PEM fuel cell and PEM electrolyzer for electric power generation. *Heliyon* **2021**, *7*, e06506. [CrossRef]

20. Alshehri, F.; Suárez, V.G.; Torres, J.L.R.; Perilla, A.; van der Meijden, M. Modelling and evaluation of PEM hydrogen technologies for frequency ancillary services in future multi-energy sustainable power systems. *Heliyon* **2019**, *5*, e01396. [CrossRef]

21. Shivaprasad, K.; Raviteja, S.; Chitragar, P.; Kumar, G. Experimental Investigation of the Effect of Hydrogen Addition on Combustion Performance and Emissions Characteristics of a Spark Ignition High Speed Gasoline Engine. *Procedia Technol.* **2014**, *14*, 141–148. [CrossRef]

22. Ismail, T.M.; Ramzy, K.; Elnaghi, B.E.; Mansour, T.; Abelwhab, M.; El-Salam, M.A.; Ismail, M. Modelling and simulation of electrochemical analysis of hybrid spark-ignition engine using hydroxy (HHO) dry cell. *Energy Convers. Manag.* **2019**, *181*, 1–14. [CrossRef]

23. Yilmaz, A.C.; Uludamar, E.; Aydin, K. Effect of hydroxy (HHO) gas addition on performance and exhaust emissions in compression ignition engines. *Int. J. Hydrogen Energy* **2010**, *35*, 11366–11372. [CrossRef]

24. Carl, J.; Fedor, D. Tracking global carbon revenues: A survey of carbon taxes versus cap-and-trade in the real world. *Energy Policy* **2016**, *96*, 50–77. [CrossRef]

25. Mendoza-Casseres, D.; Valencia-Ochoa, G.; Duarte-Forero, J. Experimental assessment of combustion performance in low-displacement stationary engines operating with biodiesel blends and hydroxy. *Therm. Sci. Eng. Prog.* **2021**, *23*, 100883. [CrossRef]

26. Dhinesh, B.; Raj, Y.M.A.; Kalaiselvan, C.; KrishnaMoorthy, R. A numerical and experimental assessment of a coated diesel engine powered by high-performance nano biofuel. *Energy Convers. Manag.* **2018**, *171*, 815–824. [CrossRef]

27. van Wyk, S.; van der Ham, A.; Kersten, S. Pervaporative separation and intensification of downstream recovery of acetone-butanol-ethanol (ABE). *Chem. Eng. Process. Process. Intensif.* **2018**, *130*, 148–159. [CrossRef]

28. Anderhofstadt, B.; Spinler, S. Preferences for autonomous and alternative fuel-powered heavy-duty trucks in Germany. *Transp. Res. Part D Transp. Environ.* **2020**, *79*, 102232. [CrossRef]

29. Ismail, T.M.; Ramzy, K.; Abelwhab, M.; Elnaghi, B.E.; El-Salam, M.A.; Ismail, M. Performance of hybrid compression ignition engine using hydroxy (HHO) from dry cell. *Energy Convers. Manag.* **2018**, *155*, 287–300. [CrossRef]

30. Ferguson, C.R.; Kirkpatrick, A.T. *Internal Combustion Engines—Applied Thermosciences*; John Wiley & Sons: Hoboken, NJ, USA, 2001.

31. Williams, F.A. *Combustion Theory Benjamin/Cummings*; Westview Press: Menlo Park, CA, USA, 1985.

32. Ochoa, G.V.; Isaza-Roldan, C.; Forero, J.D. A phenomenological base semi-physical thermodynamic model for the cylinder and exhaust manifold of a natural gas 2-megawatt four-stroke internal combustion engine. *Heliyon* **2019**, *5*, e02700. [CrossRef]

33. Pain, J. Gas Dynamics. *Nat. Cell Biol.* **1967**, *213*, 1182. [CrossRef]

34. Hernández-Comas, B.; Maestre-Cambronel, D.; Pardo-García, C.; Fonseca-Vigoya, M.; Pabón-León, J. Influence of Compression Rings on the Dynamic Characteristics and Sealing Capacity of the Combustion Chamber in Diesel Engines. *Lubricants* **2021**, *9*, 25. [CrossRef]

35. Irimescu, A.; Di Iorio, S.; Merola, S.S.; Sementa, P.; Vaglieco, B.M. Evaluation of compression ratio and blow-by rates for spark ignition engines based on in-cylinder pressure trace analysis. *Energy Convers. Manag.* **2018**, *162*, 98–108. [CrossRef]

36. Woschni, G. A Universally Applicable Equation for the Instantaneous Heat Transfer Coefficient in the Internal Combustion Engine. *SAE Tech. Pap. Ser.* **1967**. [CrossRef]

37. Consuegra, F.; Bula, A.; Guillín, W.; Sánchez, J.; Forero, J.D. Instantaneous in-Cylinder Volume Considering Deformation and Clearance due to Lubricating Film in Reciprocating Internal Combustion Engines. *Energies* **2019**, *12*, 1437. [CrossRef]

38. Ağbulut, Ü.; Sarıdemir, S.; Albayrak, S. Experimental investigation of combustion, performance and emission characteristics of a diesel engine fuelled with diesel–biodiesel–alcohol blends. *J. Braz. Soc. Mech. Sci. Eng.* **2019**, *41*, 389. [CrossRef]

39. Heseding, M.; Daskalopoulos, P. *Exhaust Emission Legislation-Diesel-and Gas Engines*; VDMA: Frankfurt Am Main, Germany, 2006.

40. Musthafa, M.M.; Kumar, T.A.; Mohanraj, T.; Chandramouli, R. A comparative study on performance, combustion and emission characteristics of diesel engine fuelled by biodiesel blends with and without an additive. *Fuel* **2018**, *225*, 343–348. [CrossRef]

41. Ma, F.; Wang, M.; Jiang, L.; Deng, J.; Chen, R.; Naeve, N.; Zhao, S. Performance and emission characteristics of a turbocharged spark-ignition hydrogen-enriched compressed natural gas engine under wide open throttle operating conditions. *Int. J. Hydrogen Energy* **2010**, *35*, 12502–12509. [CrossRef]

42. Senthilkumar, S.; Sivakumar, G.; Manoharan, S. Investigation of palm methyl-ester bio-diesel with additive on performance and emission characteristics of a diesel engine under 8-mode testing cycle. *Alex. Eng. J.* **2015**, *54*, 423–428. [CrossRef]

43. Zhang, J.; Nithyanandan, K.; Li, Y.; Lee, C.-F.; Huang, Z. Comparative Study of High-Alcohol-Content Gasoline Blends in an SI Engine. *SAE Tech. Pap. Ser.* **2015**, *1*. [CrossRef]

44. Duarte, J.; Amador, G.; Garcia, J.; Fontalvo, A.; Padilla, R.V.; Sanjuan, M.; Quiroga, A.G. Auto-ignition control in turbocharged internal combustion engines operating with gaseous fuels. *Energy* **2014**, *71*, 137–147. [CrossRef]

45. Duarte, J.; Garcia, J.; Jiménez, J.; Sanjuan, M.E.; Bula, A.; González, J. Auto-Ignition Control in Spark-Ignition Engines Using Internal Model Control Structure. *J. Energy Resour. Technol.* **2017**, *139*, 022201. [CrossRef]

46. Li, Y.; Nithyanandan, K.; Zhang, J.; Lee, C.-F.; Liao, S. Combustion and Emissions Performance of a Spark Ignition Engine Fueled with Water Containing Acetone-Butanol-Ethanol and Gasoline Blends. *SAE Tech. Pap. Ser.* **2015**. [CrossRef]

47. Kul, B.S.; Kahraman, A. Energy and Exergy Analyses of a Diesel Engine Fuelled with Biodiesel-Diesel Blends Containing 5% Bioethanol. *Entropy* **2016**, *18*, 387. [CrossRef]

48. Monsalve-Serrano, J.; Belgiorno, G.; Di Blasio, G.; Guzmán-Mendoza, M. 1D Simulation and Experimental Analysis on the Effects of the Injection Parameters in Methane–Diesel Dual-Fuel Combustion. *Energies* **2020**, *13*, 3734. [CrossRef]

 *applied sciences*

Article

Experimental Study of the Effect of Hydrotreated Vegetable Oil and Oxymethylene Ethers on Main Spray and Combustion Characteristics under Engine Combustion Network Spray A Conditions

José V. Pastor [1], José M. García-Oliver [1], Carlos Micó [1,*], Alba A. García-Carrero [1] and Arantzazu Gómez [2]

[1]   CMT–Motores Térmicos/Universitat Politècnica de València, Camino de Vera s/n. 46022 Valencia, Spain; jpastor@mot.upv.es (J.V.P.); jgarciao@mot.upv.es (J.M.G.-O.); algar6@mot.upv.es (A.A.G.-C.)
[2]   Universidad de Castilla-La Mancha, Campus de Excelencia Internacional en Energía y Medioambiente, Escuela de Ingeniería Industrial y Aeroespacial de Toledo, Real Fábrica de Armas. Edificio Sabatini. Av. Carlos III, s/n, 45071 Toledo, Spain; Aranzazu.Gomez@uclm.es
*   Correspondence: carmirec@mot.upv.es

Received: 14 July 2020; Accepted: 2 August 2020; Published: 7 August 2020

Featured Application: This work contributes to the understanding of the macroscopic characteristics of the spray as well as to the evolution of the combustion process for alternative fuels. All these fuels have been studied under the same operating conditions than diesel therefore the comparison can be made directly, leaving in evidence that some fuels can achieve a similar behavior to diesel in terms of auto ignition but avoiding one of the biggest disadvantages of diesel such as the soot formation. Moreover, the quantification of characteristic parameters such as ignition delay, liquid length, vapor penetration and flame lift-off length represent the most important data to adjust and subsequently validate the computational models that simulate the spray evolution and combustion development of these alternative fuels inside the combustion chamber.

Abstract: The stringent emission regulations have motivated the development of cleaner fuels as diesel surrogates. However, their different physical-chemical properties make the study of their behavior in compression ignition engines essential. In this sense, optical techniques are a very effective tool for determining the spray evolution and combustion characteristics occurring in the combustion chamber. In this work, quantitative parameters describing the evolution of diesel-like sprays such as liquid length, spray penetration, ignition delay, lift-off length and flame penetration as well as the soot formation were tested in a constant high pressure and high temperature installation using schlieren, OH* chemiluminescence and diffused back-illumination extinction imaging techniques. Boundary conditions such as rail pressure, chamber density and temperature were defined using guidelines from the Engine Combustion Network (ECN). Two paraffinic fuels (dodecane and a renewable hydrotreated vegetable oil (HVO)) and two oxygenated fuels (methylal identified as OME_1 and a blend of oxymethylene ethers, identified as OME_x) were tested and compared to a conventional diesel fuel used as reference. Results showed that paraffinic fuels and OME_x sprays have similar behavior in terms of global combustion metrics. In the case of OME_1, a shorter liquid length, but longer ignition delay time and flame lift-off length were observed. However, in terms of soot formation, a big difference between paraffinic and oxygenated fuels could be appreciated. While paraffinic fuels did not show any significant decrease of soot formation when compared to diesel fuel, soot formed by OME_1 and OME_x was below the detection threshold in all tested conditions.

Keywords: hydrotreated vegetal oil; oxymethylene ethers; ignition delay; liquid length; lift-off length; soot

---

1. Introduction

Pollutant emission regulations are becoming more stringent every year and strategies to reduce them in compression ignition (CI) engines are being constantly investigated. Among them, active and passive solutions can be found. These latter strategies or after-treatment systems, such as diesel particle filters (DPF), catalytic oxidizers and selective catalytic reduction systems (SCR), are focused on retaining the pollutant emissions prior being expelled through the tail pipe [1]. However, the efficiency and life cycle of the after-treatment systems are deeply related to the fuel behavior inside the combustion chamber. In this sense, active strategies to avoid pollutant formation, as the redesign of the combustion chamber or the study of new combustion concepts [2], the improvement of the mixing formation and injection systems [3], the study of new exhaust gas recirculation schemes as well as the development of cleaner alternative fuels, have great interest and importance to the automotive manufacturers [4].

Biodiesel is the most used biofuel in conventional compression ignition engines because of its potential for reducing the particulate matter due to its oxygenated nature without aromatic compounds [5]. However, lower heating value and higher viscosity, as well as a negative effect on nitrogen oxides (NO_x) are important drawbacks compared to conventional diesel fuel [6,7]. These have promoted the study of more appropriate alternative fuels. Among them, hydrotreated vegetable oil (HVO), gas-to-liquid (GTL) and new oxygenated fuels like the oxymethylene ethers with the general structure $CH_3–O–(CH_2–O)_n–CH_3$, are the most promising alternatives to replace conventional diesel [8–10].

HVO and GTL are paraffinic fuels without oxygen on their composition and similar chain length. They are obtained by a hydrotreatment of vegetable oils at controlled temperature or by a Fischer-Tropsch process of natural gas or gasified biomass, respectively. Due to their analogous molecular structure, their physical-chemical properties are also similar. However, while viscosity and heating value are comparable to conventional diesel fuel, their higher cetane number, lower cold filter plugging point (CFPP) and the absence of aromatic compounds make these paraffinic fuels very attractive. Moreover, several studies in CI engines agree on their potential for reducing the particulate matter in the exhaust without NOx penalization [11–13].

Just as HVO or GTL, OME_x presents a molecular structure without double bonds and can be generated from methanol or formaldehyde by a synthetic process that consumes CO_2 and water [14]. Depending on the chain length their physical-chemical properties differ, increasing the density, viscosity, oxygen content and cetane number as carbon number increases from 1 to 5 [15]. Although with different molecular structure, HVO and OME_x coincide in their higher cetane number and exhaust emissions benefits, even better than paraffinic fuels in terms of soot emissions [16]. However, the low viscosity and boiling point of the simplest OME (methylal, usually called OME_1) limits its use in diesel engines without injection system and storage modifications [15,17].

Even though, these alternative fuels have proved to be environmentally friendly, the study of their spray characteristics and combustion behavior is of importance for a complete understanding of their effects on the exhaust emissions [18]. In this sense, optical techniques are very helpful and effective tools to evaluate combustion progress and soot formation and oxidation [7,19]. Therefore, in this work, an exhaustive study of spray and combustion behavior of alternative fuels has been carried out in a high pressure and high temperature combustion chamber under diesel-like conditions.

The objective of this study was to define the differences on combustion behavior of four alternative fuels in comparison to conventional diesel fuel: on the one hand, dodecane and HVO with paraffinic structures and, on the other hand, OME_1 and OME_x as oxygenated fuels. The experiments were carried out under an operating condition known as "Spray A conditions" settled as standard by the Engine

Combustion Network (ECN) [20]. Experimental data about the fundamental behavior of these fuels under these conditions is a novelty of this study, due to the lack of this type of information in the current literature.

Quantitative parameters describing the evolution of diesel-like sprays such as liquid length, spray penetration, ignition delay, lift-off length and flame length as well as the soot formation were determined using schlieren, OH^* chemiluminescence and diffused back-illumination extinction high-speed imaging techniques. The results can provide a useful information to better understanding of factors that dominate the combustion of these alternative fuels and therefore contributing to the upgrade of combustion models. Furthermore, the results will expand the database available in the ECN [20].

2. Experimental and Theoretical Tools

2.1. High Pressure and High Temperature Rig

The experiments were carried out in a high pressure and high temperature (HPHT) rig. Parameters such as the composition of the gas, the pressure and the temperature of the ambient gas can be controlled independently to obtain oxygen concentration between 0 and 21%, pressures up to 15 MPa and temperatures up to 1100 K. Thereby, it is possible to replicate the thermodynamic conditions of the cylinder of an internal combustion engine when the fuel is injected. Furthermore, as the temperature field is homogeneous and constant in the area of interest, this reduces the uncertainties that could be associated to engine transients. The HPHT rig has wide optical accesses, which allow the application of different visualization techniques. The installation can operate as an open circuit with air or, in order to reduce O_2 concentration, as a closed loop circuit with a mixture of air and Nitrogen. The regulation and control system ensures steady thermodynamic conditions during long time periods with the aim of getting reliable statistical results of many injections and combustion events. Additionally, one injection is performed every 4 s to avoid any temperature transients. The boundary conditions have been detailed widely in [21] where a full description of the facility is given.

The common-rail used is capable to achieve injection pressures up to 230 MPa and is equipped with a solenoid-activated single-hole nozzle injector. It is worth mentioning that the injection system pump used is made by polymerizing tetrafluoroethylene material. This material has shown compatibility with oxymethylene ether fuels according to the investigation reported in [22].

2.2. Fuel and Test Matrix

2.2.1. Fuel Description and Properties

A set of two different groups of pure alternative fuels were tested and compared to a low sulfur diesel fuel without biodiesel, which was used as reference. First, two paraffinic fuels with different molecular structure: a renewable hydrotreated vegetable oil (HVO) and a single component *n*-paraffinic fuel (dodecane) were tested. The last is used as standard surrogate for diesel by the engine combustion network (ECN) [20] and, in this work, dodecane is tested to compare the results with ECN database and validate the current experiments. Then, two oxymethylene ethers, with the general structure $CH_3–O–(CH_2–O)_n–CH_3$ were tested under the same thermodynamic conditions as HVO and dodecane. The first one was a single component one corresponding to the shortest carbon chain (n = 1) of the family, which will be denoted here as OME_1. The second one is a multi-component fuel, which will be denoted here as OME_x, which is a blend of components of different chain lengths. Table 1 compiles the main properties for all five fuels and Table 2 shows the OMEx actual composition.

Table 1. Fuel properties.

Characteristic [unit]	Diesel	Dodecane	HVO	OME_1	OME_x
Density [kg/m^3] (T= 15 °C)	835.20	751.20	779.10	866.70	1057.10
Viscosity [mm^2/s] (T= 40 °C)	2.80	1.44	2.70	0.36	1.08
Cetane number [–]	54.18	74	75.5	28	68.6
Lubricity [μm]	386	563	316	747	320
Flash point [°C]	-	83	70	<40	65
Lower heating value [MJ/kg]	39.79	44.20	43.90	19.25	19.21
Initial Boiling Point [°C]	155.10	214.00	185.50	37.40	144.90
Final Boiling Point [°C]	363.1	218	302	38	242.4
Total contamination [mg/kg]	<24	-	6.0	<1	<1
Carbon content [% m/m]	85.3	84	85.7	48.4	44.2
Hydrogen Content [% m/m]	13.4	16	14.3	10.4	8.8
Oxygen content [% m/m]	0	0	0	42.1	45
$(A/F)_{st}$ at 21% of O_2	14.39	14.92:1	14.55:1	7.22:1	5.89:1
$(A/F)_{st}$ at 15% of O_2	19.98	20.72:1	20.20:1-	10.03:1	8.18:1

Table 2. Composition of OME_x used in this study.

Molecule	Content (wt%)
OME_1	0.01
OME_2	<0.01
OME_3	57.90
OME_4	28.87
OME_5	10.08
OME_6	1.91

2.2.2. Operating Conditions

For the five tested fuels, a set of parametric studies was performed. Oxygen concentration, ambient temperature and injection pressure were defined as variables. Table 3 shows the test matrix.

Table 3. Variation of parameters to evaluate for each fuel.

Oxygen Concentration (%)	Temperature [K]	Injection Pressure [bar]
15	800	500
21	900	1000
	1000	1500

The nominal condition is based on Spray A specification from the ECN [20]. It corresponds to 900 K as ambient temperature, 15% of oxygen concentration and 1500 bar as injection pressure and air density of 22.8 kg/m^3. The whole test matrix is shown in Table 3.

A single-hole nozzle was used, which has been extensively studied in previous ECN studies [23,24]. The injector serial number is 210,675, with an actual nozzle diameter of 89.4 μm. A single injection was used as injection strategy. The energizing time was 2000 μs, which provides an injection event long enough to study the spray evolution and flame development under stabilized mixing-controlled combustion.

2.3. Diagnostic Techniques

Aiming to analyze the spray development, the ignition delay, the combustion behavior and the soot formation for each fuel, four visualization techniques have been implemented using the two opposed optical access windows in the installation. The measurements were carried out in two sets with the slightly different setups shown in Figure 1. The change from one setup to the other was made in few seconds, thanks to the use of an automated translational stage. Consequently, it can be assumed that measurements will all the techniques were performed nearly simultaneously for any given condition of the test matrix.

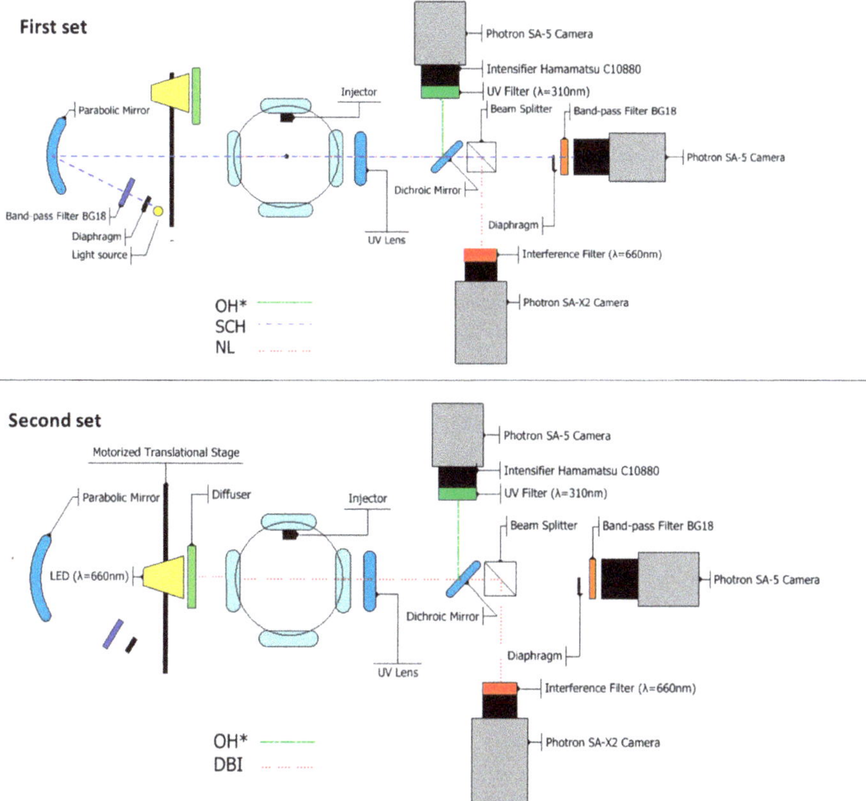

Figure 1. Optical arrangement for the first set up (**top**) and second setup (**bottom**).

In the first set of measurements, the sketch shown at the top of Figure 1 was used for simultaneous recording of images from schlieren, and OH* chemiluminescence optical techniques. A minimum of 20 injection events were recorded with each camera in this first set of measurements. Immediately after recording and saving these images for each operating conditions, an illumination device based on LEDs was placed in front of the vessel's window in the illumination side using a motorized translational stage (sketch at the bottom of Figure 1). Then, in this second set, images from OH* chemiluminescence and diffused back illumination (DBI) optical technique were recorded for 100 injections in the nominal case and 40 injections for the other operating conditions. The OH* images were recorded in both cases and these results were used to verify consistency between both measurement sets.

The configuration used for the application of each optical technique will be described below.

2.3.1. Schlieren Imaging

The schlieren technique is based on the fact that when a light ray travels through a medium with refractive index gradients, it suffers a deflection due to the refraction phenomenon [25]. Accordingly, any variations of refractive index such as those produced by density variations at the injection of fuel can be recorded as different gray levels in an image. Consequently, this technique allows to observe the local density variations that the fuel air mixing, auto ignition or flame development provoke. In this experiments, a high-speed single-pass schlieren imaging configuration was implemented to visualize the spray and flame boundaries at any operating conditions, with an optical arrangement similar as that described in [26].

On the illumination side, light from a xenon lamp is driven with a liquid light guide into an iris diaphragm, to generate a point light source at the focal length plane of a parabolic mirror (f = 610 mm, D = 150 mm) so that the measurement area is illuminated with a collimated beam. In addition, to avoid interference and restrict the spectrum of the xenon lamp in the DBI and NL images commented later, a BG18 band pass filter was used. On the other side of the chamber, a spherical lens (f = 750 mm) was placed very close to optical access. This lens focusses the light onto the Fourier plane where an iris diaphragm with a cut-off diameter of around 3 mm was located. Additionally, another BG18 band pass filter was placed just before the Photron Fastcam SA-5 camera (Photron, San Diego, CA, USA) to minimize undesired light from sources other than the schlieren illumination lamp. The camera was equipped with a Carl Zeiss Makro-Planar T 100 mm f/2 ZF2 camera lens (Carl Zeiss, Aalen, Germany). Images were recorded at 25 kfps. The shutter time was 4.24 µs and it was kept constant throughout all the experiments. The resolution was 800 × 320 pixel with a total magnification of 6.9 pixel/mm. 20 injection cycles per test were recorded in all cases. The schlieren images have been used to describe the spray tip penetration under nonreactive, as well as the penetration under reactive conditions and the ignition delay. Image segmentation from the background has been performed by using the standard methodology developed by ECN [20].

In this paper, ignition delay (ID) is obtained from schlieren images as the time at which the derivative of the accumulated pixel intensity within the spray boundaries is 50% of range between the minimum and maximum ones. This criterion was defined and explained in [27].

2.3.2. High-Speed OH* Chemiluminescence (OH*) Imaging

Excited hydroxyl radicals (OH") are good tracers of high temperature combustion regions in a flame [28]; therefore, visualization of OH* chemiluminescence at the base of the flame allows to quantifying the lift-off length (LOL). Moreover, if an intensified high-speed camera is used, the ID can be measured too. In these experiments, a high-speed image intensifier (Hamamatsu C10880 by Hamamatsu, Hamamatsu city, Japan) was used, with gain set between 700 and 850 depending on fuel and operating condition. It was coupled to a Photron Fastcam SA5 camera with a 1:1 relay lens. The system was equipped with a UV f/4 100mm focal length lens. An interference filter centered at 310 nm (10 nm FWHM) was placed in front of the camera to remove most of the radiation of the flame while keeping OH^* chemiluminescence. As shown in Figure 1, a dichroic mirror, which reflects UV spectrum and transmit the visible light spectrum, is used to reflect the UV flame radiation to the intensified camera.

Images were taken at 25 kfps. The resolution was 704 × 416 pixels and the magnification was 5.43 pixel/ mm. The shutter time was adapted between 19.97 µs and 39.75 µs to accommodate the camera dynamic range to the flame radiation intensity as a function of the fuel and operating condition. Again, 20 injection cycles per test were recorded in any case. ID was obtained determining the first frame with detectable light intensity, and it was used as a check on the results obtained with the schlieren technique.

The algorithm used to obtain the LOL is based on the procedure described in [29] and recommended by the engine combustion network (ECN). Thus, the LOL was as the distance between the injector tip and the first axial locations above and below the spray centerline with intensity greater than 50% of the intensity peak of that zone.

2.3.3. Diffused Back Illumination (DBI) Extinction Imaging

Diffused back-illumination (DBI) is a technique based on measuring the amount of light attenuated by liquid droplets or by soot particles within the flame, which is related to the liquid and soot concentration, respectively. A red LED (λ = 660 nm) was used in these experiments as the light source to create short flashes synchronized with the camera frames. An engineered diffuser (EDC-20 by RPC Photonics, Henrietta, NY, USA) was placed in front of the LED to create a diffused Lambertian intensity profile [30]. On the collection side, the transmitted light from the LED and the flame radiation went through a beam splitter with a 50% reflection rate. Then, half of the light was collected by the Photron SA-X2 camera. The exposure time was 1 μs and the resolution was 896x384 with a magnification of 6.85 pixel/mm. The sampling frequency was 25 kfps.

The images taken were analyzed considering that the total light registered by the camera has two contributions: the transmitted LED light intensity and the flame radiation. Due to the use of a bandpass filter centered at 660 nm (FWHM = 10 nm), the crosstalk of flame radiation into the DBI signal is minimized. However, the flashing frequency of the LED was set as quarter of the camera frame rate to capture a LED image between every three consecutive dark images. This configuration was used to deal with the non-ideal CMOS sensor behavior, that occurs when the camera is exposed to a sudden change in light intensity between two consecutive frames [18]. The flame luminosity from dark image before the LED image (third dark image) was quantified and used to isolate the transmitted LED light from the total registered radiation. Then, the light attenuation can be related with the optical properties of the soot cloud by means of Lambert-Beer's law, as described in Equation (1):

$$\frac{I_{on} - I_{off}}{I_o} = e^{-KL} \tag{1}$$

where I_{on} is the light intensity recorded by the camera when the LED is on, i.e., the sum of the transmitted LED intensity and the flame luminosity. I_{off} is the intensity of the flame acquired when the LED is off. I_O is the LED light intensity obtained from images recorded before the start of injection (SOI). K is the soot dimensional extinction coefficient and L is the light beam path length through the soot cloud. Thus, the product KL represents the integral value of the soot extinction coefficient along the light path, which is related with the soot concentration [24].

Besides determining soot production through the KL factor, DBI is also suggested as an experimental standard to measure the liquid length (LL) of fuels by ECN [20]. The method uses the extinction produced by the spray droplets to provide a quantitative parameter related to the liquid volume fraction along the path of the light. ECN recommends also taking care with vapor phase beam steering from temperature gradients, which could disturb the measurement. In the current work, the liquid length has been determined only at 800 K and 900 K. It was observed that at higher temperature the measurement is not reliable because the flame lift-off zone is close to the liquid jet zone and more intense beam steering exists there.

The image processing method to get KL was based on creating a mask from the dark image I_{off} with the aim of collecting the information only corresponding to flame radiation. Based upon this mask, attenuation ($I_{on} - I_{Off}$) is calculated, from which Equation (1) is applied. Figure 2 shows a KL profile along the spray axis at a time instant where the flame is well developed (3500 μs), within the quasi- steady period chosen in this study (between 3000 μs and 4000 μs). The confidence interval at 95% (red shadow) for the measurement of the ensemble averaged KL value and the standard deviation (blue shadow) have been represented. The operating condition shown is the nominal case and the fuel chosen is the diesel.

Similarly, the total soot mass (s_{mass}) at a given time was determined from Equation (2) as the sum of the values of over all the pixels of the average image taken at that time, and corrected with the other factors indicated. In this equation, ρ_{soot} corresponds to the soot density defined as 1.8 g/cm^3 by Choi [31], λ is the wavelength used in the current work (660 nm), r is the pixel-mm ratio (6.85 in this work) and ke is the dimensionless extinction coefficient equal to 7.27 determined in this study through the ratio of scattering and absorption cross-sections which is used in small particle Mie theory:

$$s_{mass} = \frac{\sum KL \cdot \rho_{soot} \cdot \lambda}{ke \cdot r^2} \tag{2}$$

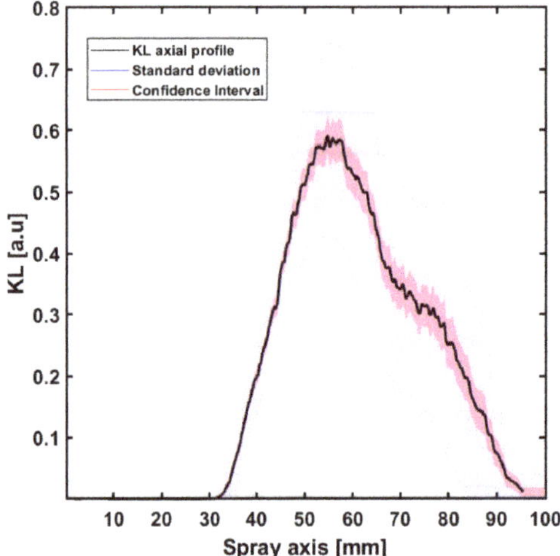

Figure 2. KL profile on the spray axis for diesel at nominal case in the time instant 3500 µs.

In Figure 3 the temporal evolution of total soot mass for diesel at nominal condition is shown. The standard deviation (blue shadow) and confidence interval at 95% for the ensemble average soot mass value (red shadow) have been represented too. In the results section, these values will be shown for the other fuels and operating conditions tested.

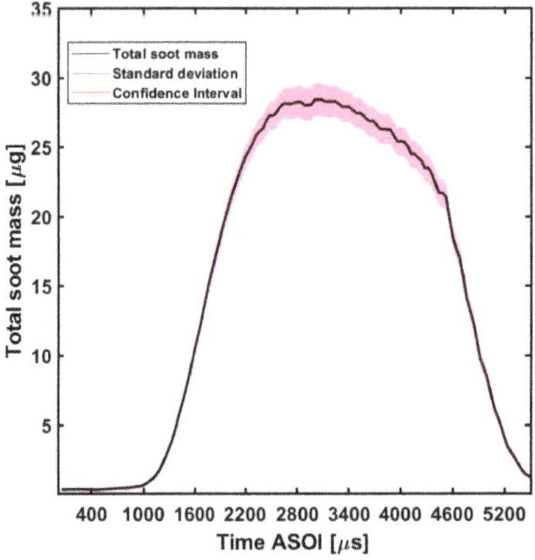

Figure 3. Total soot mass for the diesel at nominal condition.

3. Results and Discussion

Before entering the quantification and detailed analysis on the effect that fuel has upon the spray characteristics, the combustion process and soot formation, a comparison between results for the dodecane fuel from current study and from similar studies available in the ECN database [20] is presented. Table 4 summarizes the mean values obtained under the Spray A condition (15% of Oxygen concentration, 900 K of temperature and 1500 bar of injection pressure) by different research centers members of the ECN. It can be observed that results obtained for liquid length (LL), ignition delay (ID), and flame lift-off length (LOL) in this work are very close to those previously obtained. This makes possible to confirm the reliability of current results and extending the methods in the current work for the experiments with the other fuels tested.

Table 4. Comparison of main spray parameters between database available in the ECN [20] and current work for dodecane under Spray A conditions (900 K, 1500 bar and 15% O_2).

Parameter	Current Work	CMT (2012)	SNL (2017)	TU/e (2012)	IFPEn (2012)	CAT (2010)
LL [mm]	10.0	10.8	9.60	-	11	8.67
ID [ms]	0.405	0.435	0.428	0.41	0.4	-
LOL [mm]	18.89	17.73	17.66	15.8	14.5	16.1

3.1. Maximum Liquid Length

Spray liquid length is the maximum axial penetration of the liquid phase fuel. Previous studies have demonstrated that a shorter spray liquid length leads to a better air-fuel mixing, because the fuel vaporization is completed before the fuel reaches the combustion region [32]. On the other hand, too long liquid length leads to fuel wall impingement on the cylinder, which could produce large soot emission with reduced engine efficiency [33]. Figure 4 shows the liquid length (LL) at 800 K and 900 K for the tested fuels. The injection pressure is 1500 bar and the oxygen concentration is 15%. The LL is constant in time; therefore, the values correspond to an average between 800 µs and 4000 µs. From this figure, it can be seen that the spray liquid length of OME_1 is shorter, which could be due to the lower distillation temperature of OME_1 (37.40 °C) compared to the other fuels. Additionally, as Kook indicated in a previous study [34], low viscosities and high densities lead to shorter liquid lengths. This could explain why OME_x (υ = 1.082 mm^2/s; ρ = 1057.1 kg/m^3) shows shorter LL than diesel and HVO (υ = 2.7 mm^2/s; ρ = 779.1 kg/m^3).

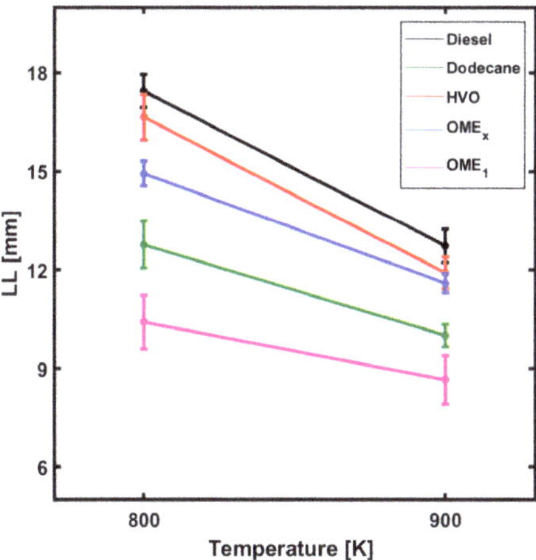

Figure 4. Liquid length for 15% O_2 and 1500 bar of injection pressure.

3.2. Ignition Delay

Ignition delay (ID) is defined as the time elapsed from the start of injection (SOI) to the start of combustion. Figure 5 shows the ignition delay comparison obtained from schlieren technique and from OH* chemiluminescence imaging. It can be seen that ID values for both techniques are almost identical, therefore, any ID value used in the current work is valid. Hereafter, the represented ID values correspond to those from the schlieren technique.

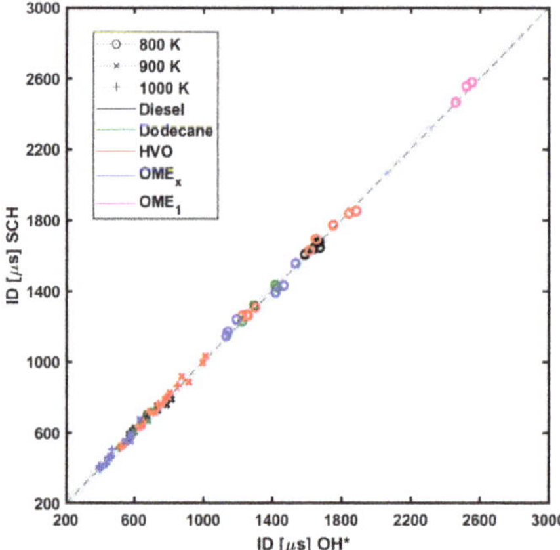

Figure 5. Comparison between ignition delay measured from schlieren and from OH* chemiluminescence images.

In Figure 6, ID for dodecane, HVO, OME$_x$ and OME$_1$ have been compared against diesel. All the operation conditions have been presented with their standard deviation. The results attend to the trend found in previous studies [23,35]. When the air temperature, injection pressure, and ambient oxygen concentration are increased, the ID values decrease. It is a common behavior for all the tested fuels.

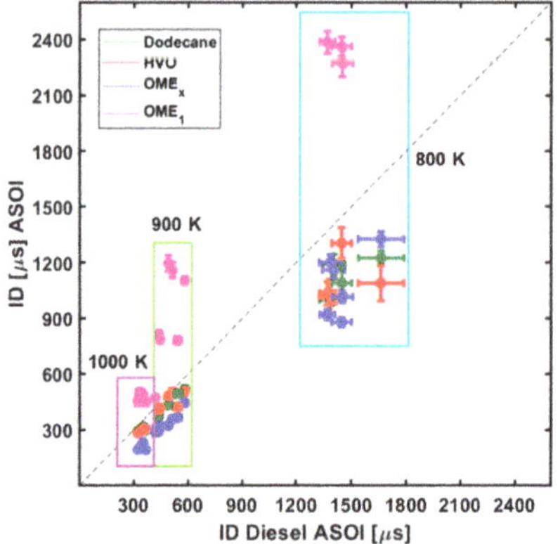

Figure 6. Comparison between ignition delay of diesel and the ignition delay for the other fuels tested.

Furthermore, Figure 6 shows that dodecane, HVO, and OME$_x$ ignite before diesel does. The three fuels have a higher cetane number than diesel (see Table 1), and the results confirm the relevance of this parameter on autoignition. Based on this, one could expect OME$_x$ to ignite later than HVO and dodecane. However, its molecular structure has high oxygen content, which makes it more reactive. This explains that OME$_x$ ignites earlier than HVO and dodecane, despite its lower cetane number. For OME$_1$, although its molecular structure also contains oxygen, its cetane number is much lower when compared to the other fuels tested. Thus, it is the last one to ignite as shown in Figure 6. It is important to mention that the OME$_1$ at 800 K and 15% of O$_2$ does not ignite and at 21% of O$_2$, the ignition occurs very late (around 2400 μs).

3.3. Lift-Off Length

An important parameter in the combustion and soot production processes is the flame lift-off length (LOL). This is strongly related with the amount of fuel–air mixing process upstream the combustion region. Enough air entrained reduces the average equivalence ratio at the LOL [29] and this result in less soot production. Consequently, flame LOL is worth to be studied for the different fuels tested in current work.

As previously done with the ignition delay, the LOL for the diesel compared to the other fuels is shown in the Figure 7, for the whole test matrix. The interval time to average the LOL corresponds to that in which the flame is "quasi steady" (between 3000 µs and 4000 µs). The HVO and dodecane present a very close behavior, showing smaller LOL than diesel. However, the tendency is not so clear for OME_x since at 800 K its LOL is slightly longer that diesel, at 900 K it is similar but at 1000 K is smaller, close to HVO and dodecane. It is also possible to observe that at 800 K and 15% of O_2 the LOL of OME_x has the highest standard deviation, indicating a lower combustion stability. The OME_1 cases show longer LOL than the other fuels because of its lower reactivity (higher ignition delay). However, at 1000 K the difference between LOL of diesel and LOL of OME_1 is smaller compared with the other operating conditions.

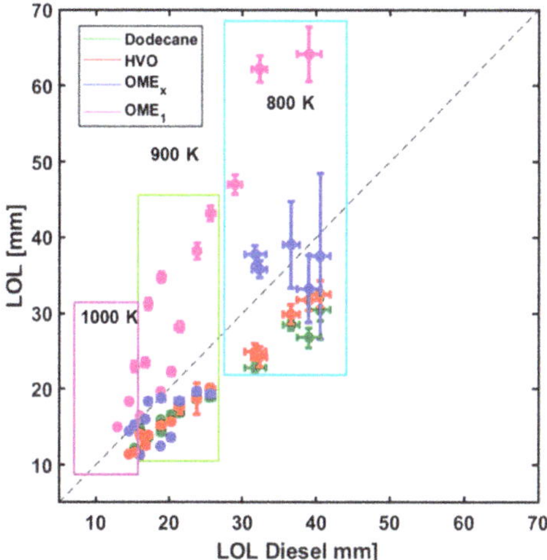

Figure 7. Comparison between lift-off length (LOL) of diesel and the other fuels tested.

3.4. Spray Tip and Flame Penetration

The temporal evolutions of vapor penetration and LOL, as well as the ID for all the fuels at nominal condition have been shown in Figure 8. The vapor penetration and ID have been measured from schlieren images and LOL from OH* chemiluminescence, as was mentioned previously. The standard deviation for each parameter is represented by shaded areas. A first stage from start of injection until ignition can be observed, where vapor penetration is identical for all the fuels. Only at certain time after ignition differences appear. This first stage corresponds to the ignition delay phase, which extends for each fuel until combustion starts. The ignition delay time has been marked in the figure as a dashed vertical line for each fuel. During the period until ignition, there are not effects on the spray behavior, as expected because the penetration is governed by the momentum flux at the nozzle [34] which only depends on the pressure drop across the nozzle and on the orifice area, fuel effects are negligible. In addition, in Figure 8 the effect of the different fuels upon the indicated parameters can be observed. The dashed-dot horizontal lines correspond to the temporal evolution of LOL. After the ignition, the fuel that ignites first penetrates faster than the others do. In general, it is possible to see that while OME$_x$ has similar behavior to the paraffinic fuels, OME$_1$ is quite far from them.

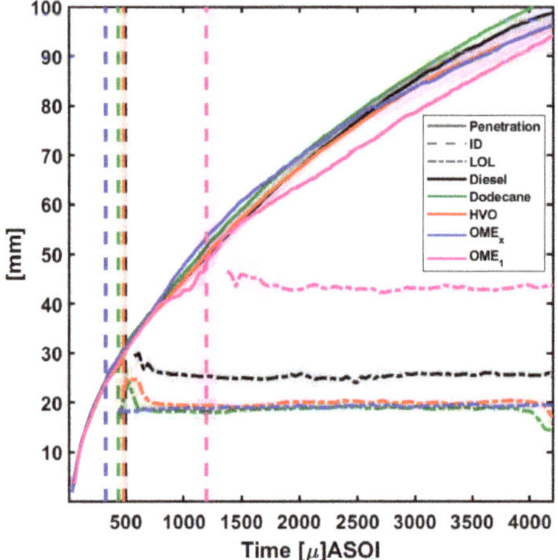

Figure 8. Spray tip penetration and lift-off length at 900 K 15% O2 and 1500 bar (Spray A condition by ECN). Vertical dashed lines represent ignition delay.

On the other hand, from OH* images, the flame tip penetration can be measured also from the high-speed OH* chemiluminescence imaging as the axial distance from tip injector until the flame front. In Figure 9, the temporal evolution of the flame penetration for all fuels at 900 K and 1500 bar and both oxygen concentrations has been represented. The solid lines correspond to the case of 15% of oxygen concentration and the dotted lines to 21% case

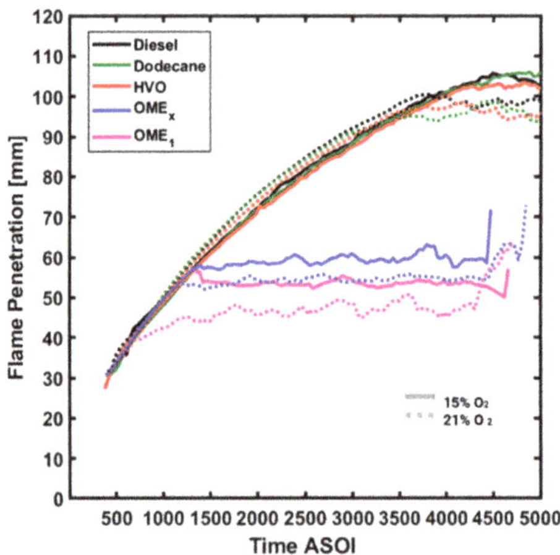

Figure 9. Flame penetration at 900 K and 1500 bar. The solid lines correspond to 15% of Oxygen and dotted lines to 21%.

From this figure, flame tip penetration is seen to undergo a similar time evolution as spray tip penetration during most of the injection period. However, for all cases flame tip becomes constant after a given period, which depends on fuel type and oxygen concentration. For oxygenated fuels, this occurs between 1000 to 1500 µs, while for all other paraffinic fuels it happens close to the end of injection event (around 4200 µs). During that period, the stoichiometric reacting surface stabilizes, and hence the maximum distance where OH* chemiluminescence is recorded does not change with time. For constant ambient and injection conditions this distance has been shown to scale inversely with the stoichiometric mixture fraction, i.e., directly with the stoichiometric A/F ratio [36,37], which depends on oxygen concentration and fuel composition. Table 1 shows stoichiometric air-fuel ratios, from which one can observe that paraffinic fuels have similar values, resulting in the similar long stabilization distance, while the oxygenated fuels approximately a 50% lower $(A/F)_{st}$, resulting in a much shorter flame length.

Furthermore, differences among Diesel/HVO/Dodecane are small, in agreement with the small $(A/F)_{st}$ differences, OME_1 shows a consistently roughly 10% shorter penetration compared to OME_x. Figure 10 has been used to depict the differences in flame structure for each fuel at the nominal operating condition in an instant of time where the combustion is quasi steady (3500 µs). Such images confirm the previously plotted trends among fuels in stabilized flame length, with paraffinic fuels having a similar length, and a shorter one is observed for oxygenated fuels. On the other hand, observed differences in lift-off length also point out at interesting features. As shown above, this quantity is quite similar for all paraffinic fuels and OME_x, while it is far longer for OME_1. The equivalence ratio at the lift-off length (Φ_{LOL}) can be estimated for all fuels using Equation (3), and the values are shown in the Figure 10:

$$\Phi_{LOL} = \frac{f_{LOL}}{1 - f_{LOL}} \cdot A/F_{st} \tag{3}$$

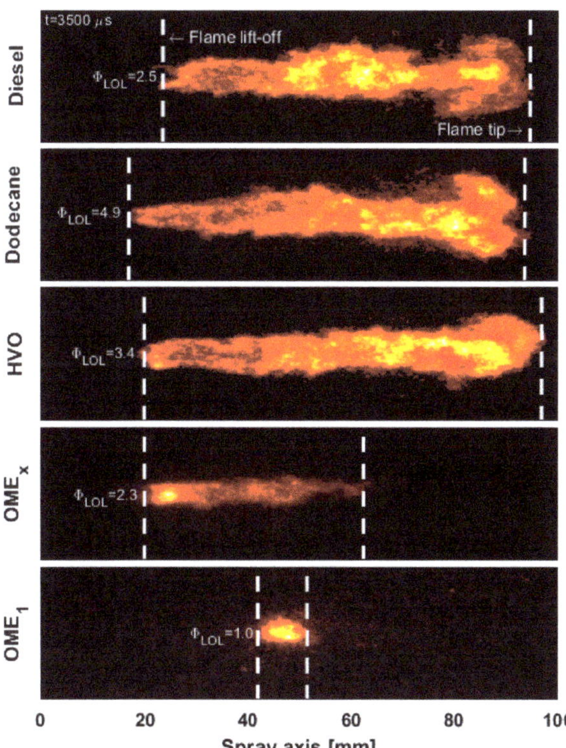

Figure 10. Flame images from OH* chemiluminescence for all fuels at the nominal operating condition at 3500 µs ASOI.

In the Equation (3), the term f_{LOL} represents the fuel mixture fraction along spray axis and it is calculated using the equation (4), where K is a constant equal to 7, d_o is the nozzle diameter and ρ_f and ρ_a correspond to the density of the fuel and the ambient respectively.

$$f_{LOL} = \frac{K \cdot d_0 \cdot \sqrt{\frac{\rho_f}{\rho_a}}}{LOL} \tag{4}$$

Results show that the flame is rich at the lift-off for all fuels except for the OME_1, and hence a typical lifted diffusion flame stabilized at stoichiometric conditions is observed for all fuels except for the latter one. OME_1 flame stabilizes at lean conditions, where all needed air is already available, which explains the previously observed differences in flame length between OME_x and OME_1. Similar lean stabilized flames were observed for an oxygenated fuel (70% tetraethoxypropane 30% heptamethylnonane) with a 100 um orifice at 21% O_2, 850 K and 14.8 kg/m^3 ambient conditions in [38].

3.5. Soot Production

As was indicated in methodology section, the DBI technique was used to determine the soot production. The DBI images were collected after schlieren. However, OH* chemiluminescence was recorded with the two sets. To verify that both sets are consistent and there are no experimental discrepancies among them, the results obtained for ID from OH* images measured with the first and in the second set are compared in Figure 11. The grey shadow in the bisector represents the uncertainty associated to the time interval between consecutive frames. It can be observed that ID differences between both sets are always less than the time elapsed between two consecutives frames (40 μs). Thus, it is possible to conclude that both sets are consistent and results can be analyzed together.

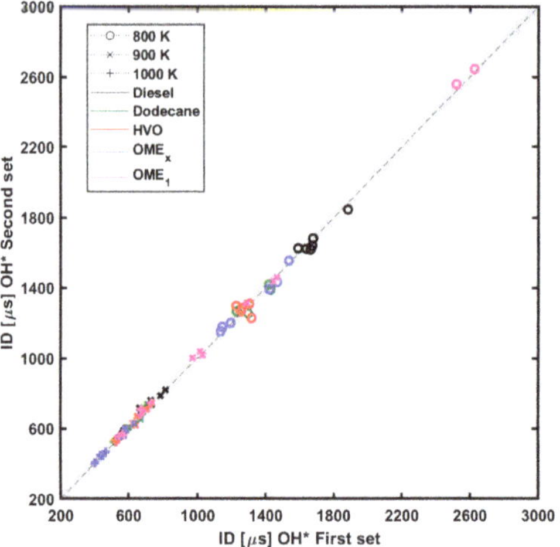

Figure 11. Comparison of ignition delay (ID) obtained in the first set and in the second one.

DBI images were processed to calculate the flame soot in terms of the optical thickness (KL). KL maps were constructed to depict soot evolution throughout the combustion event, by calculating the KL for each axial distance and time-step. A good way to simultaneously evaluate the KL evolution in time and space is transforming the KL map at any given instant into a 1D vector where KL values at any given cross section from the nozzle are accumulated into a single value. Thus, each KL map at any instant is converted into a vector of "accumulated KL" values along the flame axis. By compiling those vectors at every time step (i.e., at every frame), a plot such as those of Figure 12 can be obtained. In any of those plots, the abscissa axis reads for time, the ordinates axis indicates axial distance from the nozzle and the color at any point (X,Y) in the plot indicates the accumulated KL at a cross section at distance Y from the nozzle, of frames taken at time X after SOI.

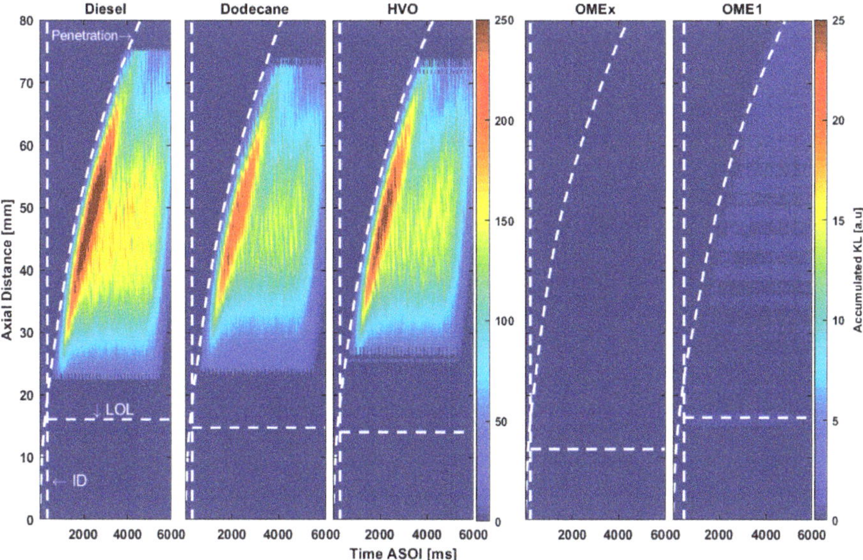

Figure 12. Accumulated KL map at the operating condition with the greatest tendency to soot formation (1000 K of temperature, 500 bar of injection pressure and 15% of O_2).

In Figure 12, the accumulated KL has been represented for the operating condition with greater tendency to soot formation, that is, at the highest temperature (1000 K), the lowest injection pressure (500 bar) and the lowest oxygen concentration (15%). Dashed lines have been used to depict ID, vapor penetration and the flame lift off length. In this figure, it is possible to observe that soot production for OME_1 and OME_x is below the detection threshold. The color scale for these two fuels has been modified in other to enhance this fact. Therefore, a sootless flame seems to be established. This holds for all operating conditions and, consequently, KL values for those fuels cannot be analyzed further. Dodecane, HVO and OME_x show smaller KL values than diesel. As for the time evolution, in all three cases it can be observed that initially the largest soot amount is located at the spray tip, in the head vortex area, but when reaching a distance around 70 mm, this soot disappears due to the establishment of the quasi-steady flame The highest KL region appears for the three fuels from 40 to 50 mm from the nozzle.

For a given fuel composition, the soot production is related to the amount of air entrainment that occurs upstream of the lift-off [29]. Longer LOL suggests less soot formation, but this depends also on fuel composition. This can be observed in Figure 12, where diesel has the longest lift-off length but also the largest soot formation. The stoichiometric air/fuel ratio (ma/mf) for dodecane and OME_1 is quite different. At 15% of oxygen concentration, stoichiometric air-fuel ratio for dodecane is 20.72 versus 10.03 for OME_1 indicating that oxygenated fuels require less air to oxidize. This explains that OME_x, although with LOL closer to diesel, does not produce soot. This statement is supported by the equivalence ratio (Φ) at LOL. For dodecane, Φ is 4.9 at baseline operating conditions, while for the OME_1 is 1.0. Furthermore, OME fuels do not have carbon-carbon (C-C) bonds, which contribute to the absence of soot.

In the case of HVO, its aromatic free composition suggests that less polycyclic aromatic hydrocarbons (PAH) are formed [39]. This could explain the lower soot production for HVO compared to diesel, as they can be considered as the building blocks for particulates in flames [40].

The total soot mass production was calculated from instantaneous KL, following the procedure described in Section 2.3.3. Figure 3 depicts the total soot mass for diesel, dodecane and HVO at different operating conditions with the aim of showing the effect of decreasing temperature, increasing injection pressure and increasing the ambient oxygen content. The shown temperatures are 1000 K and 900 K, from top to bottom; the injection pressures are 500 bar and 1000 bar, from left to right; and 15% and 21% of oxygen have been represented for 900 K and 500 bar at the bottom of the figure.

It can be seen that when decreasing the temperature from 1000 K to 900 K (from top to bottom, at the left side of the Figure 13), the maximum value of soot production is reduced 40% for diesel and 30% for dodecane and HVO. When the injection pressure is increased from 500 bar to 1000 bar (from left to right at the top of figure), the soot reduction is around 30% for all fuels. When the oxygen concentration is increased from 15% to 21% (from left to right at the bottom of figure) the total average soot production is reduced 30% for diesel, 40% for HVO and 50% for dodecane. Results for all conditions show that the parametric variation in terms of soot tendency always holds, with diesel showing the largest soot production, while HVO and dodecane present similar soot production. In fact, considering all results presented in this work regarding combustion and soot formation characteristics, it can be concluded that n-dodecane is a good surrogate to model the HVO due to its similarities.

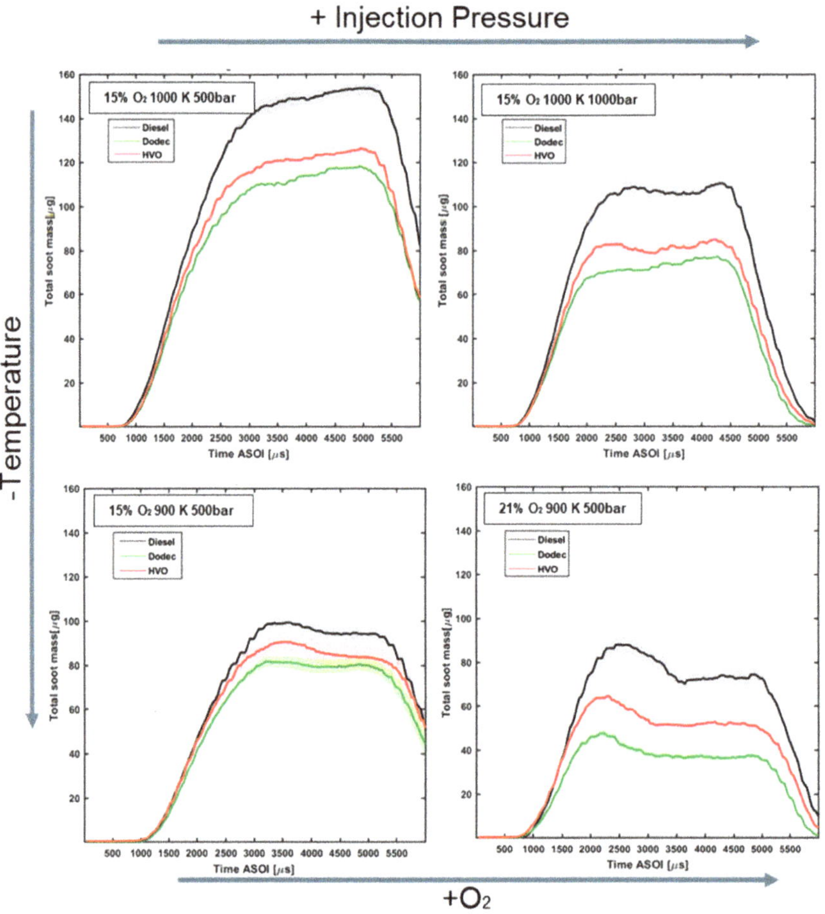

Figure 13. Influence of the variation of injection pressure, temperature and oxygen concentration on the total soot mass.

4. Conclusions

In this study, the main spray parameters and combustion characteristics for diesel, HVO, dodecane, OME_x and OME_1 were evaluated in an ambient at High Pressure and High Temperature installation, with controlled oxygen concentration. The parameters studied were liquid and spray tip penetration, ignition delay, flame lift-off length, flame length, flame tip penetration and soot production. These were assessed using four high-speed imaging based optical techniques: schlieren, OH* chemiluminescence and diffused back illumination. The operating conditions were defined using as baseline the "Spray A" targets recommended by the Engine Combustion Network (ECN). The main conclusions of this study are summarized below.

Results obtained for the dodecane at the baseline condition (900 K 1500 bar and 15% of O_2) for ignition delay, liquid length and flame lift-off length are consistent with those found in the ECN database, which provides confidence in the reliability of the results of this study:

-   The maximum liquid length of all five fuels correlates with the corresponding distillation temperature, stratified in increasing order: OME_1-dodecane-OME_x-HVO-diesel.
-   ID trend matches the cetane number rating of paraffinic fuels. HVO and dodecane ignite very close and earlier than diesel. However, OME_x ignites before the other fuels although its cetane number is not the highest, but its shorter ID is due to the oxygen in its molecular structure, which improves its ignition. The only fuel that ignites after diesel does is OME_1. However, at 1000 K, OME_1 shows a similar behavior than diesel.
-   The longest flame lift-off length is for OME_1 and at 800 K and 900 K, which is quite far from OME_x and paraffinic fuels. Conversely, OME_x has a closer behavior to paraffinic fuels in all operating conditions.
-   During the period until ignition, all fuels show the same behavior in terms of penetration, since the spray tip penetration is known to be governed by the momentum flux conservation at the nozzle orifice, which only depends on the pressure drop across the nozzle and the orifice area.
-   The difference of flame penetration between paraffinic fuel and oxygenated fuel is very marked. Oxygenated fuels stabilize early while paraffinic fuels reach the flame stabilization later as a consequence of the fuel's stoichiometry. Furthermore, an important difference between OME_1 and OME_x at nominal operating condition exists due to the OME_1 flame stabilizes at lean air-fuel ratio.
-   Regarding soot production, the oxygenated fuels (OME_x and OME_1) did not produce detectable soot at any operating condition tested. This can be explained by the very low equivalence ratio at lift-off owing to the oxygen content in the molecule as well as by the absence of C-C bonds.
-   HVO produces less soot than diesel, but more than dodecane. This behavior can be rel0ated with the absence of aromatic compounds in HVO formulation.

Author Contributions: Conceptualization, J.M.G.-O.; Data curation, A.A.G.-C.; Formal analysis, A.A.G.-C. and A.G.; Methodology, J.V.P. and C.M.; Project administration, J.M.G.-O.; Resources, J.V.P.; Supervision, J.M.G.-O.; Writing–original draft, A.A.G.-C. and A.G.; Writing–review & editing, J.V.P., J.M.G.-O., C.M., A.A.G.-C. and A.G. All authors have read and agree to the published version of the manuscript.

Funding: This research has been partly funded by the European Union's Horizon 2020 Programme through the ENERXICO project, grant agreement n° 828947, and from the Mexican Department of Energy, CONACYT-SENER Hidrocarburos grant agreement n° B-S-69926 and by Universitat Politècnica de València through the Programa de Ayudas de Investigación y Desarrollo (PAID-01-18).

Conflicts of Interest: The authors declare no conflict of interest.

Abbreviations

$(A/F)_{st}$	Stoichiometric air–fuel ratio
ASOE	After Start of Energizing
ASOI	After Start of Injection
CAT	Caterpillar
CFPP	Cold Filter Plugging Point
CI	Compression Ignition
CMT	CMT-Motores Térmicos
D	Diameter of lens
DPF	Diesel Particle Filter
ECN	Engine Combustion Network
F	Focal length
FWHM	Full Width at Half Maximum
GTL	Gas to Liquid
HVO	Hydrotreated Vegetable Oil
ICCD	Intensified Charge –Coupled Device
IFPEn	Institut Français du Pétrole Énergies Nouvelles
KL	Optical Thickness
LED	Light-Emitting Diode
NO_x	Nitrogen Oxides
OH*	Excited state of hydroxyl radical
OME_1	Methylal
OME_x	A blend of oxymethylene ethers
PM	Particle Matter
SCR	Selective Catalytic Reduction
SNL	Sandia National Laboratory
SOI	Start of Injection
TU/e	Eindhoven University of Technology

References

1. Reşitoğlu, İ.A.; Altinişik, K.; Keskin, A. The pollutant emissions from diesel-engine vehicles and exhaust aftertreatment systems. *Clean Technol. Environ Policy* **2015**, *17*, 15–27. [CrossRef]
2. Pundir, B.P. *IC Engines: Combustion and Emissions*; Alpha Science International: Oxford, UK, 2010.
3. Mohan, B.; Yang, W.; Chou, S. kiang Fuel injection strategies for performance improvement and emissions reduction in compression ignition engines—A review. *Renew. Sustain. Energy Rev.* **2013**, *28*, 664–676. [CrossRef]
4. Leach, F.; Kalghatgi, G.; Stone, R.; Miles, P. The scope for improving the efficiency and environmental impact of internal combustion engines. *Transp. Eng.* **2020**, *1*, 100005. [CrossRef]
5. Kim, H.Y.; Ge, J.C.; Choi, N.J. Application of Palm Oil Biodiesel Blends under Idle Operating Conditions in a Common-Rail Direct-Injection Diesel Engine. *Appl. Sci.* **2018**, *8*, 2665. [CrossRef]
6. Tziourtzioumis, D.N.; Stamatelos, A.M. Experimental Investigation of the Effect of Biodiesel Blends on a DI Diesel Engine's Injection and Combustion. *Energies* **2017**, *10*, 970. [CrossRef]
7. Merola, S.S.; Tornatore, C.; Iannuzzi, S.E.; Marchitto, L.; Valentino, G. Combustion process investigation in a high speed diesel engine fuelled with n-butanol diesel blend by conventional methods and optical diagnostics. *Renew. Energy* **2014**, *64*, 225–237. [CrossRef]
8. Choi, K.; Park, S.; Roh, H.G.; Lee, C.S. Combustion and Emission Reduction Characteristics of GTL-Biodiesel Fuel in a Single-Cylinder Diesel Engine. *Energies* **2019**, *12*, 2201. [CrossRef]
9. Dimitriadis, A.; Seljak, T.; Vihar, R.; Žvar Baškovič, U.; Dimaratos, A.; Bezergianni, S.; Samaras, Z.; Katrašnik, T. Improving PM-NOx trade-off with paraffinic fuels: A study towards diesel engine optimization with HVO. *Fuel* **2020**, *265*, 116921. [CrossRef]
10. Pastor, J.V.; García, A.; Micó, C.; Lewiski, F. An optical investigation of Fischer-Tropsch diesel and Oxymethylene dimethyl ether impact on combustion process for CI engines. *Appl. Energy* **2020**, *260*, 114238. [CrossRef]

11. Bergthorson, J.M.; Thomson, M.J. A review of the combustion and emissions properties of advanced transportation biofuels and their impact on existing and future engines. *Renew. Sustain. Energy Rev.* **2015**, *42*, 1393–1417. [CrossRef]

12. Yehliu, K.; Boehman, A.L.; Armas, O. Emissions from different alternative diesel fuels operating with single and split fuel injection. *Fuel* **2010**, *89*, 423–437. [CrossRef]

13. Gómez, A.; Soriano, J.A.; Armas, O. Evaluation of sooting tendency of different oxygenated and paraffinic fuels blended with diesel fuel. *Fuel* **2016**, *184*, 536–543. [CrossRef]

14. Benajes, J.; García, A.; Monsalve-Serrano, J.; Martínez-Boggio, S. Potential of using OMEx as substitute of diesel in the dual-fuel combustion mode to reduce the global CO2 emissions. *Trans. Eng.* **2020**, *1*, 100001. [CrossRef]

15. Burger, J.; Siegert, M.; Ströfer, E.; Hasse, H. Poly(oxymethylene) dimethyl ethers as components of tailored diesel fuel: Properties, synthesis and purification concepts. *Fuel* **2010**, *89*, 3315–3319. [CrossRef]

16. Iannuzzi, S.E.; Barro, C.; Boulouchos, K.; Burger, J. POMDME-diesel blends: Evaluation of performance and exhaust emissions in a single cylinder heavy-duty diesel engine. *Fuel* **2017**, *203*, 57–67. [CrossRef]

17. Omari, A.; Heuser, B.; Pischinger, S. Potential of oxymethylenether-diesel blends for ultra-low emission engines. *Fuel* **2017**, *209*, 232–237. [CrossRef]

18. Bjørgen, K.O.P.; Emberson, D.R.; Løvås, T. Combustion and soot characteristics of hydrotreated vegetable oil compression-ignited spray flames. *Fuel* **2020**, *266*, 116942. [CrossRef]

19. Marchitto, L.; Merola, S.S.; Tornatore, C.; Valentino, G. *An Experimental Investigation of Alcohol/Diesel Fuel Blends on Combustion and Emissions in a Single-Cylinder Compression Ignition Engine*; SAE Technical Paper 2016-01-0738; SAE: Warrendale, PA, USA, 2016. [CrossRef]

20. Engine Combustion Network | Engine Combustion Network Website. Available online: https://ecn.sandia.gov/ (accessed on 14 July 2020).

21. Payri, R.; Gimeno, J.; Bardi, M.; Plazas, A.H. Study liquid length penetration results obtained with a direct acting piezo electric injector. *Appl. Energy* **2013**, *106*, 152–162. [CrossRef]

22. Crusius, S.; Müller, M.; Stein, H.; Goral, T. Oxy-methylen-di-methylether (OMEx) as an alternative for diesel fuel and blend compound: Properties, additizing and compatibility with fossil and renewable fuels. In Proceedings of the 12th International Colloquium Fuels-Conventional and Future Energy for Automobiles, Esslingen, Germany, 25 June 2019; p. 8.

23. Benajes, J.; Payri, R.; Bardi, M.; Martí-Aldaraví, P. Experimental characterization of diesel ignition and lift-off length using a single-hole ECN injector. *Appl. Ther. Eng.* **2013**, *58*, 554–563. [CrossRef]

24. Xuan, T.; Desantes, J.M.; Pastor, J.V.; Garcia-Oliver, J.M. Soot temperature characterization of spray a flames by combined extinction and radiation methodology. *Combust. Flame* **2019**, *204*, 290–303. [CrossRef]

25. Settles, G.S. *Schlieren and Shadowgraph Techniques: Visualizing Phenomena in Transparent Media*; Springer Science & Business Media: Berlin/Heidelberg, Germany, 2012; ISBN 978-3-642-56640-0.

26. Pastor, J.V.; Payri, R.; Garcia-Oliver, J.M.; Briceño, F.J. Schlieren Methodology for the Analysis of Transient Diesel Flame Evolution. *SAE Int. J. Engines* **2013**, *6*, 1661–1676. [CrossRef]

27. Pastor, J.V.; García, A.; Micó, C.; García-Carrero, A.A. Experimental study of influence of Liquefied Petroleum Gas addition in Hydrotreated Vegetable Oil fuel on ignition delay, flame lift off length and soot emission under diesel-like conditions. *Fuel* **2020**, *260*, 116377. [CrossRef]

28. Reyes, M.; Tinaut, F.V.; Giménez, B.; Pastor, J.V. Effect of hydrogen addition on the OH* and CH* chemiluminescence emissions of premixed combustion of methane-air mixtures. *Int. J. Hydrogen Energy* **2018**, *43*, 19778–19791. [CrossRef]

29. Siebers, D.; Higgins, B. Flame Lift-Off on Direct-Injection Diesel Sprays Under Quiescent Conditions. *SAE Trans.* **2001**, *110*, 400–421.

30. Xuan, T.; Pastor, J.V.; García-Oliver, J.M.; García, A.; He, Z.; Wang, Q.; Reyes, M. In-flame soot quantification of diesel sprays under sooting/non-sooting critical conditions in an optical engine. *Appl. Ther. Eng.* **2019**, *149*, 1–10. [CrossRef]

31. Choi, M.Y.; Mulholland, G.W.; Hamins, A.; Kashiwagi, T. Comparisons of the soot volume fraction using gravimetric and light extinction techniques. *Combust. Flame* **1995**, *102*, 161–169. [CrossRef]

32. Li, D.; He, Z.; Xuan, T.; Zhong, W.; Cao, J.; Wang, Q.; Wang, P. Simultaneous capture of liquid length of spray and flame lift-off length for second-generation biodiesel/diesel blended fuel in a constant volume combustion chamber. *Fuel* **2017**, *189*, 260–269. [CrossRef]

33. Lequien, G.; Berrocal, E.; Gallo, Y.; Themudo e Mello, A.; Andersson, O.; Johansson, B. *Effect of Jet-Jet Interactions on the Liquid Fuel Penetration in an Optical Heavy-Duty DI Diesel Engine*; SAE Technical Paper 2013-01-1615; SAE: Warrendale, PA, USA, 2013. [CrossRef]

34. Kook, S.; Pickett, L.M. Liquid length and vapor penetration of conventional, Fischer–Tropsch, coal-derived, and surrogate fuel sprays at high-temperature and high-pressure ambient conditions. *Fuel* **2012**, *93*, 539–548. [CrossRef]

35. Payri, R.; Salvador, F.J.; Manin, J.; Viera, A. Diesel ignition delay and lift-off length through different methodologies using a multi-hole injector. *Appl. Energy* **2016**, *162*, 541–550. [CrossRef]

36. Pickett, L.M.; Siebers, D.L. Orifice Diameter Effects on Diesel Fuel Jet Flame Structure. *J. Eng. Gas Turbines Power* **2005**, *127*, 187–196. [CrossRef]

37. Pastor, J.V.; García-Oliver, J.M.; López, J.J.; Vera-Tudela, W. An experimental study of the effects of fuel properties on reactive spray evolution using Primary Reference Fuels. *Fuel* **2016**, *163*, 260–270. [CrossRef]

38. Pickett, L.M.; Siebers, D.L. *Non-Sooting, Low Flame Temperature Mixing-Controlled DI Diesel Combustion*; SAE Technical Paper 2004-01-1399; SAE: Warrendale, PA, USA, 2004. [CrossRef]

39. Aatola, H.; Larmi, M.; Sarjovaara, T.; Mikkonen, S. Hydrotreated Vegetable Oil (HVO) as a Renewable Diesel Fuel: Trade-off between NOx, Particulate Emission, and Fuel Consumption of a Heavy Duty Engine. *SAE Int. J. Engines* **2008**, *1*, 1251–1262. [CrossRef]

40. Marinov, N.M.; Pitz, W.J.; Westbrook, C.K.; Vincitore, A.M.; Castaldi, M.J.; Senkan, S.M.; Melius, C.F. Aromatic and Polycyclic Aromatic Hydrocarbon Formation in a Laminar Premixed n-Butane Flame. *Combust. Flame* **1998**, *114*, 192–213. [CrossRef]

applied sciences

MDPI

Article

Effect of Cylinder-by-Cylinder Variation on Performance and Gaseous Emissions of a PFI Spark Ignition Engine: Experimental and 1D Numerical Study

Luigi Teodosio [1,*], Luca Marchitto [1], Cinzia Tornatore [1], Fabio Bozza [2] and Gerardo Valentino [1]

[1]   STEMS, Italian National Research Council, Via Marconi, 4, 80125 Naples, Italy; luca.marchitto@stems.cnr.it (L.M.); cinzia.tornatore@stems.cnr.it (C.T.); gerardo.valentino@stems.cnr.it (G.V.)
[2]   Department of Industrial Engineering, University of Naples "Federico II", Via Claudio, 21, 80125 Naples, Italy; fabio.bozza@unina.it
*   Correspondence: luigi.teodosio@stems.cnr.it

Abstract: Combustion stability, engine efficiency and emissions in a multi-cylinder spark-ignition internal combustion engines can be improved through the advanced control and optimization of individual cylinder operation. In this work, experimental and numerical analyses were carried out on a twin-cylinder turbocharged port fuel injection (PFI) spark-ignition engine to evaluate the influence of cylinder-by-cylinder variation on performance and pollutant emissions. In a first stage, experimental tests are performed on the engine at different speed/load points and exhaust gas recirculation (EGR) rates, covering operating conditions typical of Worldwide harmonized Light-duty vehicles Test Cycle (WLTC). Measurements highlighted relevant differences in combustion evolution between cylinders, mainly due to non-uniform effective in-cylinder air/fuel ratio. Experimental data are utilized to validate a one-dimensional (1D) engine model, enhanced with user-defined sub-models of turbulence, combustion, heat transfer and noxious emissions. The model shows a satisfactory accuracy in reproducing the combustion evolution in each cylinder and the temperature of exhaust gases at turbine inlet. The pollutant species (HC, CO and NOx) predicted by the model show a good agreement with the ones measured at engine exhaust. Furthermore, the impact of cylinder-by-cylinder variation on gaseous emissions is also satisfactorily reproduced. The novel contribution of present work mainly consists in the extended numerical/experimental analysis on the effects of cylinder-by-cylinder variation on performance and emissions of spark-ignition engines. The proposed numerical methodology represents a valuable tool to support the engine design and calibration, with the aim to improve both performance and emissions.

Keywords: exhaust emissions; spark ignition engine; experiments; 1D model; cylinder-by-cylinder variation; exhaust gas recirculation

Citation: Teodosio, L.; Marchitto, L.; Tornatore, C.; Bozza, F.; Valentino, G. Effect of Cylinder-by-Cylinder Variation on Performance and Gaseous Emissions of a PFI Spark Ignition Engine: Experimental and 1D Numerical Study. *Appl. Sci.* **2021**, *11*, 6035. https://doi.org/10.3390/app11136035

Academic Editor: Ramin Rahmani

Received: 29 April 2021
Accepted: 26 June 2021
Published: 29 June 2021

Publisher's Note: MDPI stays neutral with regard to jurisdictional claims in published maps and institutional affiliations.

1. Introduction

Modern internal combustion engines (ICEs) are designed with the aim to reduce the pollutant and CO_2 emissions, while delivering the required torque performance, complying with the binding legislations for vehicle homologation [1]. Referring to spark ignition (SI) engines, car manufacturers are facing the challenging path towards the low-emissions vehicles through the development of innovative, and sometimes very complex, engine architectures. In the present-day scenario, various technical solutions, characterized by a different cost/effectiveness compromise, have been successfully implemented in SI engines for the control and the abatement of noxious species at the exhaust. The most common technology still consists in the adoption of the three-way catalyst (TWC) along the exhaust line. TWC device requires a close-to-stoichiometric air/fuel (A/F) mixture to guarantee a high efficiency, with significant performance losses at the engine cold start operations. To overcome this issue, a greater interest is also devoted to the emerging techniques to limit the in-cylinder production of pollutant emissions for SI engines, such as the adoption of innovative combustion modes moving towards the Low-Temperature Combustion (LTC) concept: the homogeneous charge compression ignition (HCCI), the spark-assisted compression ignition (SACI) and the turbulent jet injection [2]. Turbulent jet injection (TJI) demonstrates to be a promising technique to reduce the exhaust emissions of SI engines [3], especially in the case of an active pre-chamber thanks to the ultra-lean combustion; on the other side, HCCI and SACI combustion modes allow significant improvements in NOx emissions, while some penalties on the HC and CO are obtained [4].

In addition, well-established technologies, such as the employment of cooled EGR [5], direct or port water injections [6] and alternative fuels, including ethanol/gasoline blends [7] or methanol/gasoline blends [8], allow certain benefits on the main exhaust emissions.

In addition to the above-discussed techniques to improve pollutant species, a particular attention has to be devoted to the control of cylinder-by-cylinder variation, since it could lead to a combustion deterioration with consequent increase in the exhaust emissions, especially of HC and CO species.

Different factors may play a role on the onset of cylinder-by-cylinder variation in SI engines. Indeed, the increasing complexity of engine subsystems, the high number of mechanical components, the manufacturing tolerance, the components aging can be considered as examples inducing the cylinder-by-cylinder variation and leading to a worsening in both efficiency and exhaust emissions. Of course, this aspect cannot be overlooked and should be taken into account during the engine development phase. An individual control and optimization of combustion for cylinders in a multi-cylinder engine can further contribute to suppress the cylinder-by-cylinder variation and to improve emissions. To this aim, both experimental and modeling approaches are employed. Experimental and numerical methods are increasingly combined to merge their relative advantages and to offer the availability of validated numerical tools capable to reproduce the behavior of both engine and single cylinders under various operating conditions.

Despite the relevant effects of cylinder-by-cylinder variation on SI engine performance and emissions, few technical papers are available in the literature. In addition, a reduced attention is devoted to understanding the causes that originates this phenomenon. As an example, Czarnigowski [9] investigated the effect of cylinder-by-cylinder variation on indicated mean effective pressure (IMEP) in a radial nine-piston engine, founding differences in single cylinder IMEP up to 40%. Zhou et al. [10] conducted an experimental and numerical study on a four-cylinder spark ignition engine to evaluate the influence of cylinder-by-cylinder variation on the performance. They attribute this phenomenon to the non-uniformity of gas exchange between cylinders and estimated a relative deviation of individual cylinder IMEP larger than 30%. Recently, Xu et al. [11] proposed a combustion variation control strategy, optimizing the thermal efficiency of a lean burn spark ignition engine by means of the reduction in the cylinder-by-cylinder variability. They reduced the combustion variation up to 28% with a maximum and an average increase in the brake thermal efficiency of 0.32% and 0.13%, respectively.

Referring to the cylinder-out emissions, Einewall et al. [12] carried out individual cylinder measurements of emissions and pressure cycles on a six-cylinder lean burn natural gas engine. Mixture quality variations between cylinders were confirmed by the analysis of heat release and A/F ratio. Emission measurements in each cylinder were performed only at high/medium loads and low speeds. No clear trend between A/F ratio and HC emissions of single cylinders was observed, while lower NOx emissions were detected with the air/fuel mixture leaning.

Concerning the numerical approach, a number of methodologies were adopted by worldwide researchers to explore the potentials of the models in reproducing both engine and cylinder performance. In the case of exhaust emissions, regression methods, Artificial Neural Network (ANN) models, Adaptive Neuro-Fuzzy Systems (ANFIS) and various phenomenological sub-models integrated into the fluid-dynamic codes are used. As an example, Sayin et al. [13] utilized an ANN approach to predict the overall performance, HC and CO emissions at the exhaust of an SI engine fueled with gasoline at different octane numbers. It was observed that the ANN model was able to reproduce the engine behavior at different speed/load points with very low root mean square errors. Zschutschke et al. [14] coupled a 1D code to a detailed chemical kinetics solver for the estimation of the engine-out NOx emissions of a direct injection spark ignition engine. The model was validated in a single operating point against the experimental findings and the outcomes of a developed 3D CFD model. Then, it was extended for the prediction of NOx emissions in the entire engine operating map, denoting a good agreement with the experimental data.

In a previous work [15], the authors studied a turbocharged gasoline engine through a 1D code to predict the combustion and the emissions (HC, CO and NO) of a single cylinder under a limited set of operating points. A certain inaccuracy was found in reproducing the experimental trend of HC emission at part load and increasing the EGR content.

Liu et al. [16] analyzed the combustion process and the emissions of an original compression-ignition (CI) engine converted to an SI natural gas (NG) engine using 3D G-equation based RANS simulations. According to a unique set of model tuning parameters, 3D model was able to qualitatively predict the effect of NG composition on emissions over a reduced range of operating conditions.

In the light of the above-discussed literature works, it emerges a lack of combined numerical/experimental studies on the main pollutant emissions of SI engines over different operating conditions (variation in speed/load point and EGR rate), also including the effects of cylinder-by-cylinder variation.

The main topic of the present paper is represented by the combined experimental and 1D numerical analyses of a small turbocharged PFI spark ignition engine in order to provide fast and accurate predictions for individual cylinder-out emissions at different operations and with an apparent cylinder-by-cylinder variation. A previous dedicated experimental study on the examined engine has shown a certain difference in the injected fuel quantity by the port-injectors, mainly ascribed to the fuel rail geometry [15].

In this work, in a first stage, the original engine test bench was modified with the aim to measure both the cylinder-out exhaust emissions and the overall emissions at the engine exhaust, just upstream of the TWC. An extensive experimental campaign was carried out: at each operating condition, the individual cylinder behavior was characterized both in terms of performance, combustion evolution and stability, knock occurrence and exhaust emissions.

Tests were performed at various speed/load points of the engine domain, including operations under different external EGR rates to accurately explore the emission variations for the engine and cylinders. In particular, part-load points typical of the engine Worldwide harmonized Light-duty vehicles Test Cycle (WLTC) were investigated under stoichiometric A/F ratio and increased residual contents.

The experimental outcomes were employed to validate a 1D model of the entire engine, developed within the GT-Power^{TM} code. Refined sub-models of turbulence, combustion, heat transfer and pollutant emissions were utilized and integrated within the commercial code. The propagation of the cylinder-out noxious species within the exhaust system up to the TWC was also considered. The main innovative aspect of present work is represented by the adopted modeling approach which allows to easily forecast the effects of cylinder-by-cylinder variation on both combustion and exhaust emissions of a SI engine and in a wide range of operating conditions. An additional novelty of work, compared to the ones reported in the current literature, is represented by the first-attempt prediction of the single cylinder emission characteristics. Once validated, the model is applied to reproduce the improvements in terms of Indicated Specific Fuel Consumption (ISFC) and pollutants resulting from the suppression of A/F ratio imbalance between engine cylinders.

Summarizing, the proposed numerical procedure can be considered a valuable tool to control the cylinder-by-cylinder variation with the aim to optimize the individual cylinder behavior for improved engine stability, fuel economy and pollutant emissions.

2. Engine System

The engine used for the experimental activity is a twin-cylinder turbocharged spark ignition engine, equipped with two port injectors, one for each cylinder, to supply the gasoline just upstream of the intake valves. Engine's main characteristics are reported in the following Table 1. It is provided by a conventional pent-roof combustion chamber, a centered spark plug and a standard ignition system. Each cylinder presents 4 valves, two intake valves and two exhaust valves. An electro-hydraulic Variable Valve Actuation (VVA) module is mounted on the intake side, allowing for a flexible control of the lift profile. This device provides the actuation of both Early Intake Valve Closure (EIVC) and Full Lift valve strategies. On the exhaust side, a fixed valve lift strategy is employed. Engine boosting is realized by a small waste-gated turbocharger. The compressor operating domain bounds the engine performance because of surge and choke phenomena and of a maximum allowable rotational speed (255,000 rpm). In particular, at low speeds, the boost pressure is limited to avoid the compressor surging occurrence [17]. An additional constraint is imposed by the manufacturer to the maximum boost level, achieved at high speeds, to guarantee the mechanical integrity of the intake plenum.

Table 1. Engine characteristics.

Model	2 Cylinders, Turbocharged PFI
Compression ratio, -	9.9
Displacement, cm^3	875
Valve number, -	4 per cylinder
Bore/Stroke, mm	80.5/86
Connecting rod length, mm	136.85
Max Brake Power, kW	64.6@5500
Max Brake Torque, Nm	146.1@2500
IVO-IVC at 2 mm lift, CAD AFTDC	342/356–420/624
EVO-EVC at 2 mm lift, CAD AFTDC	134–382

The engine is designed following the so-called "downsizing" concept. At high loads the engine works under knock-limited conditions and this requires to delay the combustion phasing (50% of mass fraction burned—MFB 50%), especially at lower speeds where a higher knock tendency usually occurs. On the other hand, at high speeds the A/F mixture has to be particularly enriched to maintain the Temperature at Turbine Inlet below a certain maximum allowable level. These control strategies, although mandatory for the examined engine, greatly penalize the fuel consumption at high loads. For this reason, the base engine configuration is properly modified for research purpose through the installation of an external low-pressure (LP) exhaust gas recirculation (EGR) circuit as depicted in Figure 1.

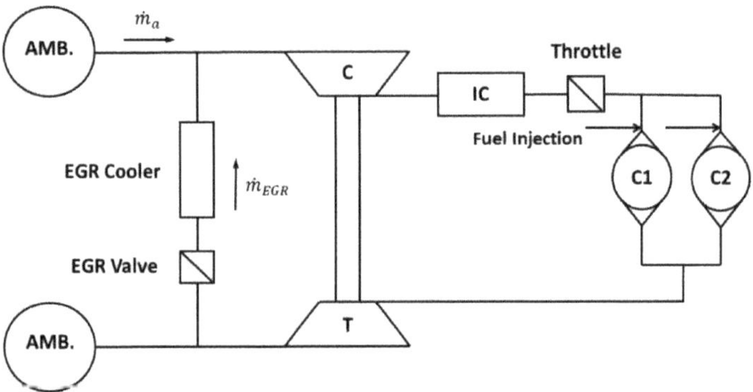

Figure 1. Engine layout with LP EGR circuit.

EGR rate and temperature are controlled through the throttle EGR valve and the cooler device, respectively. An enhanced cooling system, requiring a water cooled heat exchanger, is utilized to cool down the recycled gas. The LP EGR system was preferred over a high-pressure (HP) device, because it allows to avoid the pressure fluctuations and the counter-flow within the EGR circuit, due to the turbocharger dumping effect.

As known, EGR is essentially considered for knock control purposes at high loads, although certain EGR-related efficiency advantages are also obtained at part loads, due to the engine de-throttling.

The engine at test bench is also equipped along the exhaust line with a Three-way Catalytic converter (TWC) to guarantee the pollutant emissions (HC, CO and NOx) abatement, provided that the air/fuel ratio window very close to the stoichiometric value is maintained.

The exhaust system was opportunely modified to extract a portion of the exhaust gas from a single cylinder. In particular, a thin metallic tube is inserted inside of the exhaust manifold, through a properly realized hole, and oriented towards the exhaust side of one cylinder. The internal end of the above tube is located very close the exhaust valves. In this way, the exhaust gases of the selected cylinder (Cyl #1 in the following) are extracted and externally derived to measure the individual cylinder-out noxious emissions.

3. Experimental Setup and Test Procedure

As aforementioned, the tested engine is provided by an electro-hydraulic variable valve actuation (VVA) device. In spite of the VVA potentials, in the experimental tests the load is adjusted only acting on the throttle valve opening or waste-gate valve opening, without modifying the intake valve lift profile which is set to the "Full Lift" configuration. The intake air is constantly supplied to the engine at 293 ± 1 K by an air conditioning unit. Each engine cylinder is equipped with piezo-quartz pressure transducer (accuracy of $\pm 0.1\%$) to detect the in-cylinder pressure signal.

The instantaneous pressure signals are acquired by the AVL INDICOM over 270 consecutive pressure cycles for the combustion analysis, assuming a polytropic thermodynamic process. A resolution of 0.1 CAD within the angular window between -90 and 90 CAD AFTDC is chosen, while outside this angular interval the sampling resolution is set at 1 CAD.

An automatic post-processing tool, based on a thermodynamic model, computes the ensemble averages of in-cylinder pressure and burn rate profiles and the combustion characteristics data. Furthermore, the boost pressure and the upstream turbine pressure are acquired through piezo-resistive low pressure indicating sensors located at the compressor outlet and at the outlet of the exhaust manifold, respectively. The engine is also equipped with thermocouples to monitor intake and exhaust temperatures, with particular attention to the control of the turbine inlet temperature in order to avoid unacceptable levels for the turbine blades.

A prototype driver, managed within LabView environment, is capable to switch from the commercial ECU to an external user control of the main engine variables: fuel injection timing and duration, boost pressure, spark timing, external EGR flow and temperature. EGR-related variables are managed acting on valve opening and cooler device within the EGR circuit.

Exhaust gas emissions are sampled upstream of the three-way catalyst. An ultra-violet gas analyzer (ABB UV Limas 11) measures NO_x. A cold extractive IR gas analyzer (ABB URAS) detects CO, CO_2 and O_2 while a FID analyzer (Siemens, Milano, Italy) is used for THC. A non-dispersive infrared sensor measures the CO_2 concentration at the inlet, downstream of the throttle valve to monitor the EGR rate. The sensor error is set at 0.07% for the measured CO_2 concentration.

The engine is tested in steady-state conditions at two different speeds (1800 and 3000 rpm) and various loads (from low to medium/high levels), also including increasing external EGR rates. In particular, EGR is modified by keeping constant the engine IMEP and the relative A/F ratio, λ. This means that at increasing the EGR rate the IMEP is restored to the nominal value with engine boosting; in a similar way, λ is maintained to the stoichiometric value by a modulation of gasoline injection duration. These operating points are selected because they are representative of the WLTC cycle for a segment A vehicle equipped with the engine under investigation. Performance in these points, such as fuel consumption and CO_2 emissions, are relevant for the vehicle homologation Various operating parameters are collected during the experiments, including torque/power, air flow rate, fuel flow rate, boost and exhaust pressures and temperatures, exhaust emissions, etc.

The overall considered operating conditions are reported in Table 2. Test grid collects 31 points which were gathered into 7 groups, each one characterized by the same nominal load level and rotational speed. For a single group, a label is defined (Table 2), referring to the nominal net IMEP and speed, which is utilized in the following figures. The table also shows the measured values of EGR rate, relative A/F ratio, λ, and the spark advance (SA). This last is selected to realize the Maximum Brake Torque (MBT) condition at knock free operations, while at high/medium loads the SA is chosen to operate at knock borderline (knock limited spark advance, KLSA). In particular, knocking is monitored by a knock sensor, installed between the two cylinders on the block, which automatically cuts out the power output and avoids the engine operation under severe knock conditions.

Table 2. Overall set of experimental engine points.

Load	Speed, rpm	IMEP, bar	λ, -	EGR Rate,%	SA, CAD AFTDC	Label
Low	3000	6.96	1.01	0.0	−30	3000@7
	3000	6.93	1.01	5.8	−38	
	3000	6.92	0.99	9.0	−40	
	3000	7.00	1.00	11.2	−45	
	3000	7.02	1.01	14.9	−48	
Med	1800	7.98	1.00	0.0	−20	1800@8
	1800	8.00	1.00	5.6	−27	
	1800	8.07	1.00	10.0	−31	
	1800	8.04	1.00	13.5	−31	
	3000	9.05	1.00	0.0	−25	3000@9
	3000	9.04	1.00	5.4	−25	
	3000	9.05	0.99	8.6	−30	
	3000	9.00	0.99	16.2	−45	
	3000	9.16	1.00	19.3	−50	
	1800	10.08	1.00	0.0	−10	1800@10
	1800	10.04	1.00	5.3	−15	
	1800	10.03	1.00	9.4	−21	
	1800	9.96	1.00	14.4	−31	
	3000	10.02	1.00	0.0	−16	3000@10
	3000	10.05	1.00	5.6	−21	
	3000	10.16	1.00	14.6	−36	
Med	3000	10.99	1.00	0.0	−14	3000@11
	3000	11.09	1.00	4.9	−18	
	3000	11.08	1.00	7.7	−22	
	3000	11.11	1.00	10.7	−26	
	3000	11.04	1.00	15.7	−35	
	3000	11.01	1.00	19.8	−50	
Med/High	3000	13.10	1.00	0.0	−19	3000@13
	3000	13.12	1.01	6.4	−22	
	3000	13.02	0.99	10.0	−26	
	3000	13.07	0.99	14.6	−33	

The measured net IMEP (Table 2) is properly derived from the acquired in-cylinder pressure traces over the entire engine cycle and represents the difference between the gross IMEP (in-cylinder pressure over compression and expansion strokes) and the pumping mean effective pressure (PMEP), evaluated over the intake and exhaust strokes.

Further parameters are monitored to preserve the safety of the turbocharger and of the entire engine. In particular, the boost pressure is controlled at high loads, acting on the waste-gate (WG) valve opening, to provide proper operation for the port injectors and to avoid the mechanical failures for the intake manifold. The average maximum in-cylinder pressure is constrained to limit the engine mechanical stresses. The mechanical and thermal safety of the turbine also obliges to control the turbocharger speed and the turbine inlet temperature (TIT). Summarizing, the following constraints are taken into account:

- Boost pressure: below 2.4 bar;
- TIT: below 950 °C;
- In-cylinder pressure peak: below 85 bar;
- Turbocharger speed: below 255,000 rpm.

4. Experimental Results

As discussed in the introduction section, experiments carried out on the examined engine have highlighted significant differences in the combustion evolution between cylinders, mainly ascribed to a non-uniform effective in-cylinder air/fuel (A/F) ratio [15]. Cylinder-by-cylinder variations are apparent in Figure 2a which shows the experimental in-cylinder pressure trace and the rate of heat release (ROHR) for two representative operating points (3000@9 and 3000@13, 0% EGR). As aforementioned, the pressure data refer to the ensemble average over 270 consecutive cycles. For both operating conditions, the pressure curves of two cylinders are overlapped in the compression stage, denoting the same air volumetric efficiency, while Cyl #2 provides higher pressure peak and combustion rate than Cyl #1. The same behavior is found at each investigated operating point, suggesting a systematic difference in pressure cycles. Further experimental tests on the adopted fuel injection system have highlighted a different fuel supply to cylinders, resulting from a variation in port injectors fuel rates (Figure 2b). The results plotted in Figure 2b were collected through injection tests realized at am-bent air pressure with fuel injection system removed from the engine. A drive signal to injectors was used to simulate the injection timing; a certain number of injection timings were considered to measure the gasoline mass per stroke delivered by injectors. In particular, consistently with the in-cylinder pressure traces shown in Figure 2a, the injector corresponding to Cyl #2 provides a higher fuel mass flow rate. Interestingly, no difference was found also when the relative position of two injectors was switched, indicating the fuel rail geometry as responsible of the cylinder-by-cylinder variation. Indeed, the gasoline rail mounted on the examined engine at test bench represents a non-optimized prototype geometry. The optimized version of rail, mounted on the commercial SI engine, does not exhibit differences in the gasoline mass flow rates between port injectors. All the previous considerations underline that rail geometry is main responsible for the experimental cylinder-by-cylinder variation analyzed in this research activity.

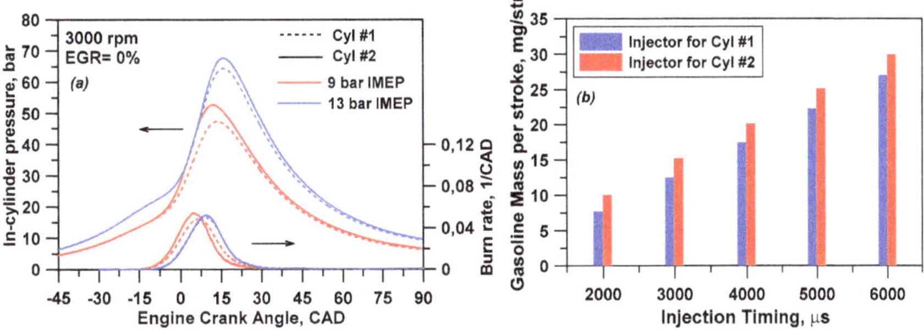

Figure 2. Comparison of experimental in-cylinder pressure traces and burn rates profiles at 3000 rpm, two IMEP levels (9 and 13 bar) and EGR = 0% (**a**). Differences of the injected gasoline mass per stroke between port injectors (**b**).

Figure 3 shows the relative air–fuel ratio (λ) in Cyl #1, Cyl #2 and the engine exhaust at different engine operating points and EGR rates. λ of Cyl #1 was estimated from the composition of the exhaust gases of that cylinder through the carbon balance of species. Similarly, the overall engine λ was verified from the overall exhaust gas composition. This allows to indirectly evaluate the λ value of Cyl #2 under the hypothesis of an equal volumetric efficiency for both cylinders, as suggested by the comparison of pressure traces in the compression stage. Even though the engine conditions are close to stoichiometric, lean (λ between 1.03 and 1.08) and rich (λ between 0.96 and 0.98) mixtures are obtained in Cyl #1 and Cyl #2, respectively. Considering that the engine does not provide an individual set of injection and combustion phasing for each cylinder, the same spark advance and injection parameters are chosen for both cylinders, optimizing the engine load. As a consequence, a difference in the single cylinder load is found at each operating condition.

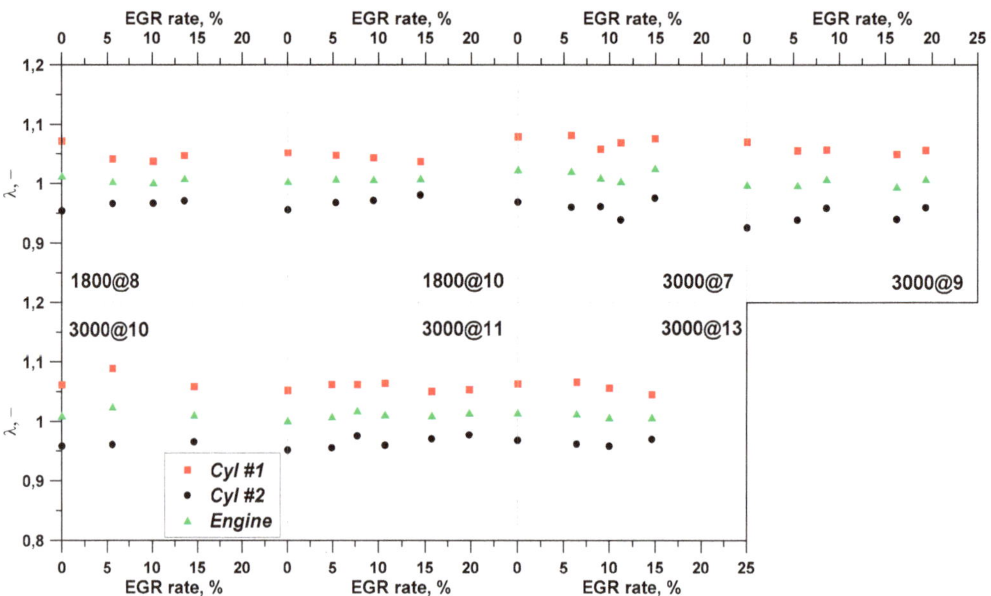

Figure 3. Relative air–fuel ratio (λ) in Cyl #1, Cyl #2 and overall engine at different operating points as a function of EGR rate.

The individual IMEP levels for both cylinders at different operating points are shown in Figure 4 as a function of the EGR rate. As expected, the larger fuel amount introduced in the Cyl #2 results in a higher IMEP at each investigated condition. Furthermore, the adoption of the same spark advance for both cylinders penalizes the combustion phasing of Cyl #1 which results slightly late compared to the MBT. As concerns the trend versus the EGR rate, it is worth noting that the measurements are performed keeping constant the overall engine load; hence, the single cylinder IMEPs are almost constant at varying the exhaust-gas recirculation.

In Figure 5, the EGR-related trend of the engine ISFC is presented for the measured operating points. As expected, a reduced ISFC is observed at low and medium/high IMEPs by increasing the EGR content. At low loads, the ISFC benefits are ascribed to the EGR-induced reduction in the pumping losses promoted by the engine de-throttling; at medium/high loads, the ISFC advantages are obtained thanks to the EGR capability in reducing the knock tendency [18].

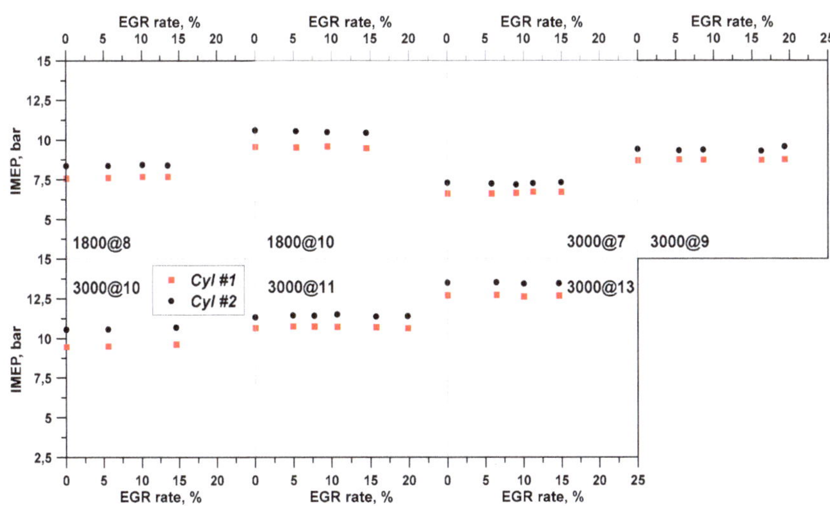

Figure 4. Indicated load in Cyl #1 and Cyl #2 at different engine operating points as a function of EGR rate.

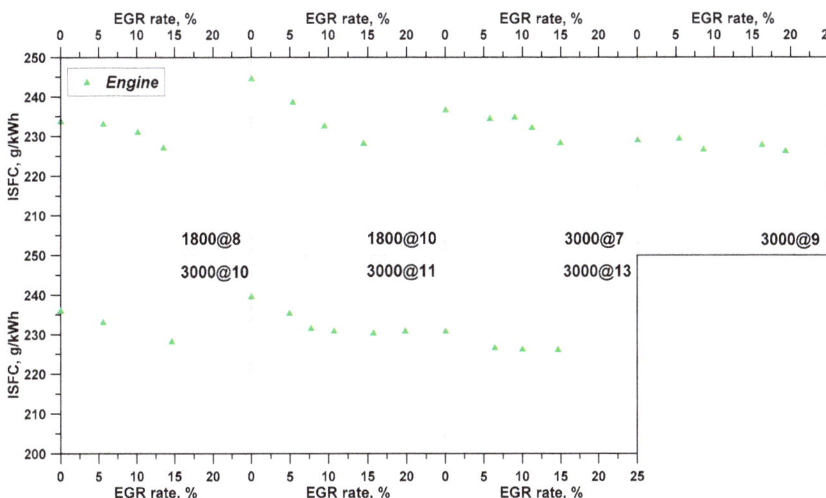

Figure 5. Engine Indicated Specific Fuel Consumption at different operating points as a function of EGR rate.

Figures 6 and 7 show the characteristic angle of MFB 10% and the combustion core duration (MFB 10–50%) for both cylinders at different operating conditions. Looking at the early stage of combustion, at each operating condition, a slight delay in MFB 10% (Figure 6) of Cyl #1 compared to Cyl #2 is evident: the richer mixture in Cyl #2 provides a higher flame speed, advancing the MFB 10%. As shown in Figure 6, this gap increases at higher EGR rates; as is well-known, the charge dilution slows down the combustion rate, enhancing the effect of the equivalence ratio on the flame speed. The cylinder-to-cylinder variation in MFB 10% does not provide significant differences in the core combustion duration as reported in Figure 7. Even though the combustion duration in Cyl #1 results always prolonged compared to Cyl #2, the maximum difference in MFB 10–50% is less than 1.5 CAD at 1800@10 with 14.4% of external EGR. Regarding the effect of charge dilution on the combustion duration, the EGR prolongs the MFB 10–50% at each operating condition. This trend is more evident at higher engine speed, due to the reduction in cycle length, leaving a shorter period available to complete the combustion process. Further increase in combustion duration against the EGR is found at lower engine load, because of the greater impact of internal EGR which contributes to increase the overall in-cylinder residual content at decreasing the load.

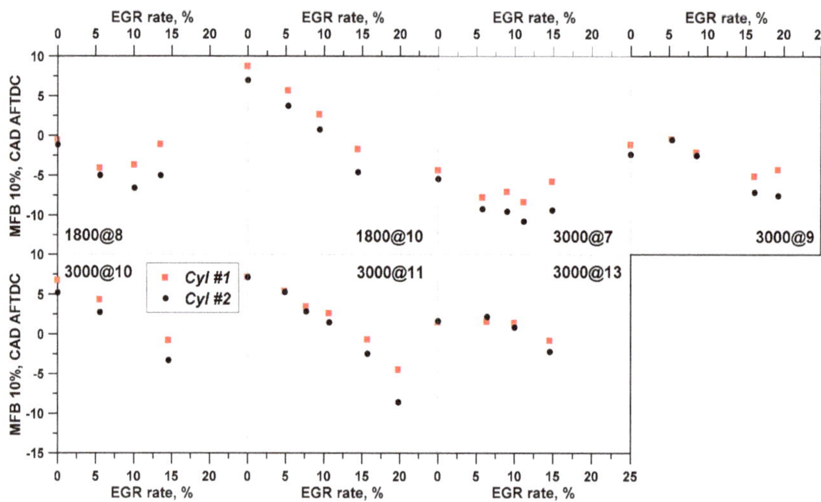

Figure 6. MFB 10% in Cyl #1 and Cyl #2 at different operating conditions as a function of EGR rate.

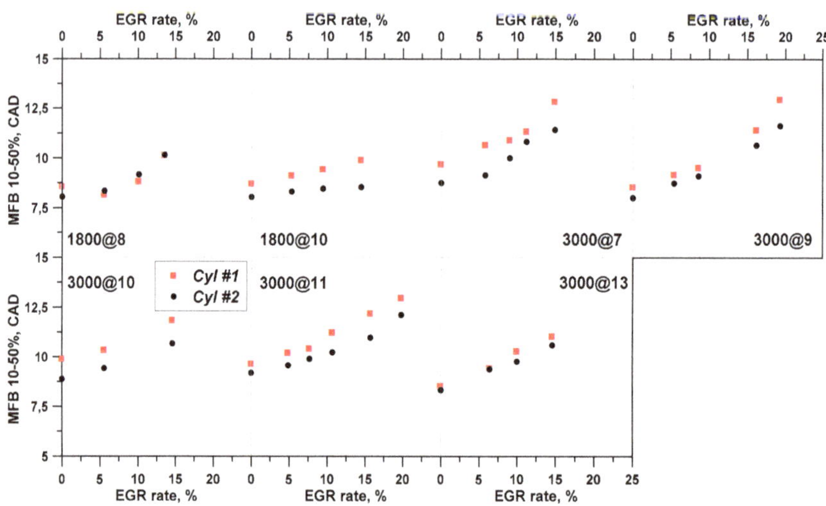

Figure 7. MFB 10–50% in Cyl #1 and Cyl #2 at different operating conditions as a function of EGR rate.

5. Description of Modeling Approach

The engine used in this paper is outlined in a 1D model by employing the GT-Power commercial code. The whole engine is represented in sub-components, including cylinders, intake/exhaust pipes and turbocharging system. In particular, the flow inside the intake/exhaust ducts is reproduced by solving the conservation equations of mass (1), energy (2) and momentum (3) reported below:

$$dm/dt = \Sigma \dot{m}, \tag{1}$$

$$d(m \cdot e)/dt = -p \cdot dV/dt + \Sigma(\dot{m}h) - dQ_w/dt, \tag{2}$$

$$dm/dt = \{-dp \cdot A + \Sigma(\dot{m} \cdot u) - 4 \cdot C_f \cdot \rho \cdot u \cdot |u| \cdot A \cdot dx/2 \cdot D - A \cdot C_p \cdot \rho \cdot u \cdot |u|/2\}/dx, \tag{3}$$

where $\dot{m}$, m, V, p, e, h and u are mass flow rate, mass, volume, pressure, internal energy per unit mass, enthalpy per unit mass and velocity at boundary. Q_w is the heat exchanged through the walls, A is the flow area, D is the equivalent diameter. C_p and C_f are the pressure and friction losses coefficients, while dx and dp represent the discretization length and the pressure differential across dx. Each cylinder is treated as a 0D volume. In this case, the scalar variables (e.g., pressure, temperature, internal energy, etc.) are assumed to be uniform over the volume. However, during the combustion process the cylinder volume is divided in two zones (burned and unburned regions), where the mass and energy conservation equations are solved. In the closed valve period of single cylinder (i.e., from IVC up to EVO), the energy equation is detailed for burned and unburned zones as reported by relations (4) and (5):

$$d(m_u \cdot e_u)/dt = -p \cdot dV_u/dt - dQ_{w,u}/dt - h_u \cdot dm_b/dt, \tag{4}$$

$$d(m_b \cdot e_b)/dt = -p \cdot dV_b/dt - dQ_{w,b}/dt + h_u \cdot dm_b/dt, \tag{5}$$

where dm_b/dt represents the burning rate term evaluated through the turbulent combustion sub-model discussed in the following.

The turbocharging operation is simulated by a "map-based" approach. Turbine and compressor maps, supplied by the manufacturer, are employed within the standard turbine and compressor objects, respectively. Flow permeability of the single cylinder head is taken into account through the measured steady flow coefficients, both under forward and reverse flow conditions. Standard injector objects, available in the 1D code, are adopted for gasoline injection. During port injections, it is assumed that 30% of the total liquid mass vaporizes immediately upon the injection event, as advised for a typical liquid injector in GT-Power user manual; details about the spray evolution and wall film formation are not taken into account. In order to enhance the model capability, it is integrated with refined phenomenological in-cylinder sub-models of turbulence, combustion, heat transfer and pollutant emissions. A detailed description of these sub-models is reported in the next subsection.

5.1. In-Cylinder Sub-Models and Tuning: Combustion, Turbulence and Pollutant Emissions

In-cylinder 0D sub-models were integrated within the 1D engine model to refine the prediction of turbulent combustion and of the exhaust emissions. The combustion process is modeled by the fractal approach [19–21], where the burning rate is written as follows:

$$dm_b/dt = \rho_u \cdot S_L \cdot A_T = \rho_u \cdot S_L \cdot A_L \cdot (L_{max}/L_{min})^{(D_3 - 2)}, \tag{6}$$

where ρ_u is the unburned gas density, A_L and A_T the area of the laminar and turbulent flame fronts, respectively, and S_L the laminar flame speed (LFS). L_{min} and L_{max} are the minimum and maximum flame wrinkling scales and D_3 is the fractal dimension. L_{min} represents the Kolmogorov scale evaluated under the hypothesis of isotropic turbulence, while L_{max} is proportional to a characteristic dimension of the flame front, i.e., the flame radius in the present study. D_3 is computed by the equation proposed in Reference [22], as a function of the ratio between the in-cylinder turbulence intensity and LFS. The latter is evaluated by using an "in-house" developed correlation, based on 1D LFS computations via chemical kinetics solver (CANTERA), accounting for the in-cylinder thermodynamic state (pressure p and temperature T), equivalence ratio, Φ, and EGR dilution. In particular, the adopted correlation reported in Equation (7) presents the well-known expression of the power law:

$$S_L = S_{L0} \cdot (T/T_{ref})^\alpha \cdot (p/p_{ref})^\beta \cdot EGR_{factor}, \tag{7}$$

In Equation (7), S_{L0} is the flame speed at reference conditions $T = T_{ref}$ and $p = p_{ref}$, depending on the fuel sensitivity S = RON-MON and Φ; EGR_{factor} is a reduction term for LFS that accounts for the presence of residual gas in the unburned mixture. Exponent α depends on Φ, T and p, while exponent β includes the dependencies on Φ and S. It is the case to underline that the above LFS correlation was developed considering a maximum EGR percent mass of 20%; therefore, the employed LFS correlation shows the capability to cover the range of examined EGR rates during the experiments (Table 2). The laminar flame area, A_L, is computed by an automatic procedure implemented in a CAD software and processing the actual 3D geometry of the combustion chamber. The estimation of L_{min}, and D_3 is based on the K–k–T turbulence sub-model, extensively reported in Reference [23]. The governing equations for the mean flow kinetic energy, K, the turbulent kinetic energy, k, and the tumble momentum, T_m, are as follows:

$$d(m \cdot K)/dt = (\dot{m} \cdot K)_{inc} - (\dot{m} \cdot K)_{out} - f_d \cdot m \cdot K/t_{Tum} + m \cdot K \cdot \dot{Q}/\rho - P, \tag{8}$$

$$d(mk)/dt = (\dot{m} \cdot k)_{inc} - (\dot{m} \cdot k)_{out} + (2 \cdot \dot{Q}/3 \cdot \rho) \cdot (m \cdot k - m \cdot \upsilon_t \cdot \dot{Q}/\rho) + P - m \cdot \varepsilon, \tag{9}$$

$$d(m \cdot T_m)/dt = (\dot{m} \cdot T_m)_{inc} - (\dot{m} \cdot T_m)_{out} - f_d \cdot m \cdot T_m/t_{Tum}, \tag{10}$$

In Equations (8)–(10), the first and second terms describe the incoming and the outcoming convective flows through the valves, respectively. The third term for the K and T_m equations is the decay contribution due to the shear stresses with the combustion chamber walls by means of the decay function, f_d, and the characteristic time scale, t_{Tum}. In Equation (9), third term represents an additive compressibility term, proportional to $(\dot{Q}/\rho)$; the latter is also included in the K equation as the 0fourth term. P describes the energy cascade mechanism, which causes the transfer of kinetic energy from the mean flow to the turbulent flow. Finally, the last term in Equation (9) takes into account the viscous dissipation rate of k into heat. The described turbulence sub-model proved to be able to adequately reproduce the in-cylinder turbulence evolution along the whole engine cycle, also sensing the variations in the engine operating conditions, the valve strategies and the cylinder geometrical parameters [23]. The heat-transfer model both for cylinders and exhaust subsystem is considered, applying a wall temperature solver based on a finite element (FE) approach. For the in-cylinder heat transfer (gas-to-wall), the Hohenberg correlation is implemented into the 1D code while convective, conductive and radiative heat transfer modes are considered for the exhaust pipes. The Hohenberg correlation here utilized is illustrated in Equation (11):

$$H = A \cdot V^{-0.06} \cdot p^{0.8} \cdot T^{-0.4} \cdot (v_{pm} + B)^{0.8}, \tag{11}$$

where V is the volume, p is the pressure, T is the mean gas temperature and v_{pm} is the mean piston speed. The calibration constants, A and B, are calculated by Hohenberg and here used as 130 and 1.4, respectively. The convective heat transfer coefficients for engine coolants (oil and water) are evaluated by simulations of coolant circuits. They are subsequently imposed as an input in the code, assuming a dependency on the engine speed. Cooling boundary conditions (temperatures for oil and water) are introduced in the model according to the levels recommended by the engine manufacturer. Concerning the model tuning, an integrated 3D/1D approach is adopted for turbulence sub-model [23]; model constants are identified to better reproduce the turbulence outcomes of in-cylinder 3D CFD simulations, under motored conditions and for various speeds and valve strategies [23]. Referring to the combustion model tuning, this sub-model includes three tuning constants differently affecting the distinct phases of combustion process. A single set of tuning constants was identified for the considered operating points, and tuning is kept fixed regardless the engine operating conditions. A particular attention is devoted to the modeling of the pollutant emissions. To this aim, "in-house developed" sub-models for the estimation of the main regulated cylinder-out emissions, namely HC, CO and thermal NO, are implemented into the 1D code. The propagation of the cylinder-out noxious species within the exhaust system up to the inlet section of TWC is also considered in the 1D model. In particular, CO concentrations are computed by solving, in the burned gas zone, a simplified chemical kinetic sub-model based on two-step reactions as reported in Reference [24]. The CO mechanism comprises the following high-temperature CO oxidation reactions (12) and (13):

$$CO + OH \leftrightarrow CO_2 + H, \tag{12}$$

$$CO + O_2 \leftrightarrow CO_2 + O, \tag{13}$$

For thermal NO emission, a refined approach is here adopted, where the semi-detailed chemical kinetics scheme proposed by Andrae [25] is applied to the burned gas zone to compute the NO concentrations in each cylinder. The kinetic mechanism includes 5 elements, 185 species and 937 reactions. This methodology, although computationally more expensive than the classic Zeldovich scheme, seems to be better in predicting the NO emissions of the single cylinder at varying the thermodynamic conditions, including pressure, temperature, inert content and mixture quality. HC production is simulated by the crevices model and a wall-flame quenching correlation, combined with a thermal-based HC oxidization equation during the expansion stroke. This means that the mixing-based oxidation contribution is neglected by assuming a perfect and instantaneous mixing of unburned charge from crevices with the hot burned gases. In the filling/emptying crevices model, HC species are assumed to accumulate/be released during the in-cylinder pressure rise/decrease phase in/from an arbitrary assigned constant volume, V_{crev}, as a fraction of the combustion chamber volume at TDC [26]:

$$x_{crev}, \% = 100 \cdot V_{crev} / V_{cyl\text{-}TDC}, \tag{14}$$

The volume V_{crev} schematizes the crevices in the combustion chamber where the flame front extinguishes and it is controlled in the model by the input parameter, x_{crev}. The temperature in crevices volume is considered to be the same as the cylinder wall. A constant flow coefficient is attributed to the orifice of the crevices volume both for incoming and outgoing mass flows. Therefore, the pressure evolution within the crevice volume is computed by combining the related mass and gas temperature profiles. For the flame wall-quenching contribution, a simple correlation is here assumed which furnishes an estimation of the total HC arising from the extinguished flame at cylinder walls as a function of the flame quenching distance. This last parameter is assumed inversely proportional to the laminar flame speed, since the flame reaches the wall in laminar conditions. Flame quenching distance varies during the combustion evolution, and, in this study, it is taken towards the end of combustion process (i.e., 90% of mass fraction burned). Indeed, it is quite reasonable to assume that the HC generated by the flame-quenching during the initial combustion phase rapidly diffuse in the burned zone and oxidize. The correlation employed in this study for the quenching-related unburned HC resembles the simple model proposed in Reference [27], and it is shown in Equation (15):

$$HC_{quench} = H_1 \cdot \delta_L + H_2, \tag{15}$$

where δ_L is the laminar flame thickness, while H_1 and H_2 are sub-model tuning constants. Finally, for the oxidation of the overall HC level during the expansion stroke the one-step kinetic equation proposed in Reference [28] is adopted. The rate of post-flame HC oxidation is computed according to Equation (16):

$$d[HC]/dt = -6.7 \times 10^{15} \times e^{-(18735/R \cdot Thc)} \cdot [HC] \cdot [O_2] \, (p/R \cdot T_{hc})^2, \tag{16}$$

This rate depends on the HC and molecular oxygen concentrations, the density term (p/RT_{hc}) and on the oxidation temperature, T_{hc}. The latter is computed as a weighted average between the in-cylinder mean gas temperature and the wall temperature.

5.2. Simulation Setup

Concerning the 1D model setup, at low loads, the experimental IMEP is reproduced by a Proportional-Integral-Derivative (PID) controller acting on the throttle (THR) valve opening, while the WG valve is fully opened. At high loads, the measured IMEP is obtained by another PID controller adjusting the WG valve opening and considering the fully opened throttle valve. In each operating condition, the injected fuel mass is monitored to reproduce the different experimental λ level for each cylinder. The experimental EGR rate is replicated by regulating the EGR valve opening with an additional PID controller. The spark advance is automatically modified in the cycle-by-cycle calculation to realize the measured combustion phasing (MFB 50%). The above-discussed model setup is taken into account in the numerical analysis, and the related outcomes are presented in the next section.

6. Numerical Analysis

The accuracy of the developed 1D engine model is proved for the entire set of measurements (Table 2). In a first stage, the model capability is tested in replicating the main performance variables such as the in-cylinder pressure peak and combustion core duration for both cylinders, the Indicated Specific Fuel Consumption (ISFC) and the Temperature at Turbine Inlet (TIT). A detailed analysis is also presented with the aim to demonstrate the model proficiency in reproducing the main gaseous emissions (NO, CO and HC). Particular care was paid to the numerical/experimental assessment of the single cylinder performance and emissions in order to reproduce the effect of cylinder-by-cylinder variation. In a second phase, the validated model is applied to evaluate the improvements in terms of ISFC and pollutant emissions resulting from the suppression of the cylinder-by-cylinder A/F ratio imbalance.

6.1. Model Validation

Referring to the performance parameters, the following figures, Figures 8–10, report the numerical/experimental comparisons for the overall set of measured operating points, including the variations in load level, speed, EGR rate and cylinder-related air/fuel ratio. The figures also present the average absolute or percent error between the numerical and the experimental outcomes.

The results plotted in Figure 8a,b show that the model is capable to capture with similar accuracy the experimental pressure peak of both cylinders, denoting very limited absolute percent errors of numerical predictions. These assessments also highlight the good reliability of the combustion modeling for the individual cylinders.

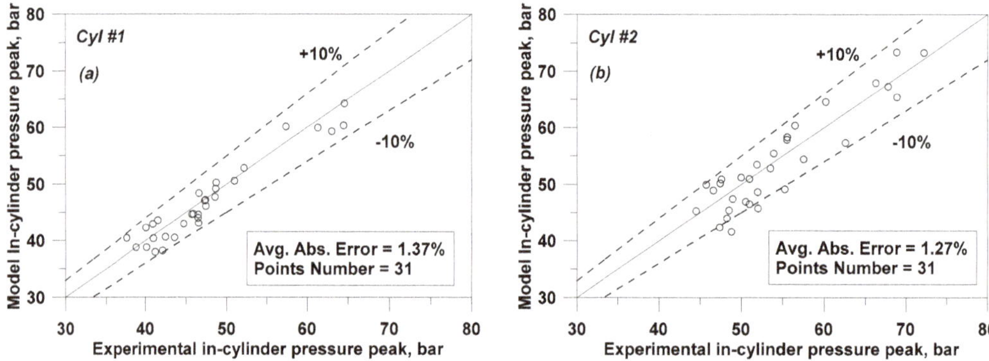

Figure 8. Numerical/experimental comparisons of in-cylinder pressure peak for Cyl #1 (**a**) and Cyl #2 (**b**) at various operating conditions.

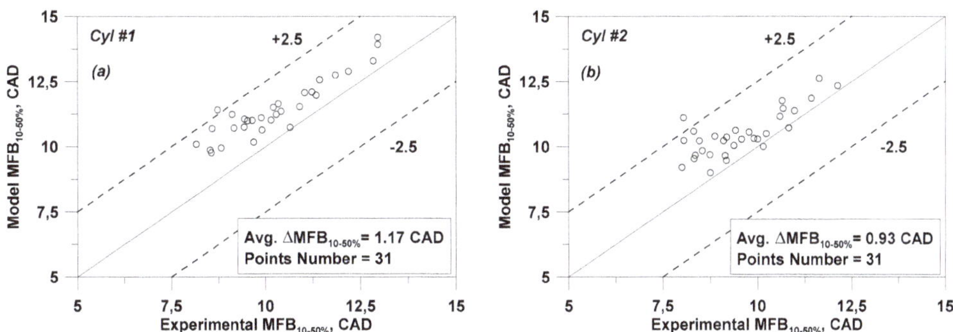

Figure 9. Numerical/experimental comparisons of combustion core duration MFB 10–50% for Cyl #1 (**a**) and Cyl #2 (**b**) at various operating conditions.

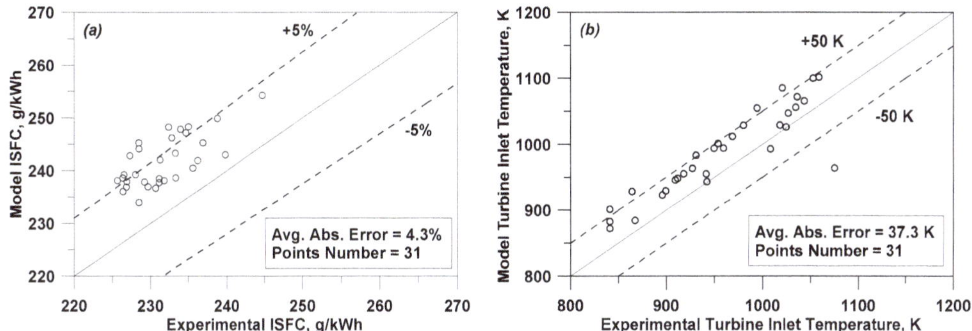

Figure 10. Numerical/experimental comparisons of ISFC (**a**) and TIT (**b**) at various operating conditions.

To further prove the model capability in reproducing the combustion process evolution, the numerical/experimental assessment of the combustion core duration (MFB 10–50%) is reported in Figure 9. Combustion duration of both engine cylinders is adequately described by the model, as demonstrated by the reduced average difference between predicted and experimental data (1.17 CAD and 0.93 CAD for Cyl #1 and Cyl #2, respectively). The above outcomes demonstrate the model ability to sense the effect of cylinder-by-cylinder differences in thermodynamic state (pressure and temperature), mixture quality and composition (A/F ratio and EGR content).

As a relevant global engine performance, Figure 10a presents the numerical/experimental assessment for the ISFC. A satisfactory correlation with the measurements is found for the overall data set with an average percent error of 4.3%. Greater ISFC differences are detected for points at speed of 1800 rpm, mainly due to the model overestimation of the air flow rate. ISFC errors are lower at speed of 3000 rpm. However, most of the computed points are included in the considered allowable error band ±5%, thus demonstrating an acceptable model capability for the ISFC prediction.

Good numerical/experimental agreements for TIT (Figure 10b) are reached, thanks to the refined thermal modeling, which also includes the FE approach for cylinders and exhaust pipes. Indeed, the computed TIT shows a good correlation with the experimental data (mean absolute temperature error of 37.3 K). Although not visible in Figure 10b, the model correctly reproduces the trend of TIT reduction at increasing the EGR rate and lowering the IMEP.

Even if not reported here for brevity, the in-cylinder pressure traces and the other global performance variables are reproduced with accuracy similar to a previous authors' work [15].

As for pollutants, Figures 11–13 show the comparison between numerical and experimental levels of regulated gaseous emissions: NO, CO and HC, respectively. Consistently with the presentation of the experimental results, the variations of considered pollutant emissions are reported as a function of the EGR rate.

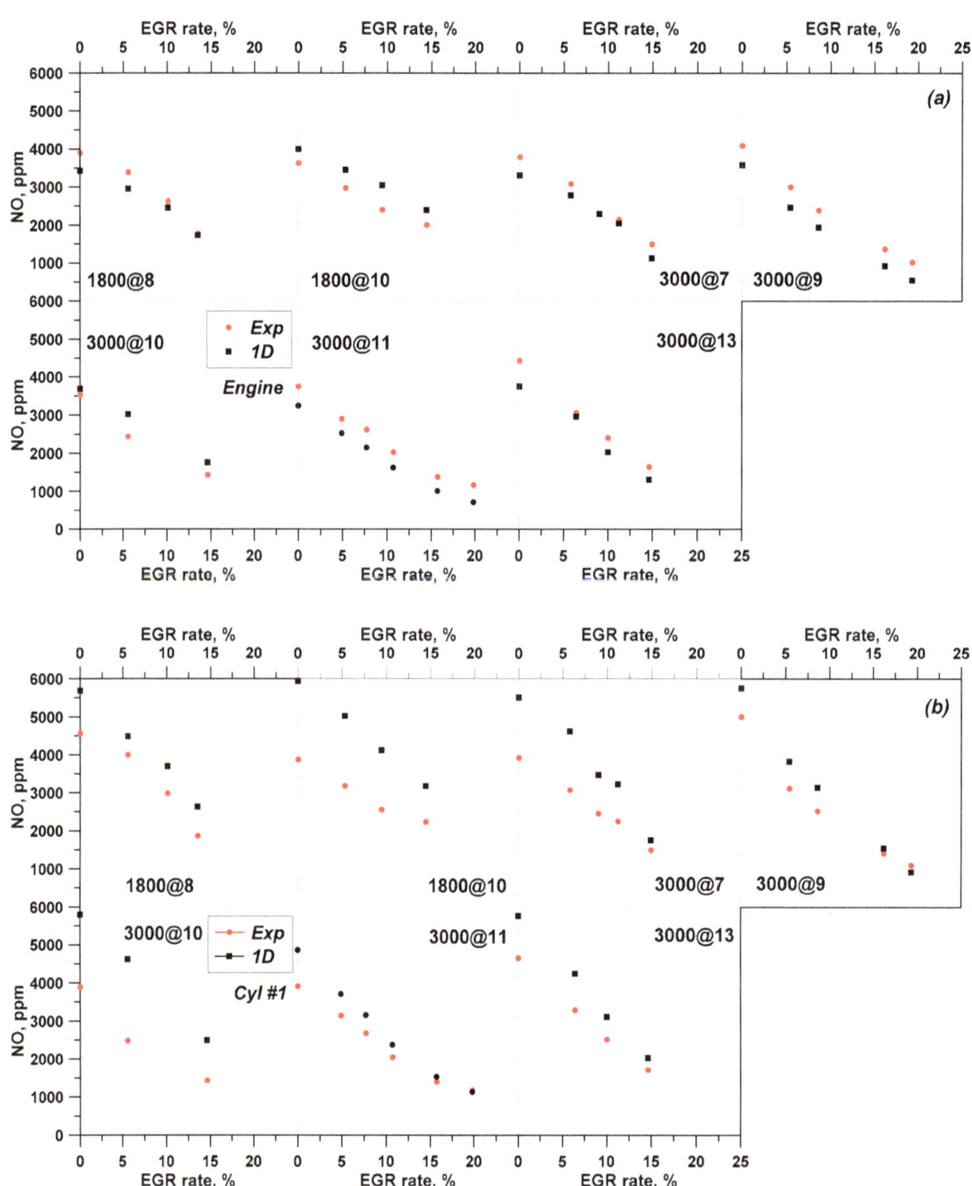

Figure 11. Numerical/experimental comparisons of nitrogen oxide (NO) for engine (**a**) and Cyl #1 (**b**) at various operating conditions.

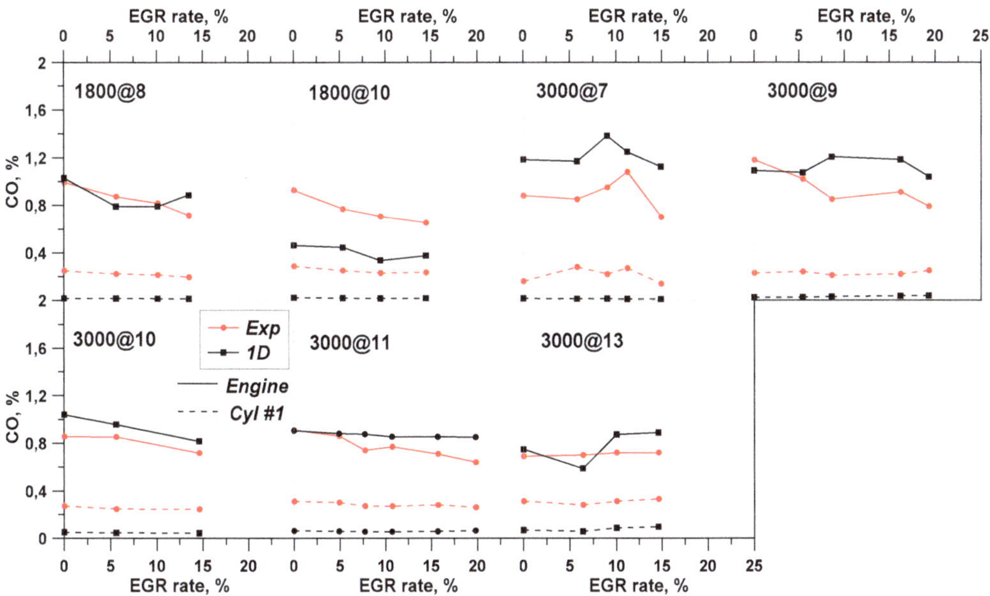

Figure 12. Numerical/experimental comparisons of percent carbon monoxide (CO%) for both engine and Cyl #1 at various operating conditions.

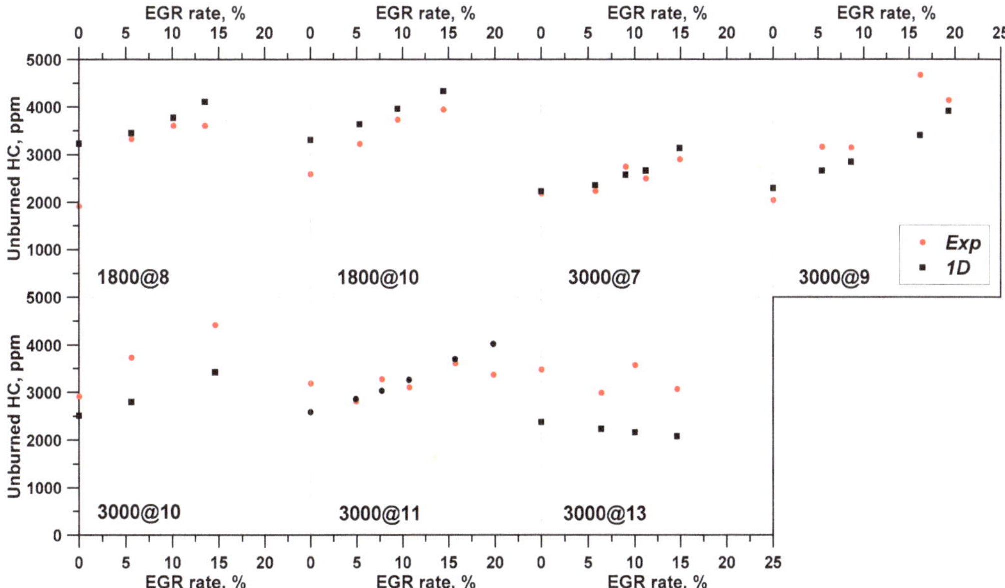

Figure 13. Numerical/experimental comparisons of engine unburned HC at various operating conditions.

Figure 11a,b show the NO emissions at the engine exhaust (Figure 11a) and at the exit of exhaust valve of Cyl #1 (Figure 11b), at different points and EGR rates. At each investigated operating condition, engine NO emission decreases at increasing the EGR rate with a reduction between ~45% and ~55% at 1800 rpm and between ~60% and ~90%

at 3000 rpm when the maximum charge dilution is adopted. Looking at Figure 11b, the Cyl #1 NO emission strongly increases compared to the overall level at the engine exhaust. This behavior can be explained considering the relevant difference in the A/F ratio of two cylinders: running the engine at stoichiometric conditions the relative A/F ratio in Cyl #1 ranges between ~1.03 and ~1.08 and the higher oxygen content speeds up the NO formation reactions. The comparison between experimental and numerical data demonstrates that the employed NO modeling approach is capable to adequately reproduce the EGR-induced trend of the NO emission reduction for both the single cylinder (Cyl #1—lean A/F ratio) and the engine exhaust (stoichiometric A/F ratio). Indeed, the NO sub-model correctly senses the formation mechanism, which is primarily affected by the in-cylinder temperature and by the presence of some species (CO_2, H_2O, N_2 and O_2) in the fresh air/fuel mixture [29].

The model demonstrates to well capture the level of the engine NO emissions (Figure 11a) even if a limited model underestimation (average error around 15%) is observed in most of the analyzed conditions. Conversely, greater errors in the prediction of NO emissions coming out of Cyl #1 are obtained (Figure 11b). In this case, the model average error in predicting the measured Cyl #1 NO emission is of about 30%. Indeed, higher differences between the computed NO emissions of Cyl #1 and of Engine are obtained by the model, when compared to the same experimental differences. This consideration highlights the prediction limitations of NO sub-model in the range of lean A/F mixtures, as in the case of Cyl #1. The numerical inaccuracies reported in Figure 11b may be probably attributed to the high sensitivity of NO sub-model to the relative air/fuel ratio variations.

CO emissions are plotted in Figure 12, where a good prediction of the experimental values for both the engine and the Cyl #1 is obtained. Engine CO is reproduced by the model with an average error of about 30% over the tested operating conditions.

Very low CO emissions are measured for Cyl #1, because of lean mixture operations. The model systematically underestimates the CO levels for Cyl #1 at varying the EGR. This could be attributed to a model limitation which does not include the CO generated by the partial oxidation of unburned HC at the cylinder walls under lean conditions.

However, the experimental/numerical correlation of CO levels for Cyl #1 is taken as satisfactory, also considering the experimental errors in recording small emission levels. The outcomes in Figure 12 suggest that the numerical CO emissions of Cyl #2 are also reliable even if they cannot be compared to any measured data.

Figure 13 proposes the numerical/experimental assessment for the sole engine unburned HC emission. Unfortunately, the single cylinder HC data cannot be proposed here, because the Cyl #1 exhaust probe is located very close the exhaust valves, and, as clearly discussed in Reference [30], a relevant instantaneous peak of unburned HC emission is realized at the event of exhaust valve opening. This phenomenon involves some difficulties to furnish a reliable measure of the HC level for Cyl #1 with the available gas-emission analyzer.

According to Figure 13, the HC sub-model demonstrates to be able to reproduce the experimental trend of the engine unburned HC emissions at varying the EGR rate. In particular, the model provides an average error of 18% in the prediction of engine unburned HC species. Concerning the set of EGR-sweep operating points, only for the highest investigated IMEP (3000@13) a slight HC reduction is observed at increasing the EGR. As known, the engine load and A/F ratio are found to be the main parameters influencing the quenching phenomenon, through the quenching distance variations [27]. Since the quenching distance is proportional at first order to the laminar flame thickness, at increasing the IMEP the wall-flame quenching furnishes a gradually reduced in-cylinder HC production. It is very likely that, at 3000@13, the quenching exerts a limited impact on the HC emission. Therefore, at 3000@13, the EGR-related HC reduction has to be ascribed to the spark timing advance (Table 2). In addition, at 3000@13, the rise in boost pressure at increasing the EGR content, required to restore the engine IMEP, is more prominent than the other load conditions. This also means that higher temperature levels are expected to occur during the expansion phase, thus improving the HC post-oxidation and contributing to reduce its overall level.

For all the other points, higher HC emissions are obtained as the EGR rises, and the model well captures this trend, mainly thanks to the contribution of the wall flame quenching.

In light of the above-discussed results, the proposed modeling approach shows the capability to satisfactorily take into account the influence of both engine operation and cylinder-by-cylinder variations on the main pollutant emissions, without the need of a case-dependent tuning.

6.2. Model Application: Suppression of A/F Ratio Imbalance between Cylinders

Once validated, the model is utilized to evaluate the engine behavior in the case of the virtual suppression of the cylinder-by-cylinder variation. To this aim, the experimental A/F ratio imbalance between cylinders for the examined engine is removed and a stoichiometric A/F ratio is imposed in each cylinder, while preserving the measured combustion phasing as input data. As a consequence, variations in the performance and emissions of the engine are expected to occur and can be easily estimated. It is the case to underline that authors focused on showing the single effect related to the suppression of A/F ratio imbalance between cylinders. Of course, a more refined engine calibration also requires a control of combustion phasing to reach the optimal levels at both low and medium/high loads. The main numerical outcomes of the above-discussed investigation are plotted in the Figures 14 and 15. In each figure, the 1D results obtained by imposing the measured A/F ratio for individual cylinder (labeled as "Unbalanced A/F ratio") are compared with the 1D outcomes resulting from the imposition of a stoichiometric A/F ratio as input data for two cylinders (labeled as "Balanced Stoichiometric A/F ratio").

Figure 14 shows the ISFC comparisons in the analyzed engine operating points. Interestingly, this figure highlights that not negligible ISFC improvements are realized and the lower ISFC levels resulting from the balanced A/F ratio (blue bars in Figure 14) mostly follow the EGR related trend. A maximum ISFC percent benefit equal to 5% (ΔISFC = 12.4 g/kWh at 3000@7, EGR% = 9%) is attained, while the average percent gain is around 2%.

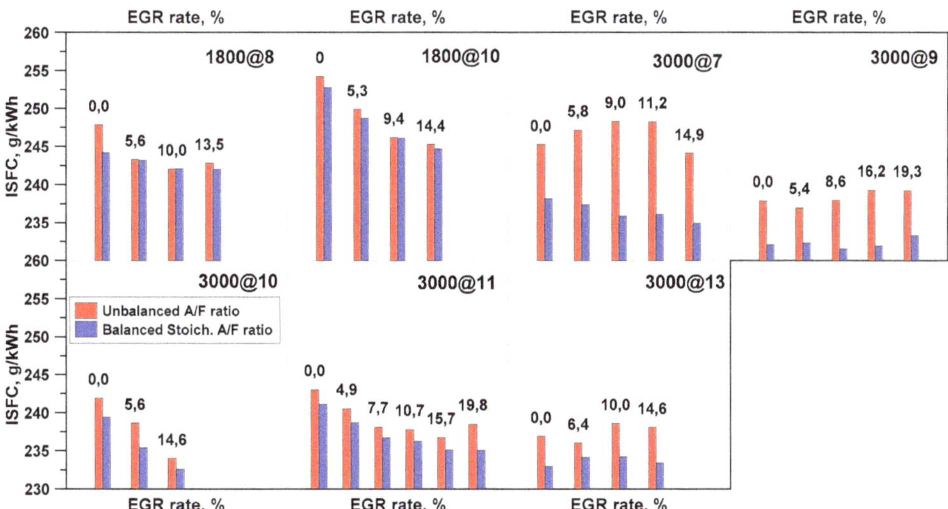

Figure 14. ISFC comparisons: Unbalanced A/F ratio vs. Balanced Stoichiometric A/F ratio at varying the engine operating condition.

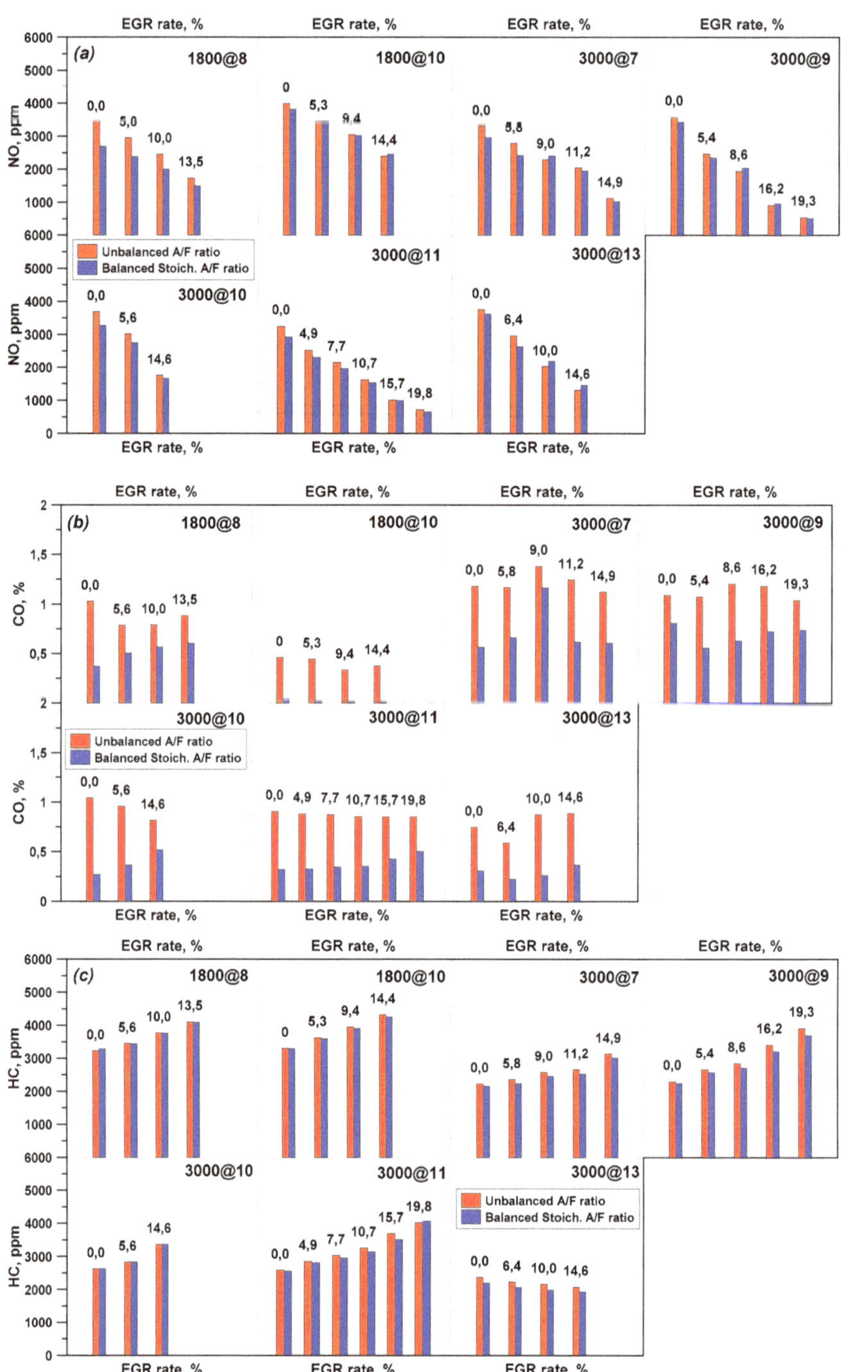

Figure 15. Comparisons of NO (**a**), CO (**b**) and HC (**c**): Unbalanced A/F ratio vs. Balanced Stoichiometric A/F ratio at varying the engine operating condition.

Referring to the main pollutants, the results plotted Figure 15a–c demonstrate that the cylinder-by-cylinder variation exerts a certain influence on both NO, CO and unburned HC emissions. CO improvements are substantially independent from the EGR variation (Figure 15b). The engine CO emission is strongly affected by the cylinder variability with a percent average reduction of about 50%. A peak of CO reduction (about 97%) is observed at 1800@10 and EGR% = 14.4% corresponding to an absolute ΔCO% of 0.36% (Figure 15b). Conversely, NO and unburned HC emissions exhibit reduced benefits (Figure 15a,c, respectively) and the optimized NO and HC species continue to follow the expected trend with EGR. The percent advantage for the HC emission reaches a peak of 8% (ΔHC = 171 ppm at 3000@13, EGR% = 10%) while the average benefit in HC emission is of 2.5%. Similar and only slightly greater benefits are observed for the NO emission: mean percent gain around 5% and peak advantage of 21% (ΔNO = 730 ppm at 1800@8 with EGR% = 0%). It is the worth to underline that the obtained advantages cannot be considered as a general case, but they should be taken as a prediction related to the analyzed engine and the measured points. Hence, the results illustrated in Figure 15 are solely related to the analyzed engine configuration and they lose their meaning in the case of mature commercial SI unit. As a final remark, a refined engine calibration which includes a control on both A/F ratio and combustion phasing can lead to further performance advantages, especially in terms of fuel consumption.

7. Conclusions

In this work, experimental and 1D numerical analyses were performed on a twin-cylinder turbocharged spark ignition engine with the aim to study the effects of cylinder-by-cylinder variation on performance, combustion and gaseous emissions.

In a first stage, experimental tests were carried out at different speed/load points and EGR rates. The experiments pointed out the following main results:

- The assessment of experimental pressure traces and the rates of heat release between cylinders suggests a systematic difference in fuel supply to cylinders, also confirmed by fuel rate difference of port injectors;
- The composition analysis of exhaust gas highlights two different mixture conditions for cylinders;
- Cylinder-by-cylinder λ variation involves a different response to EGR variation in the IMEPs and combustion evolutions between cylinders.

In a second step, a 1D engine model was realized in a commercial code and integrated with "users" in-cylinder sub-models to refine the description of turbulence, combustion, heat transfer and emissions.

Numerical study has pointed out the following main outcomes:

- Engine model satisfactorily reproduces the overall performance and combustion evolution of each cylinder at varying the operating conditions;
- Measured trends of main emissions (NO, CO and HC) for both engine and single cylinder are well captured by the model;
- Validated model is applied to quantify the improvements in terms of ISFC and main pollutants (NO, CO and HC) by suppressing the air/fuel ratio imbalance between cylinders over the examined points.

The wide numerical/experimental investigation of the influence of cylinder-by-cylinder variation on performance and emissions of a SI engine represents the novel contribution of this work compared to the existing literature analyses. The proposed modeling approach has the relevant benefit of a favorable compromise between the results accuracy and computational cost. It can be proficiently adopted to support the control and the design of a spark ignition engine for cylinder-by-cylinder variability suppression with the aim of improving performance and emissions.

Author Contributions: Conceptualization, methodology and formal analysis, L.M., C.T., L.T., F.B., G.V.; software and validation, L.T.; experimental setup, investigation, data acquisition and data curation, L.M., C.T. and L.T.; writing—original draft preparation, L.T., C.T. and L.M.; writing—review and editing, visualization and supervision, L.T., C.T., L.M., F.B. and G.V. All authors have read and agreed to the published version of the manuscript.

Funding: This research received no external funding.

Institutional Review Board Statement: Not applicable.

Informed Consent Statement: Not applicable.

Data Availability Statement: Not applicable.

Acknowledgments: The authors thanks Alfredo Mazzei for the support to the experimental campaign.

Conflicts of Interest: The authors declare no conflict of interest.

Glossary

Notations

A	area
A_L	area of the laminar flame front
A_T	area of the turbulent flame front
C_f	friction coefficient
C_p	pressure loss coefficient
Dx	discretization length
D	equivalent diameter
D_3	fractal dimension—tuning constant of combustion model
E	total internal energy per unit mass
f_d	decay function of tumble
H	total enthalpy per unit mass/heat transfer coefficient
K	turbulent kinetic energy
K	mean flow kinetic energy
L_{max}	maximum flame wrinkling scale
L_{min}	minimum flame wrinkling scale
M	mass
$\dot{m}_a$	air mass flow rate
m_b	burned mass
$\dot{m}_{EGR}$	EGR mass flow rate
N	number of boundaries
P	pressure
P	dissipation rate of mean flow kinetic energy
Q_w	heat exchanged through wall
R	gas constant
S_L	laminar flame speed
T	time
t_T	characteristic timescale of tumble
T	temperature
T_m	tumble momentum
U	velocity at boundary
V	volume
v_{pm}	mean piston speed
x_{crev}	fraction of combustion chamber volume

Greeks

A	temperature exponent in laminar flame speed formulation
B	pressure exponent in laminar flame speed formulation
δ_L	laminar flame thickness
E	dissipation rate of turbulent kinetic energy

ν_t	turbulent viscosity
ρ_u	density of unburned gas
Φ	equivalence ratio

Subscripts

B	burned zone
Crev	crevice
Cyl	cylinder
Inc	incoming
L	laminar
Max	maximum
Min	minimum
Out	outcoming
Quenc	quenching
Ref	reference
T	turbulent
Tum	tumble
U	unburned zone
W	wall

Acronyms

0D/1D/3D	zero-/one-/three-dimensional
A/F	air-to-fuel
AFTDC	After Top Dead Center
ANFIS	Adaptive Neuro-Fuzzy Systems
ANN	Artificial Neural Network
CAD	crank angle degree
CFD	computational fluid-dynamic
CI	compression ignition
CR	compression ratio
ECU	Engine control unit
EGR	exhaust gas recirculation
EIVC	Early Intake Valve Closure
EVC	Exhaust Valve Closure
EVO	Exhaust Valve Opening
FE	finite element
HCCI	homogeneous charge compression ignition
HP	high pressure
ICE	internal combustion engine
IMEP	indicated mean effective pressure
ISFC	Indicated Specific Fuel Consumption
IVC	Intake Valve Closure
IVO	Intake Valve Opening
KLSA	knock limited spark advance
LFS	laminar flame speed
LP	low pressure
LTC	Low-Temperature Combustion
MBT	maximum brake torque
MFB 10%	10% of mass fraction burned
MFB 50%	50% of mass fraction burned
MFB 10–50%	combustion core duration
MON	motor octane number
NG	natural gas
PID	proportional integral derivative
PFI	port fuel injection
PMEP	pumping mean effective pressure
RANS	Reynolds Averaged Navier Stokes
ROHR	rate of heat release

RON	research octane number
SA	spark advance
SACI	spark-assisted compression ignition
SI	spark ignition
TDC	Top Dead Center
THR	throttle valve
TIT	Temperature at Turbine Inlet
TWC	three-way catalytic converter
VVA	variable valve actuation
WG	waste-gate valve
WLTC	Worldwide harmonized Light-duty vehicles Test Cycle

References

1. *Annual European Union Greenhouse Gas Inventory 1990–2017 and Inventory Report 2019*; Report Number: 6; European Environment Agency: København, Denmark, 2019.
2. Zhou, L.; Dong, K.; Hua, J.; Wei, H.; Chen, R.; Han, Y. Effects of applying EGR with split injection strategy on combustion performance and knock resistance in a spark assisted compression ignition (SACI) engine. *Appl. Therm. Eng.* **2018**, *145*, 98–109. [CrossRef]
3. Alvarez, C.E.C.; Couto, G.E.; Roso, V.R.; Thiriet, A.B.; Valle, R.M. A review of pre-chamber ignition systems as lean combustion technology for SI engines. *Appl. Therm. Eng.* **2018**, *128*, 107–120. [CrossRef]
4. Benajes, J.; Molina, S.; García, A.; Monsalve-Serrano, J. Effects of low reactivity fuel characteristics and blending ratio on low load RCCI (reactivity controlled compression ignition) performance and emissions in a heavy-duty diesel engine. *Energy* **2015**, *90*, 1261–1271. [CrossRef]
5. Piqueras, P.; Morena, J.D.L.; Sanchis, E.J.; Pitarch, R. Impact of Exhaust Gas Recirculation on Gaseous Emissions of Turbocharged Spark-Ignition Engines. *Appl. Sci.* **2020**, *10*, 7634. [CrossRef]
6. Fan, Y.; Wu, T.; Li, X.; Xu, M.; Hung, D. *Influence of Port Water Injection on the Combustion Characteristics and Exhaust Emissions in a Spark Ignition Direct-Injection Engine*; SAE Technical Paper 2020-04-14; SAE: Warrendale, PA, USA, 2020. [CrossRef]
7. Del Pecchia, M.; Pessina, V.; Berni, F.; D'Adamo, A.; Fontanesi, S. Gasoline-ethanol blend formulation to mimic laminar flame speed and auto-ignition quality in automotive engines. *Fuel* **2020**, *264*, 116741. [CrossRef]
8. Yanju, W.; Shenghua, L.; Hongsong, L.; Rui, Y.; Jie, L.; Ying, W. Effects of Methanol/gasoline blends on a spark ignition engine performance and emissions. *Energy Fuels* **2008**, *22*, 1254–1259. [CrossRef]
9. Czarnigowski, J. Analysis of cycle-to-cycle variation and non-uniformity of energy production: Tests on individual cylinders of a radial piston engine. *Appl. Therm. Eng.* **2011**, *31*, 1816–1824. [CrossRef]
10. Zhou, F.; Fu, J.; Shu, J.; Liu, J.; Wang, S.; Feng, R. Numerical simulation coupling with experimental study on the non-uniform of each cylinder gas exchange and working processes of a multi-cylinder gasoline engine under transient conditions. *Energy Convers. Manag.* **2016**, *123*, 104–115. [CrossRef]
11. Xu, Z.; Zhang, Y.; Di, H.; Shen, T. Combustion variation control strategy with thermal efficiency optimization for lean combustion in spark-ignition engines. *Appl. Energy* **2019**, *251*, 113329. [CrossRef]
12. Einewall, P.; Johansson, B. *Cylinder to Cylinder and Cycle to Cycle Variations in a Six Cylinder Lean Burn Natural Gas Engine*; SAE Technical Paper 2000-01-1941; SAE: Warrendale, PA, USA, 2000. [CrossRef]
13. Sayin, C.; Ertunc, M.H.; Hosoz, M.; Kilicaslan, I.; Canakci, M. Performance and exhaust emissions of a gasoline engine using artificial neural network. *Appl. Therm. Eng.* **2007**, *27*, 46–54. [CrossRef]
14. Zschutschke, A.; Neumann, J.; Linsen, D.; Hasse, C. A systematic study on the applicability and limits of detailed chemistry based on NOx models for simulations of the entire operating map of spark-ignition engines. *Appl. Therm. Eng.* **2016**, *98*, 910–923. [CrossRef]
15. Marchitto, L.; Teodosio, L.; Tornatore, C.; Valentino, G.; Bozza, F. *Experimental and 1D Numerical Investigations on the Exhaust Emissions of a Small Spark Ignition Engine Considering the Cylinder-by-Cylinder Variability*; SAE Technical Paper 2020-01-0578; SAE: Warrendale, PA, USA, 2020. [CrossRef]
16. Liu, J.; Dumitrescu, C.E. 3D CFD simulation of a CI engine converted to a SI natural gas operation using the G-equation. *Fuel* **2018**, *232*, 833–844. [CrossRef]
17. Bozza, F.; De Bellis, V.; Teodosio, L.; Gimelli, A. Numerical Analysis of the transient operation of a turbocharged diesel engine including the compressor surge. *Proc. IST Mech. Eng. Part D J. Automob. Eng.* **2013**, *227*, 1503–1517. [CrossRef]
18. Božić, M.; Vučetić, A.; Sjerić, M.; Kozarac, D.; Lulić, Z. Experimental study on knock sources in spark ignition engine with exhaust gas recirculation. *Energy Convers. Manag.* **2018**, *165*, 35–44. [CrossRef]
19. Gouldin, F. An application of Fractals to Modeling Premixed Turbulent Flames. *Combust. Flame* **1987**, *68*, 249–266. [CrossRef]
20. Franke, C.; Wirth, A.; Peters, N. New Aspects of the Fractal Behaviour of Turbulent Flames. In Proceedings of the 23 Symposium International on Combustion, Orleans, France, 22–27 July 1990.
21. Gatowsky, J.; Heywood, J. Flame Photographs in a Spark-Ignition Engine. *Combust. Flame* **1984**, *56*, 71–81. [CrossRef]
22. North, G.L.; Santavicca, D.A. The fractal nature of premixed turbulent flame. *Combust. Sci. Technol.* **1990**, *72*, 215–232. [CrossRef]

23. Bozza, F.; Teodosio, L.; De Bellis, V.; Fontanesi, S.; Iorio, A. A Refined 0D Turbulence Model to Predict Tumble and Turbulence in SI Engines. *SAE Int. J. Engines* **2019**, *12*, 15–30. [CrossRef]
24. Cerri, T.; D'Errico, G.; Montenegro, G.; Onorati, A.; Koltsakis, G.; Samaras, Z.; Michos, K.; Tziolas, V.; Zingopis, N.; Eggenschwiler, P.D.; et al. *A Novel 1D Co-Simulation Framework for the Prediction of Tailpipe Emissions under Different IC Engine Operating Conditions*; SAE Technical Paper 2019-24-0147; SAE: Warrendale, PA, USA, 2019. [CrossRef]
25. Andrae, J. Comprehensive Chemical Kinetic Modeling of Tolouene Reference Fuels Oxidation. *Fuel* **2013**, *107*, 740–748. [CrossRef]
26. Kaplan, J.; Heywood, J. *Modeling the Spark Ignition Engine Warm-Up Process to Predict Component Temperatures and Hydrocarbon Emissions*; SAE Technical Paper 910302; SAE: Warrendale, PA, USA, 1991. [CrossRef]
27. Suckart, D.; Linse, D.; Schutting, E.; Eichlseder, H. Experimental and simulative investigation of flame wall interactions and quenching in spark-ignition engines. *Automot. Engine Technol.* **2017**, *2*, 25–38. [CrossRef]
28. Lavoie, G. *Correlations of Combustion Data for SI Engine Calculations-Laminar Flame Speed, Quench Distance and Global Reaction Rates*; SAE Technical Paper 780229; SAE: Warrendale, PA, USA, 1978. [CrossRef]
29. Wei, H.; Zhu, T.; Shu, G.; Tan, L.; Wang, Y. Gasoline engine exhaust gas recirculation e a review. *Appl. Energy* **2012**, *99*, 534–544. [CrossRef]
30. Esposito, S.; Mauermann, P.; Lehrheuer, B.; Günther, M.; Pischinger, S. *Effect of Engine Operating Parameters on Space- and Species-Resolved Measurements of Engine-Out Emissions from a Single-Cylinder Spark Ignition Engine*; SAE Technical Paper 2019-01-0745; SAE: Warrendale, PA, USA, 2019. [CrossRef]

Article

Optimization of Design and Technology of Injector Nozzles in Terms of Minimizing Energy Losses on Friction in Compression Ignition Engines

Jan Monieta * and Lech Kasyk

Maritime University of Szczecin, Waly Chrobrego 1–2, 70-500 Szczecin, Poland; l.kasyk@am.szczecin.pl
* Correspondence: j.monieta@am.szczecin.pl; Tel.: +48-(91)-48-09-415

Featured Application: Injector nozzles are essential elements of internal piston combustion engines. Part of the energy during fuel injection is irretrievably lost due to friction. The process of manufacturing and controlling the main features is complicated. The main design features of the injector nozzles can be optimized in order to minimize frictional energy losses. The potential application of the results obtained is the process of designing and producing injector nozzles.

Abstract: The operation of injection apparatus in self-ignition engines results from the design, manufacturing technology and wear and tear during operation. The technical state of the injector apparatus significantly affects the engine performance, fuel consumption, toxicity and smoke opacity of outlet gases. The most unreliable element of the injection apparatus is the injector nozzle, the quality of which depends on the quality of construction and production, operating conditions and the of the fuels used, etc. One of the design parameters of the injector nozzles, determining the technical state is the geometry of the nozzle holes. An attempt was made to optimize the selection of the dimensions and surface condition of the spray holes to significantly affect the flow properties of the injector nozzles and, consequently, to decide on the size and form of fuel dosed streams to individual cylinders of a self-ignition engine and the quality of fuel atomization. In work, a simulation model was developed, and the pressure losses and the mass fluid of the injected fuel were minimized for selected significant geometric features, taking into account the influence of operating conditions. With the use of Mathematica software, simulation optimization methods and methods based on evolutionary algorithms were elaborated.

Keywords: self-ignition engines; injector nozzles; spray holes; pressure losses; optimization

Citation: Monieta, J.; Kasyk, L. Optimization of Design and Technology of Injector Nozzles in Terms of Minimizing Energy Losses on Friction in Compression Ignition Engines. *Appl. Sci.* **2021**, *11*, 7341. https://doi.org/10.3390/app11167341

Academic Editors: Cinzia Tornatore and Luca Marchitto

Received: 10 June 2021
Accepted: 5 August 2021
Published: 10 August 2021

Publisher's Note: MDPI stays neutral with regard to jurisdictional claims in published maps and institutional affiliations.

1. Introduction

The fuel injection process is the primary action influencing the reach of the desired parameters of the compression ignition engine cycle. The formation of the fuel stream and its disintegration during spraying, and then the formation of the fuel -air mixture depends on the physical properties of the fuel, the pumping pressure and the opening of the injector, air swirl, as well as the advanced design of the injector nozzle and the quality of workmanship [1–4].

Incorrect operation of injection equipment in compression-ignition engines, caused by imperfect workmanship and wear, significantly affects the engine performance, increased fuel consumption, toxic emission and smoke opacity of outlet gases. The research conducted so far has shown that the injector is the most unreliable element of the injection subsystem, and the nozzle in the injector [5]. Changes in its technical state significantly affect the engine operating parameters and the emission of harmful components of the outlet gases [6,7]. The injector nozzles work in very rugged conditions, with high pressures of the fuel and the working medium in the engine's combustion chamber, and with the simultaneous interaction of the heat flux with flames and outlet gases [8].

These are elements with a complex structure, high precision of workmanship and the required high durability. The introduced design changes and modernization of the engines produced require the improvement of the manufacturing technologies used. The purpose of these changes is to increase the maximum combustion pressure and mean effective pressure, reduce specific fuel consumption and increase durability and reliability.

The requirements for injector nozzles can be divided into design and technological. Design requirements are related to the functions, operation of the injector nozzle, durability and reliability, as well as interchangeability and standardization. Technological requirements, which determine the quality and cost of execution, include the possibility of obtaining high dimensional accuracy and surface smoothness as well as small errors in shape and position.

Studies have shown that many injectors are damaged in the initial period of operation [5,9]. The reason for such a situation may be the initial unfitness or low durability, which may result from the conditions of production, transport and storage [2,10]. The analysis of the quality control requirements of the manufactured injectors has shown that they are not very credible and not uniform. The modern state of the technology of producing nozzles does not ensure their hydrodynamic similarity because the dispersion of their geometrical dimensions is difficult to determine by means of industrial quality control methods [2,9].

One of the design dimensions of nozzles, determining their technical state, and at the same time subject to intensive wear, are the geometric features of the spray hole [9,11–15]. Changing the dimensions and condition of the spray hole surface significantly affects the flow properties of injector nozzles, and this, in turn, determines the size and form of fuel jets dosed to individual cylinders of a diesel engine and the quality of fuel atomization. Many investigations have shown that the nozzle hole geometry and its internal flow characteristics play an essential role in the spray formation [2,12,16,17] and combustion process [7,8,14,18].

This article intends to present a simulation model of the injector nozzle structure of a marine engine, which is part of the fuel injection subsystem. The main task is to develop a modern model, considering the latest theoretical and numerical achievements, of a marine engine injector nozzle. In the current article, it is expected that the method of optimizing the design of the injector nozzles will contribute to the reduction of fuel flow losses and the improvement of the efficiency of the piston internal combustion engines.

The injection subsystem tests can be carried out on a real object in operation [5], on a real object in laboratory conditions [9,10] and with the use of computer simulation [19,20]. The essence of simulation studies of the fuel injection process is developing models that allow understanding the physical phenomena that determine the quality of injection.

2. Modeling and Experimental Studies of the Geometry of Injector Nozzles

2.1. Introduction

Modeling is the initial stage of the formal approach to the issues related to the analysis of the operation and the synthesis of technical objects. Modeling allows approximating the principles of organization, and principles of operation of the research object, enabling obtaining information about the modeled system. Modeling also allows reducing the cost and time of testing, especially for such complex and miniature elements as injector nozzles.

Models at the evaluation stage, design and production of objects are created for the needs of inference in simulation or experimental studies [20,21]. The model does not reflect a real object but is only a reflection of the knowledge currently possessed, cannot be treated as something permanent and is not subject to change.

An object model is a tool that allows describing an object and its behavior under various conditions through relations on a set of input and output quantities. The purpose of modeling is to obtain a reliable mathematical model that allows tracing the behavior of an object under various conditions. When building a model, the laws of physics are mainly used, which express the balance of forces, moments, flows, continuity equations and geometric relationships, as well as the results of experimental research [20–22].

The structural model of the object shows the connections and the geometric location of the distinguished elements of the object. It is a graphical representation of elements for structural analysis. These models are usually descriptive and graphic [21]. A functional model of the device is a set of functional blocks denoted by, e.g., rectangles, blocks each containing several inputs and outputs, where the injection pump output may be the injection pipe input [21,23]. Appropriate selection of the design features of this subsystem, its control system and supervision of its operation is one of the fundamental problems faced by modern designers, producers and users of engines and injection apparatus.

2.2. Influence of Geometrical Features on the Functioning of the Injector

The nozzle-hole geometry is one of the important parameters considered in the engine to alter engine's performance, combustion and emission courses. To improve the entire air-fuel mixture in self-ignition engines, swirl and turbulence characteristics need modification in the combustion chamber shape and injector nozzle geometry.

The flow characteristics at the orifice outlet for the four investigated nozzles were analyzed in terms of their stationary mass flow, effective outlet velocity and area coefficient [14]. The results showed that divergent nozzles exhibit high cavitation intensity. The geometry of injector nozzles significantly affected the characteristic spray behavior and emissions formation of diesel engines. In paper [14], a nozzle concept consisting of orifices with convergent -divergent shapes was investigated. Three nozzles, characterized by different degrees of conicity, were compared to a nozzle with cylindrical orifices, which acts as a baseline.

In the article [24], a method of calculation of the hydraulic working process of a diesel fuel feed system having an injector nozzle with different positions of its spray holes was tested. Research in self-ignition engine injector nozzle design for two groups of holes was carried out. Inlet edges of the first group of flow were in the sack volume and inlet edges of the second group were on the locking taper surface of the nozzle body. The coefficients of flow in both groups differ considerably and depend on the nozzle needle position. This makes it possible to distribute the injection quantity rationally by spray holes considering the operating conditions of the self-ignition engine and the combustion chamber zones.

Karra and Kong 2010 [11] observed the 10-hole nozzle geometry hole nozzle had the best performance in the engine at full-load conditions. The number and cone of hole variations in the nozzle injector lead to better engine performance and reduced emissions in work by Lahane and Subramanian 2014 [12]. The combined effect of cylindrical shape with the 5-hole nozzle geometry reduced NOx emissions up to 45% and slightly reduced CO emissions of the engine as compared to that of a standard shape by Shivashimpi et al. 2018 [7].

2.3. Hitherto Existing Methods of Optimizing the Geometry of Injector Nozzles

The static flow rate of high-pressure injector nozzles was characterized through measurements and computations as an optimization process of nozzle design [25]. Prototype injectors were fabricated by modifying multi-zone injection nozzles. Measurements of flow rate were compared with the computations with respect to the nozzle geometry and the needle lift. The initial cone angle was calculated and compared with measurements by the direct photographic imaging method. A parametric study was carried out to terms the flow rate and the initial cone angle for selected design parameters.

Experimental study and optimization were carried out in the layouts and the structure of the high-pressure common rail fuel injection system for a marine diesel engine [26]. A novel optimization solution to improve the steady-state performance of the common rail fuel injection system designed for a ship engine retrofitting was proposed. The tests concentrate on optimizing the hydraulic layouts and the structure parameters to manage and use the model. In tests, the modified multi-objective genetic algorithm was employed to the reduction of rail pressure fluctuation.

The nozzle needle design concept which was considered in the research project was focused on maximizing the contact area between the needle and its cartridge to reduce needle wear [27]. The injector feeding part was realized using two series of holes. The design was replenished by 3D numerical simulations which indicated the optimal feeding-hole number and geometry to obtain a maximum mass flow rate. The study's results indicated that the hole number substantially influences the flow losses along the internal flow path and the global mass flow rate.

Cavitation inside the injector nozzle was observed as a critical phenomenon, which induces a decrease in the flow capacity of the nozzle, but also an increase in the spray outlet effective velocity. Some authors [14] have also performed visualization tests on transparent nozzles, but most of these studies are performed in scaled-up or simplified injector nozzles due to difficulties in production and optical measurements in manufactured geometries.

An experimental research investigation was carried out to analyze the influence of conical and cylindrical orifice geometries on a common rail fuel injection subsystem [1]. This behavior of the injection rate in the different nozzles was characterized by using the non-dimensional parameters of cavitation number, discharge coefficient and Reynolds number. The results evidenced essential differences in the permeability of both nozzle geometries and resistance of the conical nozzle to cavitation.

In the paper, a nozzle flow model was used to design an injector nozzle and obtain initial spray conditions for the common rail injection system [28]. The injector for dimethyl ether needs nozzle holes with larger diameters and a higher sack volume for the same injection duration. Additionally, the needle lift and needle seat diameter should be increased to achieve a minimum flow area ratio.

In the conducted experimental investigation, an effort has been made to enhance the performance of a dual fuel combustion engine utilizing different nozzle orifices [29]. In work, the injector nozzle with 3, 4 and 5 holes, each having 0.2, 0.25 and 0.3 mm hole diameter, respectively, and injection pressure in the injector subsystem (varied from 210 to 240 bar in steps of 10 bar) were optimized. The configurations of re-entrant type combustion chamber and 230 bar pressure in injection subsystem, 4 hole and 0.25 mm nozzle have contributed to maximum performance.

Increasing the injection pressure course in injector space band downsizing the spray holes' orifice diameter has been the principal measures for self-ignition engine to influenced fuel—ambient gas mixture formation and combustion processes [30]. Additionally, the combination significantly accelerates the mixing of fuel and ambient gas and greatly decrease the soot formation.

2.4. Numerical Simulations Description

In paper [15], simulated performance at different load conditions for a constant speed self-ignition engine was determined using GT-Power software. The critical parameters were optimized for achieving the desired fuel nozzle hole diameter. A very close validation with less than 1% difference is achieved at full load. In comparison, a less than 10% difference is achieved at part loads. The simulation model predicts the power, torque, brake specific fuel consumption, in-cylinder pressure, outlet temperature and NOx with very good accuracy at different loads. The optimized spray hole diameter of the testing engine was validated with the experimental results and compared with the baseline model.

The work in [6] studied a new sub model of simulation model that correlated the discharge coefficient of the injector nozzle with the design variables of the nozzle and injection conditions. This solution enables the prediction of the effect of microvariations in the nozzle hole geometry as well as the variations in the main design parameters on self-ignition evolution and combustion processes. The working cycle simulation model has predicted changes in engine performance and toxic emissions due to injector nozzle design changes.

The authors of the work in [31] showed that separate measurement of these single jet flow injection quantities is necessary to evaluate the injection holes. A measuring adapter for the determination of the injection quantities for individual spray holes was elaborated here. The influence of variation in shape and geometry of the individual nozzle components on singlejet flow injection quantities can be measured and simulated with a measuring adapter. It was demonstrated that the smaller the spray hole diameter, the larger the mass flow difference between the individual nozzle holes due to the increase in variation in the manufacturing process. The macro- and micro-geometrical state of the nozzle hole surface and the transient areas of the nozzle interior into the spray hole have a strong influence on the flow conditions within the range in the space of the nozzle needle seat and within the spray holes and thus on flow stream values and flow symmetry. Design features such as the angle in which the inclination-angle of individual nozzle hole [32], and manufacturing processdriven parameters, such as the shape of the grinding radius of spray hole inlet edge substantially, influence the quantity of fuel mass flux through the individual nozzle hole.

Fuel pressurization up to 300 MPa, as required by modern common rail self-ignition engines, resulted in significant variation of the physical fuel properties (density, viscosity, heat capacity and thermal conductivity) relative to those at atmospheric pressure and room temperature conditions [33,34]. The high acceleration of the fuel at velocities reaching 700 m/s as it is flowed through the nozzle hole orifices to induce cavitation was found. The simulations were assuming variable properties reveal two processes strongly affecting the fuel injection quantity and its temperature. The intense pressure and density gradients at the central part of the spray hole induce fuel temperatures even lower than that of the inlet fuel temperature. The local values can exceed the fuel boiling point and induce reverse heat transfer from the liquid to the nozzle's metal nozzle body. Local values of the thermal conductivity and heat capacity affect the transfer of heat produced at the nozzle surface to the flowing fuel. That creates large temperature gradients within the flowing fuel, which must be considered for accurate simulations of the flow through injector nozzles.

Two different injector designs have been studied: one with sharp-inlet cylindrical holes and one with tapered holes with inlet rounding [33]. The findings indicated considerable changes in the flow development but also bulk flow values such as the volumetric efficiency of the injectors and the mean fuel injection temperature.

In this section, the main achievements of simulation modeling are described, together with a description of specific nozzle geometries studied, are presented. The analyzed state of knowledge shows that individual authors achieved partial results for various design solutions and under different conditions. The problems cannot be considered as solved but some steps have been achieved. Only a few authors have dealt with fuel flow losses resulting in pressure losses and temperature increases. The authors also took up this problem.

3. Research Results

3.1. Simulation of the Fuel Mass Stream through the Spray Holes

The fuel flow through the multi-hole injector nozzle can be compared to the flow rate when measured with a measuring orifice, e.g., a small orifice (Figure 1). The fuel mass flow q_{mi} through the spraying hole, similarly to the measuring orifice or the nozzle, is determined by the relationship [9,22,23]:

$$q_{mi} = \mu_r \varepsilon A_r \sqrt{2\Delta p \rho} \tag{1}$$

where A_r—spray hole cross-section, μ_r—flow number, Δp—pressure difference in the injector nozzle space (sack) p_i and in the combustion chamber p_c, ε—expansion number and ρ—density of the flowing fuel, determined for the conditions directly in front of the orifice.

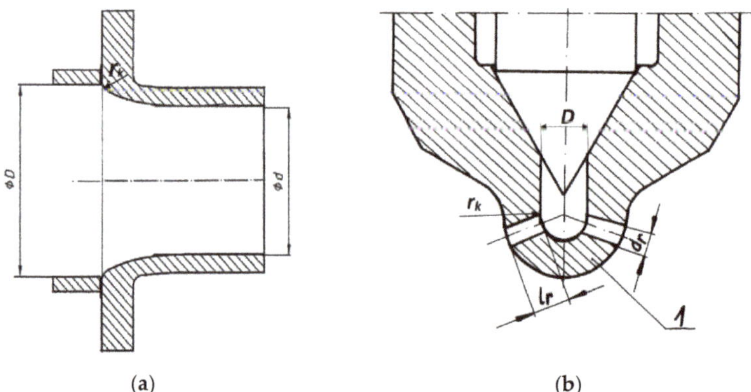

(a)   (b)

Figure 1. Cross-section of a measuring nozzle compliant with ISO 5167-1:1991 (**a**) and a cross-section of an injector nozzle (**b**): d—diameter of the throat flow opening, D—diameter of the pipeline before the measuring nozzle, r_k—radius of the nozzle edge rounding on the inlet side, d_r—diameter of the nozzle hole, l_r—nozzle hole length, 1—spherical surface of the injector nozzle.

The area of the spray hole can be calculated from the formula:

$$A_r = \frac{\pi d_{rt}^2}{4} \tag{2}$$

where d_{rt}—spray hole diameter at the temperature t.

Both cylindricity and circularity errors occur during production [35] (Figure 2). In addition, there are material des decrements operation, as well as the deposit formation. Moreover, this diameter changes with temperature; therefore, the hole diameter d_{rt} at temperature t is:

$$d_{rt}^2 = k_t d_r^2 \tag{3}$$

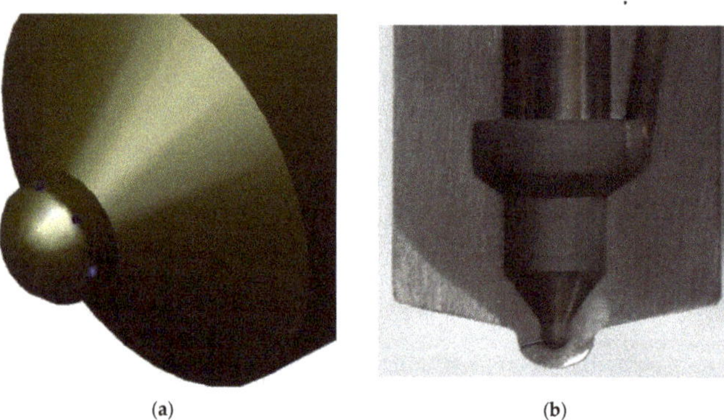

(a)   (b)

Figure 2. The solid model of the injector nozzle (**a**) and image of cross-section of the nozzle hole with rounding (**b**).

The spray hole dimensions change during the measurement with the flow of the medium at a higher temperature. The thermal expansion of the hole considers the correction factor for thermal expansion k_t:

$$k_t = 1 + 2\alpha_t(t_1 - 20°) \tag{4}$$

where $\alpha_t 10 \mid^{t_1}_{20}$ is the average coefficient of thermal expansion per 1 °C from the temperature of 20 °C to t_1.

There is also an increase in fuel temperature by 1 K at x for a pressure increase of 10 MPa [23], whereby the formula can by:

$$t_{px} = t_d + (p_{px} - p_o)/10 \tag{5}$$

where t_{px}—fuel temperature in place of x, t_d—fuel inlet temperature, p_{px}—fuel pressure at x and p_o—ambient pressure. In addition, the injector nozzle and the fuel heat up from the working medium in the combustion chamber by the Clapeyron equation [22].

The flow number μ_r can be an experimentally determined coefficient for similar flows. In general, it is calculated from the formula [9]:

$$\mu_r = k_1\, k_2\, k_3 \mu_0 \tag{6}$$

where: k_1—correction factor of viscosity taking into account the influence of Reynolds number, k_2—correction factor of roughness taking into account the influence of roughness of the internal surface of the pipeline in which the differential device is installed, k_3—correction factor of the inlet edge bluntness, μ_0—the computational number of the flow corresponding to the actual value μ determined in the smooth bore for the highest Reynolds number (Re_{max}).

The correction factor of viscosity k_1 depends on the Reynolds number and the quotient of the flow diameter of the spray hole and the inflow channel D_t, which can be written as:

$$k_1 = f\left(\frac{d_{rt}^2}{D_t^2}, R_e\right) \tag{7}$$

The correction factor for viscosity can be read from the diagram for the calculated Reynolds number and the quotient l_{rt}/d_{rt} ([23], p. 449) or a relationship can approximate it.

The correction factor k_2 depends on the quotient of the flow diameter of the nozzle hole and the inlet channel, the quotient of the surface roughness Δ and the channel diameter D_t, and the Reynolds number. Correction factor for nozzle holes k_2 was calculated from the dependence [23]:

$$k_2 = \frac{1}{\sqrt{1.5 + \dfrac{l_{rt}}{d_{rt}\left(1.14 + 2\lg\frac{d_{rt}}{\Delta}\right)^2}}} \tag{8}$$

where l_{rt}—nozzle hole length at temperature t and Δ—hole surface roughness.

The roughness of the surface of the injector nozzle sack was determined according to industry requirements and measurements using a roughness gauge for the sections of the spray channels.

The correction factor for the inlet edge defocus k_3 depends on the quotient of the flow diameter of the nozzle hole and the inlet channel, the channel diameter in front of the hole D_t and the edge radius r_k on the upstream inlet side of the spray hole. The correction multiplier k_3 was determined based on the quotient of the edge rounding radius and the diameter of the spray hole r_k/d_{rt} and the hole module m:

$$m = \frac{d_{rt}^2}{D_t^2} \tag{9}$$

This relationship was approximated by a second-order polynomial in the form:

$$k_3 = -6 \times 10^{-6}\left(\frac{r_k}{d_{rt}}\right)^2 + 0.01\frac{r_k}{d_{rt}} + 10.8 \tag{10}$$

The design number of flow μ_o corresponds to the actual value determined in a smooth pipeline for the highest Reynolds number (Re_{max}) and the quotient of the flow diameter of the spray hole and the inflow channel.

The Reynolds number can be calculated for the diameter ahead of the nozzle hole at temperature t or for the diameter d_{rt} (radius r_{rt}):

$$Re_d = \frac{c_g 4 r_{ht}}{v_{p,t}} = \frac{c_g 4 r_{ht} \rho_{p,t}}{\eta_{p,t}} \tag{11}$$

where c_g—average fuel flow velocity in the minimum cross-section, r_{ht}—hydraulic radius, $v_{p,t}$—fuel kinematic viscosity for temperature t and pressure p and $\eta_{p,t}$—absolute viscosity for temperature t and pressure p.

The kinematic viscosity can be calculated from the relationship:

$$v_{p,t} = \frac{\eta_{p,t}}{\rho_{p,t}} \tag{12}$$

The expansions number ε takes into account the change in the specific volume of the compressible fluid related to the pressure reduction as it flows through the spray hole. Fuel is a compressible liquid subordinated to Hooke's law with the modulus of elasticity [23].

The relationship between temperature and fuel density was also determined by the following relationship [3,23]:

$$\rho_{p,t} = \rho_{15} - \Delta\rho(t - 15) \tag{13}$$

where $\Delta\rho$ is the reduction of fuel density after heating by 1 °C, $\Delta\rho = 6.5$ kg/m^3.

The viscosity of the fuel is important due to the use of different liquid fuels of different viscosities for the supply of reciprocating internal combustion engines.

The pressure losses on the way from the injection pump drive to the injector nozzle were calculated for two engine types in [19,20]. The absolute pressure loss $\Delta\omega$ in the spray hole depends on the type of nozzle hole, its constriction, the flow rate and the differential pressure. For the shape compared to the orifice, it is calculated based on the formula [22]:

$$\Delta\omega = \frac{\sqrt{1 - \beta^4} - k_2 2}{\sqrt{1 - \beta} + k_2 \beta^2} \Delta p \tag{14}$$

where β—spray hole narrowing is equal. The narrowing β of the spray opening is defined by:

$$\beta = \frac{d_{rt}}{D_t} \tag{15}$$

Figure 3a shows the influence of surface roughness on pressure losses when flowing through the spray hole. This relationship has a minimum of 0.05 μm. Figure 3b shows the effect of the spray hole length on pressure losses. As the surface roughness increases and the spray hole length increases, the irreversible pressure loss increases non-linearly with related variables.

Relative pressure losses were calculated from the relationship:

$$\Delta\overline{\omega} = \frac{\Delta\omega}{\Delta p} \tag{16}$$

where $\Delta p = p_i - p_c$—pressure difference in the nozzle sack and the combustion chamber.

The fuel mass flux q_m, when flowing through the number of i spray holes can be determined by the following relationship [9]:

$$q_m = \mu_o k_1 k_2 k_3 \varepsilon i \frac{\pi d_{rt}^2}{4} \sqrt{2(p_i - p_c)\rho_{p,t}} \tag{17}$$

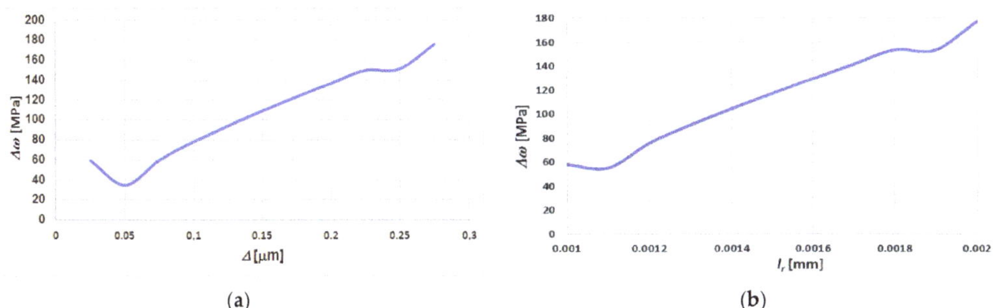

Figure 3. An example of the influence of surface roughness (**a**) and the length of the spray hole (**b**) on pressure losses $\Delta\omega$.

The value of these features is determined at the production stage. The final formula for the fuel mass flux was calculated by the following dependence:

$$q_m = \mu_r \varepsilon i \frac{\pi d_{rt}^2}{4} \sqrt{2(p_i - p_c)\rho_{p,t}} \tag{18}$$

Experimental measurements of the fuel mass flux and determination of flow characteristics for new and used injector nozzles were carried out on a flow test stand using calibration oil for many years [9,10]. These measurements made it possible to experimentally determine the flow coefficient μ_r and the spray holes' average area. The pressure courses in the combustion chamber were calculated and measured using a resistance sensor with digital processing.

The mass of the injected fuel as a result of the pressure loss is lower:

$$q_m = \mu_r \varepsilon i \frac{\pi d_{rt}^2}{4} \sqrt{2(p_i - p_c)\Delta\overline{\omega}\rho_{p,t}} \tag{19}$$

Thus, the fuel mass flux, which should be maximized due to flow losses, depends on the following geometrical values and results from the type of fuel and the design of the internal combustion engine, resulting in the parameters of the working medium:

$$q_m = f\left(i, d_{rt}, D_t, \Delta, r_k, \rho_{p,t}, \eta_{p,t}, t, \Delta p, \Delta\overline{\omega}\right) \tag{20}$$

The latter relationship will be used to optimize for surface roughness losses Δ, the radius of the spray orifices on the inlet side r_k, the influence of temperature t and pressure Δp and pressure losses $\Delta\omega$. This means that when $\Delta\overline{\omega} = 1$, then there are no friction losses, and when $\Delta\overline{\omega} = 0$, there are such large friction losses that the mass flux of dosed fuel is equal to 0. In the simulation, the change in $\Delta\overline{\omega}$ was assumed in the range from 0.3 to 1.0, in steps of 0.1. Various functions can approximate such a relationship.

From the point of view of friction losses, the value of pressure losses $\Delta\omega$ is interesting, because the quantity is irretrievably lost, depending on the characteristics of the technical state determined at the production stage and changing during operation:

$$\Delta = f\left(i, d_{rt}, D_t, l_r\Delta, \rho_{p,t}, \eta_{p,t}, t, \Delta p\right) \tag{21}$$

Therefore, there is a need to build a simulation model due to the optimization goal.

Based on the presented dependencies, the influence of selected design features on the pressure loss at the flow through the nozzle holes was presented. Figure 4a shows the effect of the number of flows through the nozzle holes on the pressure curve in the nozzle space. The diagram 4a shows that the influence of the spray hole flow coefficient μ_r on the pressure course in the nozzle space is significant. Therefore, there is a need to build a simulation model due to the optimization goal.

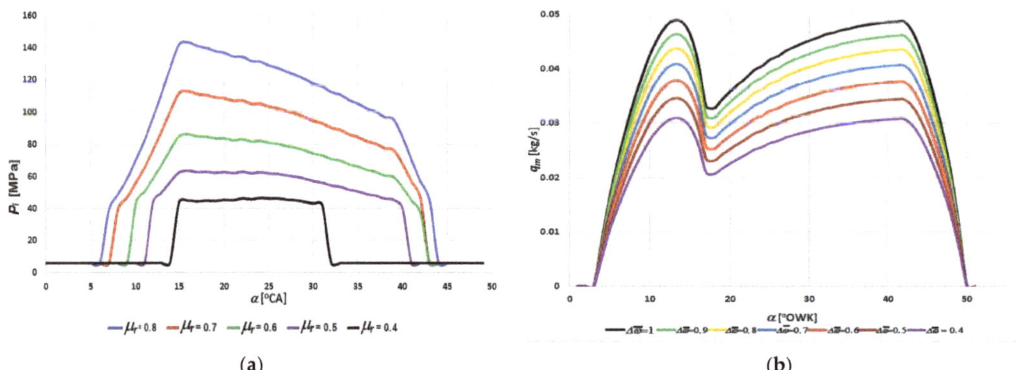

Figure 4. Influence of the value of the flow coefficient μ_r on the pressure course in the injector nozzle space p_i (**a**) and influence of the relative pressure losses $\Delta\overline{\omega}$ through the spray holes on the fuel mass flow q_m (**b**).

On the other hand, Figure 4b shows the influence of the relative pressure drop on the mass fluid of the dosage fuel. The figure also shows a significant decrease in the mass fluid of the dosed fuel with an increase in the relative pressure loss caused by friction in the geometric profiles of the spray holes. Figure 4b shows the influence of relative pressure losses through the spray holes on the fuel mass fluid.

Experimental measurements were also made in the injection subsystem outside the engine and on actual selected internal combustion engines in the Marine Power Plants Laboratory of the Maritime University of Szczecin and on sea-going ships.

The total pressure in the injector space p_i is the sum of the static pressure p_s, which is the residual pressure p_r, and the dynamic pressure p_d depending on the speed of the pumped fuel:

$$p_i = p_s + p_d = p_r + c_r^2 \frac{\rho_{p,t}}{2} \mu_r \varepsilon_s + p_c \varepsilon_s \tag{22}$$

where p_r—residual pressure, ε_s—controlling indicator, taking the values 1 (when $p_i \geq p_o$) or 0 (when $p_i < p_o$) and p_o—nozzle opening pressure. Each throttling in the injection subsystem is associated with a pressure drop, including, among others, the flow factor [23]. The results of the comparison of modeling and measurement on a real engine are shown in Figure 5. The measure of compliance of these two waveforms is the value of the correlation coefficient [5], which was 0.847, which indicates a close relationship. This compatibility is satisfactory, but at the same time, encourages further improvement of the modeling and measurement method of the pressure course with the use of the Kistler sensor.

The course and these are consistent with those presented in the literature [36]. Relative pressure losses may exceed 50%, which leads to significant weight losses of the dosage fuel.

Additionally, carried out experimental measurements of the friction force between the body and the needle of the injectors of new and used injectors, which resulted in the patent award PL 234 940 B1. The friction forces varied, and in some cases very high, expressed in tons.

3.2. Optimization of the Selection of the Geometry of Injector Nozzles

In technical or economic sciences, optimization is one of the essential research tools. The use of more and more computing power of modern computers has allowed for the development of new optimization methods. Genetic algorithms, gradient methods, evolutionary algorithms, neural networks, finite element method are examples of optimization methods widely used in scientific research [37–39]. In recent years, the application of mathematical optimization methods to practical solutions to material engineering problems [2,24,26,40] has been developed. The main task of optimization is the appropriate formulation of the analyzed problem, defining the set of searches and the optimization objective function. For this purpose, an area of permissible states S_p is introduced among the entire space of decision variables (state space S_s). If the objective function $f: S_s \rightarrow R$ and $S_p \subset S_s$, then the optimization task is to search for an element $x_{opt} \in S_p$, that:

$$\forall x \in S_p f(x_{opt}) \leq f(x), \tag{23}$$

in the case of minimizing the objective function and

$$\forall x \in S_p f(x_{opt}) \geq f(x), \tag{24}$$

in the case of maximizing the objective function.

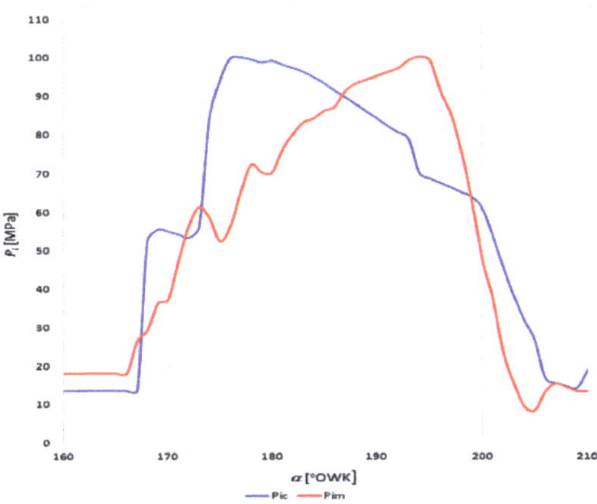

Figure 5. Comparison of the pressure course of the calculated p_{ic} in the injector nozzle space and measured before the injector valve p_{im}.

Gradient optimization methods use, in addition to the known values of the objective function, also its gradient values. They can be used only in those problems where the objective function has the form $f: f: R^n \rightarrow R$ and is differentiable in the R^n space.

The strategy of gradient methods comes down to two stages:

(1) determining the direction of searching for the minimum using the gradient of the objective function or related quantities, and
(2) selection of a step of appropriate length (constant or variable) following the found direction of the minimum search.

In the case of the conjugate gradient method used in this paper, the search direction is determined by designating conjugate directions. This method allows for the minimization of the objective function in n steps, where n is the number of variables present in the objective function [40].

3.3. Optimization of the Parameters of Multi-Hole Nozzles

The research subject is multi-hole nozzles, both conventional and with electronically controlled injection, and the aim of the research is the impact of changes in the design of nozzles on fuel flow losses. The research object is a four-stroke medium-speed marine engine. The dependencies of the pressure course in the atomizer well and the fuel stream flowing out through the spraying hole for a given fuel setting can be approximated by various functions. Based on the formulas (8), (14) and (15), the pressure loss $\Delta\omega$ can be represented as a function of six variables:

$$\Delta w = f(d_{rt}, D_t, \Delta, l_{rt}, p_i, p_c) = \frac{\sqrt{1.5\left(D_t^4 - d_{rt}^4\right) + \frac{\left(D_t^4 - d_{rt}^4\right)l_{rt}}{d_{rt}\left(1.14 + 2\log\left(\frac{d_{rt}}{\Delta}\right)\right)^2} - d_{rt}^2}}{\sqrt{1.5 D_t^3 (D_t - d_{tl}) + \frac{D_t^3 (D_t - d_{rt})l_{rt}}{d_{rt}\left(1.14 + 2\log\left(\frac{d_{rt}}{\Delta}\right)\right)^2} + d_{rt}^2}} \cdot (p_i - p_c) \tag{25}$$

four of which relate to the geometric parameters of the atomizer: hole diameter d_{rt}, the diameter of the inflow channel to the hole D_t, surface roughness Δ and channel length l_{rt}. These four variables were analyzed in terms of minimizing pressure losses Δw. As the form of the function f is too complex, the analytical solution turned out to be impossible to implement. Therefore, in order to optimize the function f, the numerical conjugate gradient method was used. The following restrictions were imposed on the values of the variables:

0.00020 m $< d_{rt} < 0.00032$ m;

0.0004 m $< D_t < 0.0010$ m;

0.001 m $< l < 0.002$ m;

$2.5 \times 10^{-8} < \Delta < 2.8 \times 10^{-7}$ m.

The pressure increase was assumed at the average level: $\Delta p = 2 \times 108$ Pa.

For the assumed limits of variability, the function f reached the minimum for the extreme values of the variability ranges of the atomizer geometric parameters: $\Delta w_{min} = 5.22858 \times 10^7$ Pa for $d_{rt} = 0.00032$ m, $D_t = 0.0004$ m, $l_{rt} = 0.001$ m and $\Delta = 2.5 \times 10^{-8}$ m.

The course of the function variability depends on the individual variables, with the values of the remaining variables fixed, witch makes it possible to determine the influence of these variables on the rate of changes in the value of the function Δw.

In the case of Δ, the change in the value of pressure loss Δw is the smallest and amounts to 1.03312×10^6 Pa, which is about 2% of the minimum value of Δw (Figure 6).

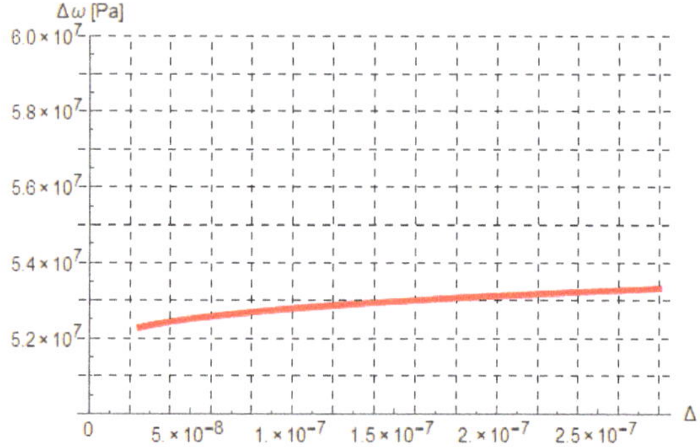

Figure 6. The course of absolute pressure loss Δw, as a function of surface roughness Δ.

In the case of l_{rt}, the change in the value of pressure loss Δw amounts to 1.55469×10^6 Pa, which is about 3% of the minimum value of Δw (Figure 7).

In the case of D_t, the change in the value of pressure loss Δw is the biggest and amounts to 1.48785×10^8 Pa, which is about 285% of the minimum value of Δw (Figure 8).

In the case of d_{rt}, the change in the value of pressure loss Δw amounts to -1.17087×10^8 Pa, which is about 240% of the minimum value of Δw (Figure 9).

In the case of the variables Δ and l_{rt}, D_t, as the variable grows, the values of Δw also increase. Only for hole diameter d_{rt}, the values of Δw decrease. The variables D_t and d_{rt} have the greatest influence on the change of the pressure loss value.

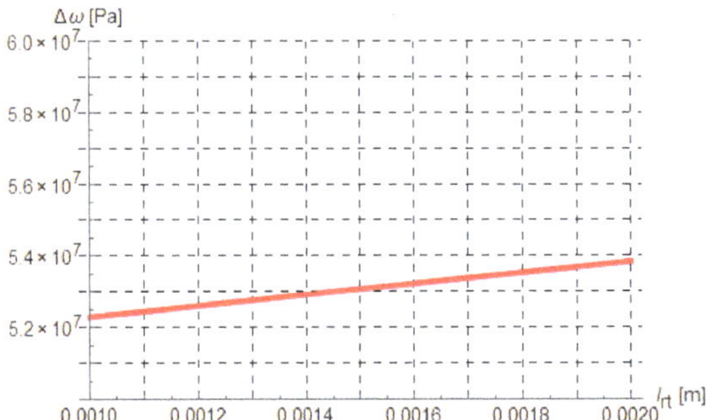

Figure 7. The course of absolute pressure loss $\Delta\omega$, as a function of channel length l_{rt}.

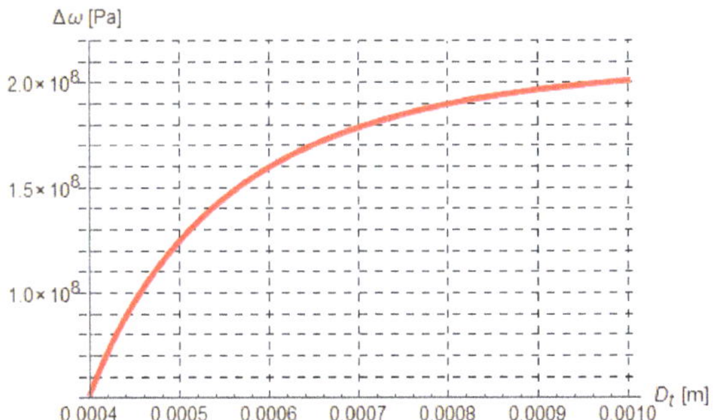

Figure 8. The course of absolute pressure loss $\Delta\omega$, as a function of diameter of the inflow channel to the hole D_t.

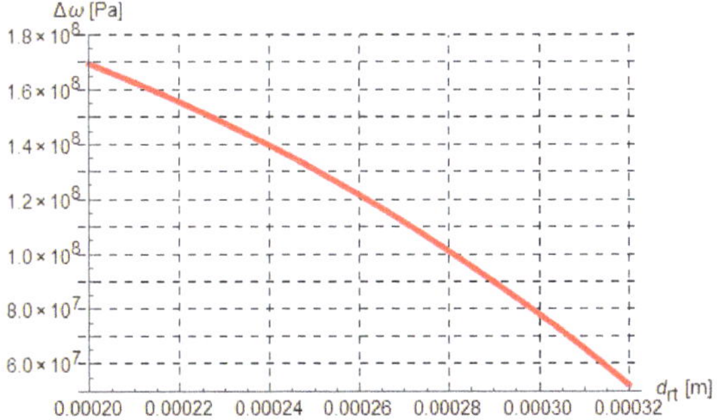

Figure 9. The course of absolute pressure loss $\Delta\omega$, as a function of hole diameter d_{rt}.

4. Conclusions

During the design and manufacture of injector nozzles, geometric features are formed that affect the performance. This is accompanied by irreversible pressure losses, which result in energy losses. For ecological and economic reasons, such losses should be minimized. Understanding the influence of geometric features allows to determine their significance and allows to determine their values to minimize pressure and energy losses. Such problems can be solved with the help of built structural and functional models and computer simulations. In addition to geometric features, pressure losses and fuel dosage are affected by the fuel quality and conditions of fuel injection. These studies focused on diesel fueling. Such problems can be solved using the constructed structural and functional models and computer simulation, and experimental tests of the internal combustion engine conditions and the injection subsystem outside the internal combustion engine. In terms of minimizing pressure losses, the optimal values of the injector nozzle geometrical features assume extreme values for the typical limitations of individual variables. In the case of the spray hole diameter (d_{rt}), hole surface roughness (Δ) and spray hole length (l_{rt}), the optimal values are the maximum permissible values of these variables. On the other hand, it is the lowest allowable value for the diameter of the inflow channel to the hole (D_t).

Based on the analysis of the variability of the pressure loss function depending on individual variables, it was found that both diameters have the most significant impact on the change $\Delta\omega$. On the other hand, roughness and the spray hole length on the pressure losses in the injector nozzles are negligible. Therefore, special attention should be paid to maintaining the appropriate dimensions of both diameters in terms of operation. Optimization of injector nozzle geometry of conventional and common rail systems could reduce manufacturing and operating costs and increase system performance significantly.

Further work may concern minimization of pressure losses in front of the injector nozzle, the influence of errors in the shape, friction losses between the needle and the injector body and position of the spray holes on pressure losses and the prediction of changes in the geometry of the nozzles during operation.

5. Patents

Monieta, J. The method and device for measuring the maximum friction force between the body and the needle of the injector in piston internal combustion engines Patent description PL 234940 B1, Patent Office of the Republic of Poland, Warsaw 2020, pp. 1–18. https://api-ewyszukiwarka.pue.uprp.gov.pl/api/collection/c2acbe35211e85ad27141555bd72d00d (accessed on 1 May 2020).

Author Contributions: Conceptualization, L.K. and J.M.; methodology, J.M.; software, L.K. and J.M.; validation, J.M.; formal analysis, L.K. and J.M.; investigation, J.M.; resources, J.M.; data curation, J.M.; writing—original draft preparation, L.K. and J.M.; writing—review and editing, L.K. and J.M.; visualization, L.K. and J.M.; supervision, J.M.; funding acquisition, L.K. and J.M. All authors have read and agreed to the published version of the manuscript.

Funding: This research outcome has been achieved under the research project "Ecological and economic aspects of the operation of selected elements of marine energy systems" No. 1/S/IESO/2014 and No. 1/S/IMFiCH/21 financed from a subsidy of the Ministry of Science and Higher Education for statutory activities.

Institutional Review Board Statement: Not applicable.

Informed Consent Statement: Not applicable.

Conflicts of Interest: The authors declare no conflict of interest. The funders had no role in the design of the study; in the collection, analyses or interpretation of data; in the writing of the manuscript, or in the decision to publish the results.

Nomenclature

d_{rt} nozzle hole diameter at temperature t

D_t diameter of the incoming fuel stream determined for the conditions directly in front of the hole at the temperature t

k_1 correction factor of viscosity taking into account the influence of Reynolds number

k_2 correction factor of roughness taking into account the influence of roughness of the internal surface of the pipeline in which the differential device is installed

k_3 correction factor of the inlet edge bluntness for the inlet edge of the spray hole

k_t the correction factor for thermal expansion of the hole

l_{rt} spray hole length at temperature t,

m spray hole module

q_m fuel mass flow as it flows through the number i of spray holes

q_{mi} fuel mass flow when flowing through the i-th nozzle hole

p_{px} fuel pressure at x

Re_d Reynolds number can be calculated for the diameter d_r in front of the nozzle hole at tempera -ture t and for the diameter

r_k radius of the nozzle edge rounding on the inlet side,

t_{px} fuel temperature in place of x by 1 K at x for a pressure increase of 10 MPa

α crankshaft rotation angle

β spray hole narrowing

μ_0 the computational number of the flow corresponding to the actual value μ_r determined in smooth bore for the highest Reynolds number (Re_{max})

Δ hole surface roughness

Δp pressure difference in the injector nozzle space (sack) p_i and in the combustion chamber p_c

E expansion number,

$\eta_{p,t}$ absolute viscosity of the fuel at atmospheric pressure p and temperature t

μ_r fuel flow coefficient through the spray hole, depending on the geometrical dimensions

$\nu_{p,t}$ fuel kinematic viscosity for temperature t and pressure p

$\rho_{p,t}$ fuel density at pressure p and temperature t

$\Delta\omega$ irreversible absolute pressure losses

$\Delta\overline{\omega}$ irreversible relative pressure losses

References

1. Benajes, J.; Pastor, J.V.; Payri, R.; Plazas, A.H. Analysis of the influence of diesel nozzle geometry in the injection rate characteristic. *Trans. ASME J. Fluids Eng.* **2004**, *126*, 63–71. [CrossRef]
2. Idzior, M. A Study of Combustion Ignition Engine Injector Nozzle Parameters in the Aspect of Operating Properties. Ph.D. Thesis, Poznan University of Technology, Poznan, Poland, 2004.
3. Piotrowski, I.; Witkowski, K. *Marine Diesel Engines*; Trademar: Gdynia, Poland, 2004; pp. 475–551.
4. Yue, Z.Y.; Battistoni, M.; Som, S. Spray characterization for engine combustion network Spray G injector using high-fidelity simulation with detailed injector geometry. *Int. J. Engine Res.* **2020**, *21*, 226–238. [CrossRef]
5. Monieta, J. Selection of diagnostic symptoms and injection subsystems of marine reciprocating internal combustion engines. *Appl. Sci.* **2019**, *9*, 1540. [CrossRef]
6. Jung, D.-H.; Assanis, D.N. A reduced quasi-dimensional model to predict the effect of nozzle geometry on diesel engine performance and emissions. *Proc. Inst. Mech. Eng. Part D-J. Automob. Eng.* **2008**, *222*, 131–141. [CrossRef]
7. Shivashimpi, M.M.; Alur, S.A.; Topannavar, S.N.; Dodamani, B.M. Combined effect of combustion chamber shapes and nozzle geometry on the performance and emission characteristics of C.I. engine operated on Pongamia. *Energy* **2018**, *154*, 17–26. [CrossRef]
8. Som, S.; Ramirez, A.I.; Longman, D.E.; Aggarwal, S.K. Effect of nozzle orifice geometry on spray, combustion, and emission characteristics under diesel engine conditions. *Fuel* **2011**, *90*, 1267–1276. [CrossRef]
9. Monieta, J.; Wasilewski, M. The utilize of mass flow for spraying holes evalution of marine diesel engines injection nozzles wear. *Tribologia* **2001**, *5*, 947–961.
10. Monieta, J.; Łukomski, M. Methods and means of estimation of technical state features of the marine diesel engines injector nozzles type Sulzer 6AL20/24. *Sci. J. Marit. Univ. Szczec.* **2005**, *5*, 383–392.
11. Karra, P.K.; Kong, S.C. Experimental Study on Effects of Nozzle Holes Geometry on Achieving Low Diesel Engine Emissions. *J. Eng. Gas Turbines Power* **2010**, *132*, 022802. [CrossRef]
12. Lahane, S.; Subramanian, K.A. Impact of Nozzle Holes Configuration on Fuel Spray, Wall Impingement and NOx Emission of a Diesel Engine for Biodiesel–Diesel Blend (B20). *Appl. Therm. Eng.* **2014**, *64*, 307–314. [CrossRef]
13. Shivashimpi, M.M.; Banapurmath, N.R.; Alur, S.A.; Dodamani, B.M. Optimization of nozzle geometry in the modified common rail direct injection biodiesel-fuelled diesel engine. *Int. J. Ambient. Energy* **2019**, *1*, 1–9. [CrossRef]
14. Salvador, F.J.; De La Morena, J.; Carreres, M.; Jaramillo-Císcar, D. Numerical analysis of flow characteristics in diesel injectors nozzles with convergent-divergent orifices. *Proc. Inst. Mech. Eng. Part D J. Automob. Eng.* **2017**, *231*, 1935–1944. [CrossRef]
15. Nerkar, A. Optimization and Validation for Injector Nozzle Hole Diameter of a Single Cylinder Diesel Engine using GT-Power Simulation Tool. *SAE Int. J. Fuels Lubr.* **2012**, *5*, 1372–1381. [CrossRef]
16. Lopez, J.J.; Salvador, F.J.; De la Garza, O.A.; Arregle, J. Characterization of the pressure losses in a common rail diesel injector. *Proc. Inst. Mech. Eng.* **2012**, *226*, 1697–1706. [CrossRef]
17. Kuenssberg Saree, C.; Kong, S.C.; Reitz, R.D. Modeling the effects of injector nozzle geometry on diesel sprays. *SAE Pap.* **1999**, *108*, 1375–1388.
18. Borkowski, T. Marine diesel engine combustion influenced by injection nozzle and primary fuel atomization. *J. KONES Powertrain Transp.* **2007**, *14*, 45–54.
19. Monieta, J. Functional model of injector of medium-speed marine diesel engine. *J. KONES Powetrain Transp.* **2007**, *14*, 423–428.
20. Monieta, J. Modelling of the fuel injection of medium speed marine diesel engines. *Combust. Engines* **2017**, *170*, 139–146. [CrossRef]

21. Żółtowski, B.; Cempel, C. Elements of the theory of technical diagnostics. In *Engineering of Diagnostics Machines*; Society of Technical Diagnostics: Bydgoszcz, Poland, 2004; pp. 86–105.
22. Bogusławski, L. *Laboratory Exercises in Fluid Mechanics*; Poznan University of Technology: Poznan, Poland, 1999; pp. 73–81.
23. Falkowski, H.; Hauser, G.; Janiszewski, T.; Jaskuła, A. *Diesel Engine Injection Systems*; Modeling; WKiŁ: Warsaw, Poland, 1989; Part 2.
24. Shatrov, M.G.; Malchuk, V.I.; Skorodelov, S.D.; Dunin, A.Y.; Sinyavski, V.V.; Yakovenko, A.L. Simulation of fuel injection through a nozzle having different position of the spray hole ransient heating effects in high pressure Diesel injector nozzles. *Period. Eng. Nat. Sci.* **2019**, *7*, 458–464.
25. Jang, C.; Bae, C.; Choi, S. Characterization of prototype high-pressure swirl injector nozzles, part II: CFD evaluation of internal flow. *At. Sprays* **2000**, *10*, 179–197. [CrossRef]
26. Hu, Y.; Yang, J.G.; Hu, N. Experimental study and optimization in the layouts and the structure of the high-pressure common-rail fuel injection system for a marine diesel engine. *Int. J. Engine Res.* **2020**, *4*, 1850–1871. [CrossRef]
27. Kolovos, K.; Koukouvinis, P.; Robert, M.; McDavid, R.M.; Gavaises, M. Transient cavitation and friction-induced heating effects of diesel fuel during the needle valve early opening stages for discharge pressures up to 450 MPa. *Energies* **2021**, *14*, 2923. [CrossRef]
28. Han, S.W.; Shin, Y.S.; Kim, H.C.; Lee, G.S. Study on the Common Rail Type Injector Nozzle Design Based on the Nozzle Flow Model. *Appl. Sci.* **2020**, *12*, 549. [CrossRef]
29. Yaliwal, V.S.; Banapurmath, N.R.; Gireesh, N.M.; Hosmath, R.S.; Donateo, T.; Tewari, P.G. Effect of Nozzle and Combustion Chamber Geometry on the Performance of a Diesel Engine Operated on Dual Fuel Mode Using Renewable Fuels. *Renew. Energy* **2016**, *93*, 483–501. [CrossRef]
30. Nishida, K.; Zhu, J.Y.; Leng, X.Y.; He, Z.X. Effects of micro-hole nozzle and ultra-high injection pressure on air entrainment, liquid penetration, flame lift-off and soot formation of diesel spray flame. *Int. J. Engine Res.* **2017**, *18*, 51–65. [CrossRef]
31. Kilic, A.; Lothar Schulze, L.; Tschöke, H. Influence of nozzle parameters on single jet flow quantities of multi-hole diesel injection nozzles. *SAE Trans. J. Engines* **2006**, *115*, 785–795.
32. Anvari, S.; Taghavifar, H.; Khalilarya, S.; Jafarmadar, S.; Shervani-Tabar, M.T. Numerical simulation of diesel injector nozzle flow and in-cylinder spray evolution. *Appl. Math. Model.* **2016**, *40*, 8617–8629. [CrossRef]
33. Theodorakakos, A.; Strotos, G.; Mitroglou, N.; Atkin, C.; Gavaises, M. Friction-induced heating in nozzle hole micro-channels under extreme fuel pressurisation. *Fuel* **2014**, *123*, 143–150. [CrossRef]
34. Theodorakakos, A.; Mitroglou, N.; Gavaises, M. Simulation of heating effects caused by extreme fuel pressurisation in cavitating flows through diesel fuel injectors. In Proceedings of the 8th International Symposium on Cavitation CAV2012, Submission no. 216, Singapore, 13–16 August 2012; pp. 1–7. [CrossRef]
35. Monieta, J. Reconstructing the complex geometry of the injection nozzle channels of marine diesel engines. *Maint. Probl.* **2016**, *3*, 65–76.
36. Xu, L.L.; Bai, X.S.; Jia, M.; Qian, Y.; Lu, X.C. Experimental and modeling study of liquid fuel injection and combustion in diesel engines with a common rail injection system. *Appl. Energy* **2018**, *230*, 287–304. [CrossRef]
37. Cichocki, A.; Unbehauen, R. *Neural Networks for Optimization and Signal Processing*; John Wiley & Sons: Toronto, ON, Canada, 1993.
38. Kusiak, J.; Danielewska-Tułecka, A.; Oprocha, P. *Optimization. Selected Methods with Examples of Applications*; PWN: Warsaw, Poland, 2019.
39. Tan, K.C.; Khor, E.F.; Lee, T.H. *Multiobjective Evolutionary Algoritms and Applications*; Springer: London, UK, 2005.
40. Kusiak, J.; Danielewska-Tułecka, A.; Żmudzki, A. Optimization of metal forming processes using a new hybrid technique. *Arch. Metall.* **2005**, *50*, 609–620.

Article

An Analysis on the Effects of the Fuel Injection Rate Shape of the Diesel Spray Mixing Process Using a Numerical Simulation

Intarat Naruemon [1], Long Liu [1,*], Dai Liu [1], Xiuzhen Ma [1] and Keiya Nishida [2]

[1]   College of Power and Energy Engineering, Harbin Engineering University, Harbin 150000, China; kniisnd@gmail.com (I.N.); dailiu@hrbeu.edu.cn (D.L.); maxiuzhen@hrbeu.edu.cn (X.M.)
[2]   Department of Mechanical System Engineering, University of Hiroshima, 1-4-1 Kagamiyama, Higashi-Hiroshima 739-8527, Japan; nishida@mec.hiroshima-u.ac.jp
*   Correspondence: liulong@hrbeu.edu.cn; Tel.: +86-451-8251-8036

Received: 26 March 2020; Accepted: 9 July 2020; Published: 20 July 2020

Abstract: In diesel engines, fuel mixing is an important process in determining the combustion efficiency and emissions level. One of the measures used to achieve fuel mixing is controlling the nature and behavior of the fuel spray by shaping the injection rate. The mechanism underlying the behavior of the spray with varying injection rates before the start of combustion is not fully understood. Therefore, in this research, the fuel injection rate shape is investigated to assess the spraying and mixing behavior. Diesel sprays with different ambient temperatures and injection pressures are modeled using the CONVERGE-CFD software. The validation is performed based on experimental data from an Engine Combustion Network (ECN). The verified models are then used to analyze the characteristics of the diesel spray before and after the end-of-injection (EOI) with four fuel injection rate shapes, including a rectangular injection rate shape (RECT), a quick increase gradual decrease injection rate shape (QIGD), a gradual increase gradual decrease injection rate shape (GIGD), and a gradual increase quick decrease injection rate shape (GIQD). The spray vapor penetrations, liquid lengths, evaporation ratios, Sauter mean diameter (SMDs), distributions of turbulence kinetic energy, temperatures, and equivalence ratios were compared under different injection rate shapes. The results show that the QIGD injection rate shape can enhance mixing during injection, while the GIQD injection rate shape can achieve better mixing after the EOI.

Keywords: diesel spray; spray mixing; varying injection rate; numerical simulation

1. Introduction

Diesel engines find widespread applications in many industries, including transportation, agriculture, and power generation, among others. Soot and NOx emissions, which are products of fuel combustion in these engines, pose a threat to the environment and the health of living organisms. Over the years, strict global regulations have been set to reduce the negative impact of such emissions. These stringent legislations require manufacturers to design cleaner and more efficient engines [1,2]. Many technologies, such as diesel particulate filters (DPFs) [1] and selective catalytic reduction (SCR) [3], have been developed to reduce emissions from diesel engines. In addition, controlling the combustion process (e.g., by using low-temperature combustion (LTC) [2], homogeneous charge compression ignition (HCCI) [4], reactivity-controlled compression ignition (RCCI) [5], and premixed charge compression ignition (PCCI)) have also attracted significant research interest [2]. Research on the fuel mixing process for preparing diesel combustion components in cylinders through effective injection rate adjustment is an important consideration. In this study, we only consider injection rate adjustment in the fuel injection system without changing the air intake and other contexts of the system; thus, this study does not modify the existing injection system too greatly. This is due to the significant influence of the air–fuel mixing process on combustion efficiency and exhaust emissions in diesel engines.

The latest developments in fuel injection strategy efficiency tend to focus on the injection rate to increase fuel mixing efficiency and reduce emissions. Notably, the combustion time will be shorter for cases with high initial injection rates. Generally, a high initial injection rate will result in better atomization and air entrainment. The duration of combustion decreases as the mixture of air and fuel improves, meaning that combustion occurs faster. Conversely, under a low initial injection rate, the initial atomization of the fuel and premixing are not good. The slow initial injection fuel droplets will combine with faster fuel droplets, which will result in a larger droplet size that yields a poor spray breakup. The combustion period will then be longer as more time is required to inject the fuel to atomize and evaporate it for combustion. In the past, researchers were interested in investigating the injection rates to improve fuel injection strategies. Juneja et al. [6] noted that increasing the injection rate after a previous injection is sufficient to increase the collision frequency and formation of large droplets, resulting in high momentum and greater penetration. Liu et al. [7] found that a higher peak injection rate yielded a higher spray tip penetration, peak entrainment rate, and entrainment rate after the end of injection (EOI). In addition, Arsie et al. [8] suggested that the start of injection is the main parameter that affects the impingement phenomenon, whereas Kun Lin Tay et al. [9] found that the start of combustion for each rate shape is different, although the injection duration and start of injection are same due to the start of pressure rise in each case. The peak in-cylinder pressures are higher when the start of combustion is advanced due to the injection rate shaping. The combustion duration will be shorter with a higher initial injection velocity. Apart from that, the results from numerical study on the effects of boot injection rate shapes by co-workers of Balaji Mohan and Kun Lin Tay [10–12], showed that NOx decreased but large soot particles occur due to low injection velocity and narrow soot distribution when the main injection velocity is higher [11]. Dezhi Zhou et al. [12] found that higher boot injection velocity and shorter boot injection duration resulted in shorter ignition delay and more fuel burning at the premixed combustion stage. This suggests that a higher injection pressure will often lead to a better spraying process, that this is one of the most effective ways to meet the efficiency requirements, and that this process has a potential benefit in diesel engine performance [13–15]. Agarwal et al. [16] found that increasing the injection pressure reduces the number and mass of particles and increases the diesel spray velocity, which improves the atomization and evaporation process. In addition, Shuai et al. [17] applied a numerical simulation to examine the effects of injection time and injection rate shape on the performance and exhaust emissions of compression ignition engines. The authors found that CO, UHC, and soot emissions can be reduced by using rectangular-type and boot-type rate shapes instead of other types. Based on these previous studies, the injection rate is clearly an important parameter that requires more attention. Injection rate parameters, such as injection velocity, injection mass quantity, and injection duration, have a significant influence on the fuel mixing and combustion process [18]. Many of these investigations considered how the diesel spray mixing process behavior can increase mixing efficiency by studying the effect of the injection rate shape. Attempts to increase fuel mixing efficiency by determining the injection rate shape remain unsatisfactory. Although there have been extensive studies in the past on the influence of injection rates, most of these studies were interested in investigating the influence of injection rates on combustion efficiency and engine emissions [9–12,17]. Few studies have investigated the influence of injection rates on spray mixing behavior.

Based on our previous study [19], we examined the spray mixing characteristics under different injection rate shapes using a modified one-dimensional spray model. This one-dimensional spray model can analyze the spray penetration, entrainment rate, and velocity over a cross-sectional area, as well as the equivalence ratio distributed along with the spray's axial distance. This model does not consider breakup and evaporation. Instead, it reveals general information on both the liquid and the vapor. The fuel and air are assumed to be immediately mixed uniformly, so turbulent mixing cannot be analyzed in detail. A 3D-CFD model is better for researching a spray that includes two-phase flow characteristics.

Creating a spray model using CFD has become an effective way to study diesel spray to analyze the mechanism of fuel–air mixing and atomization. The results of many previous studies on creating a spray model using CFD show that the various parameters of fuel injection can be efficiently monitored and predicted [20–22]. Nevertheless, there are few studies on the injection rate shape's effects on spray mixing. Unlike a quasi-steady-state spray, it is difficult to obtain clear spray images that include liquid and vapor information under varying injection rates, so there is a lack of experimental data to validate the 3D-CFD spray model with varying injection rates. Consequently, most numerical studies focus on the effects of injection rates on combustion processes and do not offer an in-depth understanding of the spray behaviors under different injection rates.

The objective of the current study is to analyze the effect of different injection rate shapes on the diesel mixing process using a numerical modeling method. In this work, the "CONVERGE" CFD code was adopted for a constant-volume combustion chamber with a single hole injector, particularly to study the spray breakup and spray mixing behavior. Validation of the model results involved a comparison of the spray shape and spray penetration with the experimental data from previous researchers of the Sandia National Laboratory, taken from the ECN website [23]. We found that the modified CFD spray model can predict spray behavior. Four injection rate shapes were used to analyze the effects of injection rate shapes on diesel spray mixing behavior to understand the mixing process in-depth, including the microscopic spray characteristics, evaporation process, and mixture properties. The results of this study are expected to provide useful insights for developing an effective fuel injection rate design for future diesel engines.

2. Numerical Modeling

In this study, numerical simulations were implemented using CONVERGE Version 2.2 [24], which was used to create constant volume models and set the turbulence model, spray model, and sub-model, as shown in Table 1. This model was created to predict the diesel spray performance by simulating the spray shape and the spray penetration distance in both a liquid and a vapor state, as well as the mixing behavior. The shape of the model is defined as a constant volume combustion chamber with a diameter of 105 mm and a length of 105 mm to reduce the grid number and increase computational efficiency, with input from the case study boundary conditions under the experimental conditions [23]. The injector was placed at the top center of the cylinder, as shown in Figure 1. The spray was designed using the Spray A condition (detailed information is shown in the ECN [23]). The $C_{12}H_{26}$ reaction mechanism was used as a diesel fuel agent like the experimental considerations since the current model estimation of the trends and evolution of vapor penetration are independent of the fuel type [25].

Table 1. Modeling and Numerical Parameters.

Modeling Tool	CONVERGE
Spray models	
Drop evaporation model	Frossling model
Collision model	No Time Counter model (NTC) collision
Collision outcome	O'Rourke collision outcomes
Drop drag model	Dynamic drop drags
Breakup	KH-RT model
Turbulence model	RANS, RNG k-ε
Grid control	
Base grid size	16 mm
Finest grid size	0.25 mm

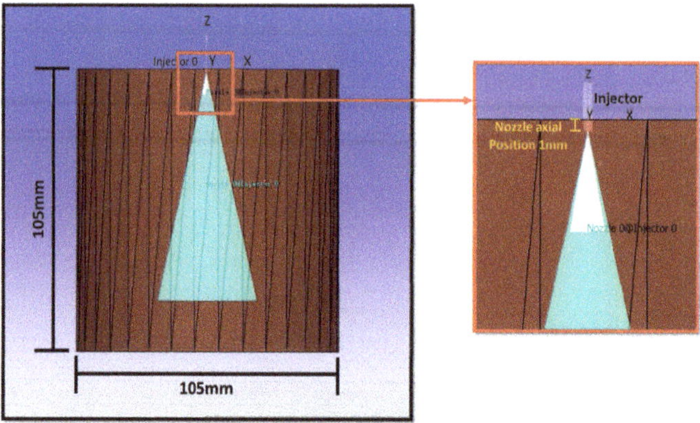

Figure 1. Model characteristics of cylinders.

The grid size has a large influence on the penetration distance [26]. When the grid size is too small, a long simulation time will be required. In this study set up, a fixed grid embedded with a minimum grid size for the spray flow field is shown in Equation (1) [24], where the "Size of cells in scaled grid" represents the sizes of the cells in the scaled grid in each axial "Size of cells in base grid" indicates the sizes of the cells in the base grid setting of each axial, and "amr_embed_vel_scale" is the level of embedding. In this study, CONVERGE can easily change the overall grid resolution prior to executing a simulation to assess grid sensitivity, which is useful for reducing the working time for constant stimulation. For simulations in areas or periods that are not of great importance, a rough grid can be used and then adjusted to consider only the important periods by setting the level of embedding for each section to obtain results that are accurate and not unnecessarily time intensive for calculations:

$$Size\ of\ cells\ in\ scaled\ grid = Size\ of\ cells\ in\ base\ grid * 2^{-amr\_embed\_vel\_scale}. \tag{1}$$

The computational fluid dynamics simulation of the in-cylinder process has many modifications in the spray modeling. The interaction of spray turbulence modeling has an important influence on the spray penetration and mixture formation prediction, which emerges at the end of the entire combustion process. A recent literature review on fuel spray modeling showed that the use of RANS guidelines is of fundamental importance to ensure sufficient performance of the spray mixing and combustion processes. In this study, we used the RANS method to describe the spray development process. The RANS model was used in conjunction with the RNG k-ε turbulence model to determine the effects of smaller movements. A drop evaporation model base on the Frossling model was also used. The liquid penetration length is defined as the maximum distance from the axial position with 99% injected fuel mass at the injection location, while the vapor penetration length is defined as the distance from the farthest location with a 0.01% fuel mass fraction to the nozzle exit. For a fast and accurate collision calculation response, the NTC model with an O'Rourke outcome was used in this study. To increase the efficiency of our droplet collision calculations, we used a dynamically simulated gas technique for spraying. For this technique to work effectively, it must be able to handle a common case, where the number of droplets in each particle varies. The configuration can work effectively with general cases under different conditions.

For the spray breakup model in this study, the Kelvin–Helmholtz instability (KH) and Rayleigh–Taylor instability (RT) models were used. The KH model size constant is defined as proposed by Reitz [27]. The wave breakup formulation was used to model the liquid breakup process. The wavelength (Λ_{KH}) and the breakup size constant (B_0) determined the child droplet sizes. The drop radius equation is calculated as follows:

$$r_{KH} = B_0 \Lambda_{KH}. \tag{2}$$

By requiring that B_0 equal 0.61 for the droplet breakup regime that has the characteristics of $25 < We < 50$, under a higher injection velocity, the B_0 will be in accordance with the applications offered by Hwang et al. [28]. The KH model velocity constant was determined to be 0.188, which is the basic value commonly used by researchers. During the breakup, the parent droplet parcel radius (r) is continuously reduced until it reaches a stable droplet radius (r_{KH}) according to the following equation:

$$\frac{dr}{dt} = \frac{r - r_{KH}}{\tau_{KH}}, \ r_{KH} \leq r \tag{3}$$

where τ_{KH} is the breakup time, given by:

$$\tau_{KH} = \frac{3.276 B_1 r}{\Lambda_{KH} \Omega_{KH}} \tag{4}$$

where Ω_{KH} is the KH wave with the maximum growth rate. The KH model breakup time constant (B_1) determines the primary breakup time. In this study, B_1 equals 21, which is more accurate than the recommended value (recommended value equals 7) in other references. The RT breakup length equation is given by:

$$L_b = C_{bl} \sqrt{\frac{\rho_l}{\rho_g}} d_0 \tag{5}$$

where C_{bl} is the RT model breakup length constant, ρ_l is the fuel density, ρ_g is the ambient gas density, and d_0 is the orifice diameter. The RT model breakup length constant (C_{bl}) generally equals 1.0.

It is necessary to switch the liquid breakup parameter from the KH model to the RT model. Using this approach, the breakup time can be determined by the RT model breakup time constant (C_τ). The value of C_τ should be less to reduce the breakup delay, as shown in the RT breakup time equation below:

$$\tau_{RT} = C_\tau \frac{1}{\Omega_{RT}} \tag{6}$$

where Ω_{RT} is the RT wave with the maximum growth rate. The RT model size constant (C_{RT}) used to determine the scaled wavelengths and the radius of the RT breakup with a higher value increases the predicted RT breakup radius size:

$$r = \pi \frac{C_{RT}}{K_{RT}} \tag{7}$$

where K_{RT} is the wavenumber. In this study, C_{RT} equals 0.1, in which the parent radius decreases continuously until it reaches the constant value in Equation (3), as in the KH model. The setting of the spray breakup model provides an additional configuration for the breakup model constants. Notably, different breakup model constants are used depending on the test conditions and the numerical calculation tools.

The spray dispersion angle is another important input variable. The spray distribution angle defines the fluctuation of air in the injected fuel due to the impulse exchange between gases and liquids. The spray distribution angle will change under different injection intervals (start of injection, the transient regime, the stable regime, and the EOI) due to changes in the nozzle sac flow, momentum, and the interactions between air and the fuel. The actual spray distribution angle detected by each method is different and cannot be compared. The differences in the spray distribution angles measured in the far-field show a higher angle variance than when measuring the angles using the near field method. Generally, the spray distribution angle can be represented in two ways: the spray angle and the spray cone angle. This study analyzed the shapes of different injection rates that have different injection pressures, making the spray distribution angle a very important variable because it is influenced by pressure. Here, the spray distribution angle is measured by taking an image of the cut-plane from the direction of the spray by specifying two lines to determine the area of the spray and then measuring the distribution angle. The spray angle is measured from the nozzle outlet to the spray penetration distance/2 and the spray cone angle is measured from the nozzle outlet to the spray penetration distance equal to the 100×hole diameter. The spray cone angles were used in this study as the input in the spray model to analyze the spray behavior. In this study, the spray cone angles changes depend on the operating conditions of the difference in rail pressure. The spray cone angle will increase noticeably when the injection pressure is higher because the pressure of the fuel mixture inside the nozzle increases and the distribution is higher. The relationship between spray cone angle and injection pressure is discussed in the fourth section.

3. Model Validation

In this section, the numerical models above were studied using a real experiment. Two different sets of operating conditions were used for predicting the spray behavior, including the spray shape, the spray penetration, and the equivalence ratio. The first set involves the model calibration using an evaporation process under experimental conditions at different ambient temperatures. The second set involves the model being calibrated at different rail pressures using the experimental conditions of the Sandia National Laboratory in the ECN [23]. The ambient density, steady flow discharge coefficient, and nozzle hole diameter are the same in all conditions (22.8 kg/m^3, 0.89 mm, and 0.084 mm, respectively) and with the other conditions different, as shown in Table 2.

Table 2. The Operating Conditions of the Test Cases.

Case No.	Ambient Composition	Ambient Temperature [K]	Ambient Pressure [MPa]	Fuel Temperature [K]	Injection Duration [ms]	Injection Mass [mg]	Rail Pressure [MPa]
1	$N_2 = 100.0$	440	2.93	363	1.54	3.46	150
2	$O_2 = 0.00$; $N_2 = 89.71$; $CO_2 = 6.52$; $H_2O = 3.77$	900	6.05	373	1.54	3.46	150
3	$O_2 = 0.00$; $N_2 = 89.71$; $CO_2 = 6.52$; $H_2O = 3.77$	900	6.07	373	5.2	9.30	100
4	$O_2 = 0.00$; $N_2 = 89.71$; $CO_2 = 6.52$; $H_2O = 3.77$	900	6.07	373	5.65	6.90	50

The injection rate of the test conditions conducted by the Sandia National Laboratory in the ECN was determined by CMT, in which the injection mass flow rate of the Spray A condition was created using the "Virtual Injection Rate Generator" model on the ECN website. The virtual injection rate generator model considers the expected hydraulic fluctuations, including the injector opening times, the pressure, the nozzle diameter, and the discharge coefficient.

3.1. Ambient Temperature Effects

In this section, the low and high ambient temperature conditions are used for model calibration. To demonstrate the predictive efficiency of the model, the spray behavior under different ambient temperatures, as well as at different ambient pressures (case No. 1 and 2 in Table 2) is used as a case study.

A comparison of the fuel spray injection in the cylinder at different ambient temperatures and ambient pressures shows the influence of evaporation and spray fuel diffusion. Figures 2 and 3 present a comparison of the spray shape and equivalence ratio between the experimental and simulation results for cases No. 1 and 2. The gray background images show the experimental data obtained from high-speed videos using the Schlieren video technique to represent the vapor and liquid for case No. 1.

For case No. 2, the experimental data image obtained from the high-speed video using the Schlieren video technique shows vapor and liquid regions, which are stored as MATLAB binary files. The white background images show the simulation result, in which the gradient color region represents the equivalence ratio data. The value of the equivalence ratio is represented by the gradient color bar in the right corner. The black particles are liquid fuel data from the simulation, which the cut-plane from the direction of the spray then rotates 90° to the left. Figures 2 and 3 illustrate the shape of the variance by using Re-Normalization Group Theory to determine the small movement effects, while the KH-ACT model calculation captures the impact of cavitation and turbulence on the primary cracks in addition to aerodynamic separation. The calculation results show a smoother distribution for the spray shape boundary simulation than for the experimental data due to the turbulence model's limitations in conjunction with the RNG k-ε turbulence model, where the RNG k-ε turbulence model is used for determining the smaller movements effects. Nonetheless, the simulation results are satisfactory, as these results indicate the effective grid resolution for this simulation. Each comparison image shows the exact thickness of the fuel mass distributed throughout and is consistent with the experimental data. The method of image analysis obtained from this model is sufficient for the study of spray behavior.

The comparison results (Figure 4 and Figure 6) show the spray penetration and spray distribution angle. Black and red represent the results under ambient temperatures of 440 and 900 K, respectively, while dots and solid lines represent the experimental and simulation results, respectively.

The results of vapor penetration are shown in the top graphs of Figures 4 and 5, while the middle and bottom graphs show the liquid penetration and injection rates, respectively.

Figure 4 shows the simulation results that are compatible with the experimental results both during injection and after the EOI. The initial injection is important to consider, as it provides the momentum to change the conditions in the combustion chamber caused by the onset of fuel injection, thereby increasing the injection rate, temperature, and pressure. As shown in Figure 5, the initial spray penetration (0.0–0.3 ms) reveals that both cases cannot properly capture the initial ramp for both vapor and liquid penetration. Due to the very short time error of the simulation results (less than 0.1 ms), the injection rate input data from the virtual injection rate generator may not be as accurate as those of the actual injection rate.

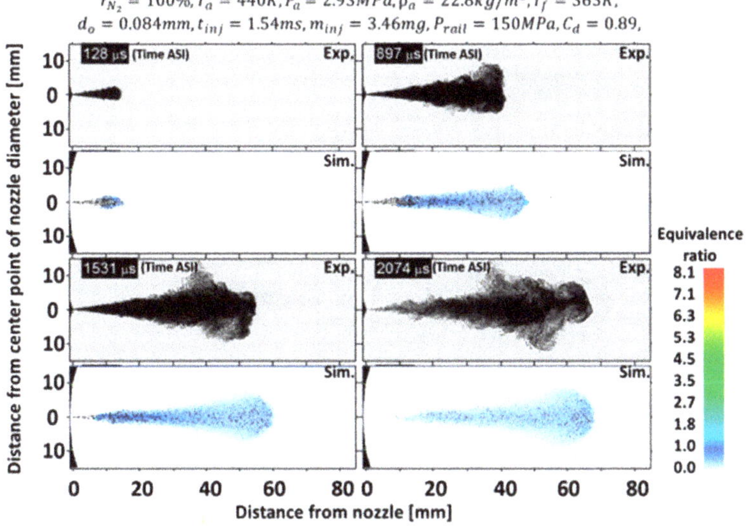

Figure 2. Comparison of the spray shapes and equivalence ratio histories of the experimental result [23] and the simulation result under an ambient temperature of 440 K.

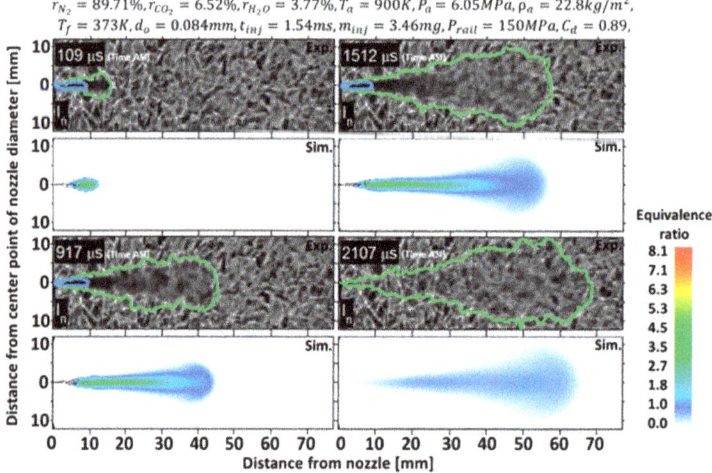

Figure 3. Comparison of the spray shapes and equivalence ratio histories of the experimental result [23] and the simulation result for an ambient temperature of 900 K.

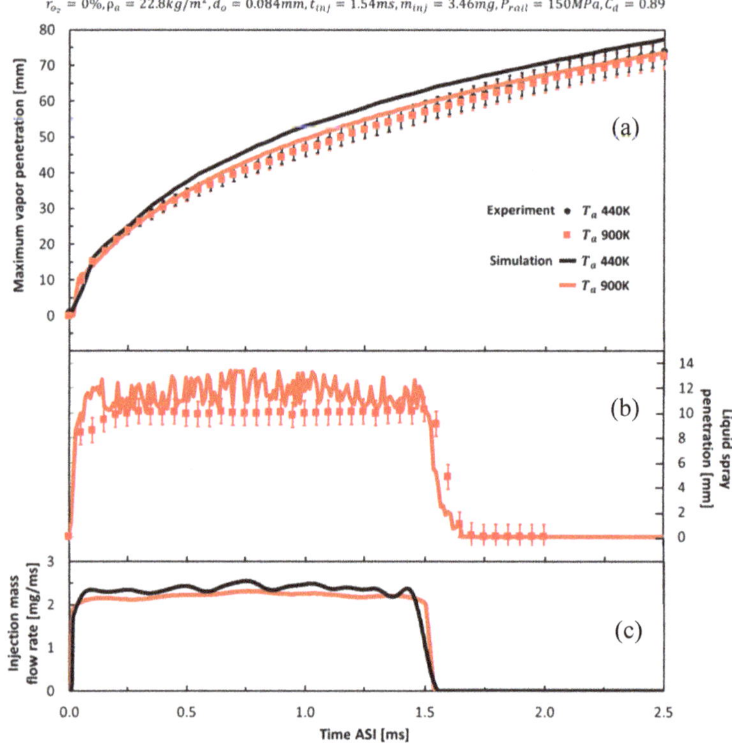

Figure 4. Comparison of the spray penetration of the experimental result [23] and simulation result at different ambient temperatures. (**a**) Maximum vapor penetration, (**b**) liquid spray penetration and (**c**) injection mass flow rate.

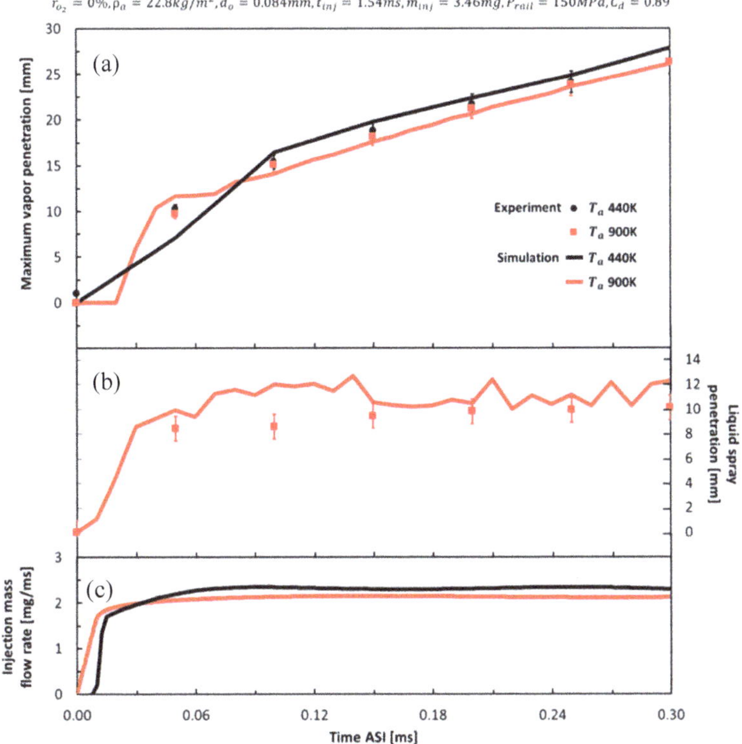

Figure 5. Comparison of the initial spray penetration (0.0–0.3 ms) of the experimental result [23] and the simulation result under different ambient temperatures. (**a**) Maximum vapor penetration, (**b**) liquid spray penetration and (**c**) injection mass flow rate.

Figure 6 shows a comparison of the spray angle (left graph) and the spray cone angle (right graph) with varying ambient temperatures. The measurement results from experimental data show that the ambient temperature influences the spray angle but not on the spray cone angle. The trend of the measurement results of the spray cone angle is the same for both cases, which means that the spray cone angle is independent of the ambient temperature and ambient pressure. In addition, the simulated spray cone angle results show the same trend as the measured experimental results. These results show that the constructed model effectively predicted spray behavior under varying conditions of ambient temperature and ambient pressure.

3.2. Rail Pressure Effects

To ensure that the effects of rail pressure changes are captured by the model when predicting the spray's behavior between the start of the injection (lamping up) and the end of the injection (lamping down), the experimental conditions from previous ECN work under operating conditions No. 2, 3, and 4 (as described in Table 2) are studied in this section.

Figures 7 and 8 show a comparison of the spray shape and equivalence ratio between the simulation results and the experimental data of the rail pressure under 100 MPa and 50 MPa conditions, respectively. The gray background images show the experimental data obtained from high-speed videos using the Schlieren video technique. The white background images show the simulation results. The gradient color region represents the equivalence ratio data, with values shown on the gradient color bar in the right corner.

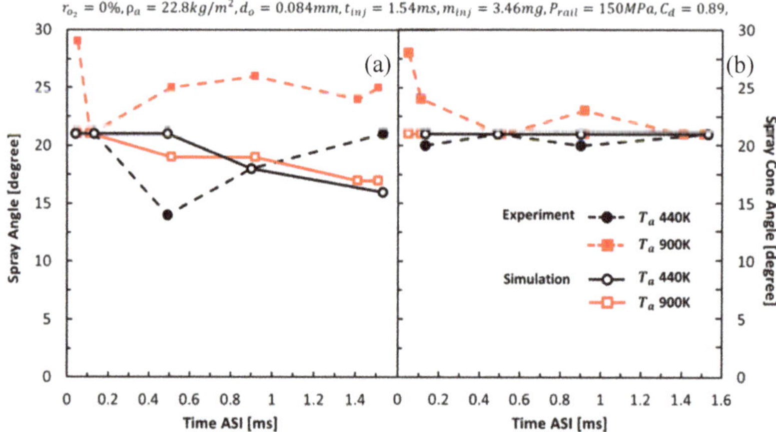

Figure 6. Comparison of the spray distribution angle for different ambient temperatures. (**a**) Spray angle and (**b**) spray cone angle.

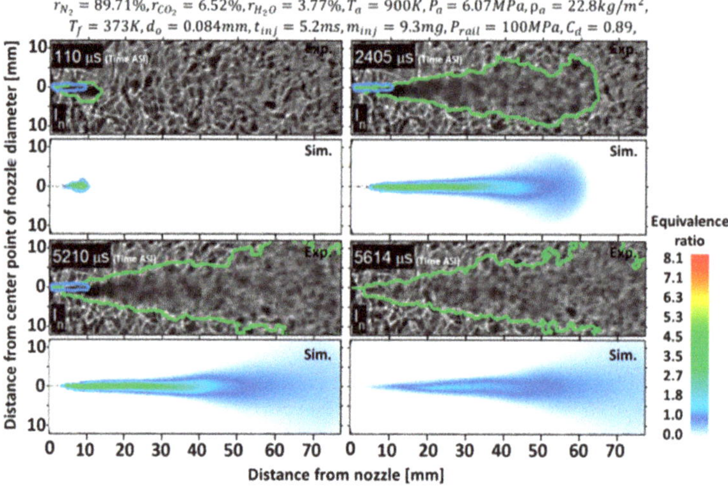

Figure 7. Comparison of the spray shapes and the equivalence ratio histories of the experimental result [23] and the simulation result under a rail pressure of 100 MPa.

The black particles are the liquid fuel data from the simulation, which the cut-plane from the direction of the spray then rotates 90° to the left (which is similar to the display in case No. 2, as mentioned in the previous section (Figure 3)). From these comparisons, the results of the simulation show a better prediction under high rail pressure conditions (150 MPa and 100 MPa) than under a low rail pressure condition (50 MPa).

The results of the spray penetration and spray distribution angle (Figures 9 and 10) shown in the red, blue, and yellow colors indicate the results for rail pressures of 150, 100, and 50 MPa, respectively. The dots and solid lines represent the experimental and simulation results, respectively. The results of vapor penetration are shown in the top graphs of Figures 9 and 10, while the middle and bottom graphs show the liquid penetration and injection rates, respectively.

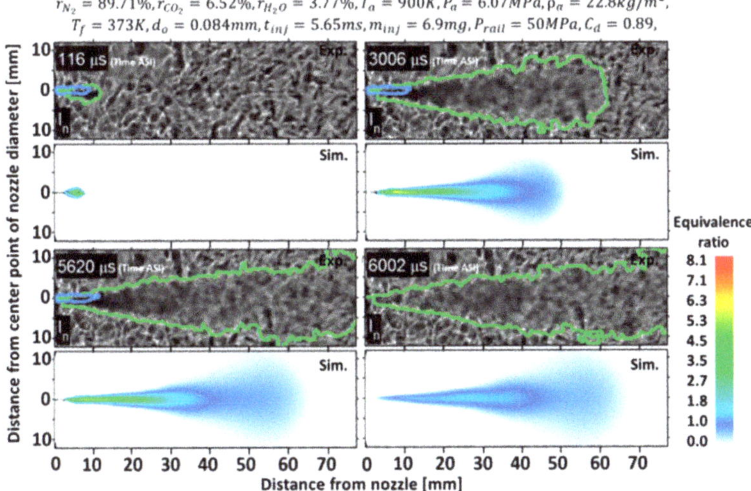

Figure 8. Comparison of the spray shapes and equivalence ratio histories of the experimental results [23] and simulation results for a rail pressure of 50 MPa.

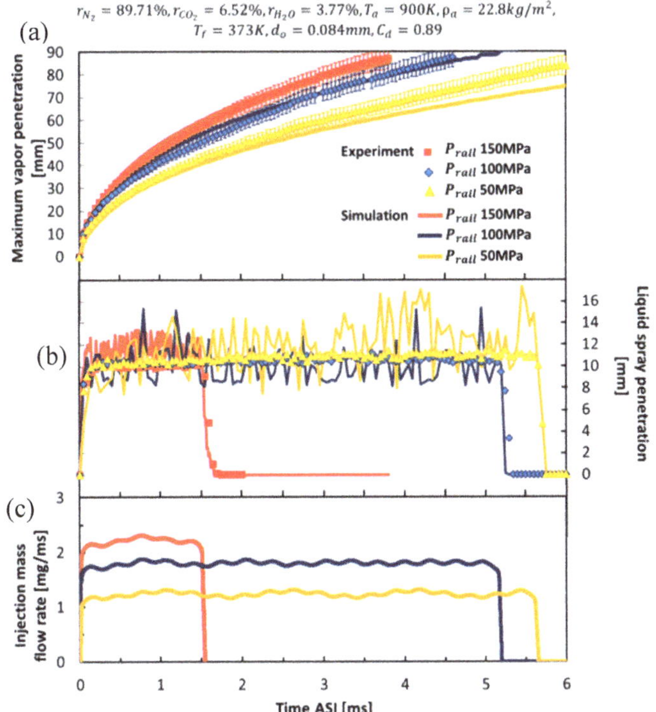

Figure 9. Comparison of the spray penetration of the experiment results [23] and the simulation results for different rail pressures. (**a**) Maximum vapor penetration, (**b**) liquid spray penetration and (**c**) injection mass flow rate.

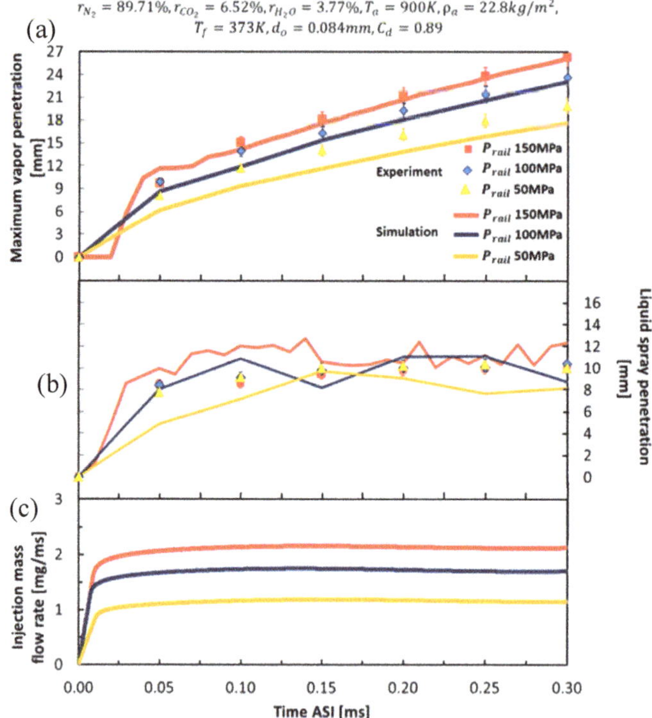

Figure 10. Comparison of the initial spray penetration (0.0–0.3 ms) for the experiment results [23] and the simulation results under different rail pressures. (**a**) Maximum vapor penetration, (**b**) liquid spray penetration and (**c**) injection mass flow rate.

Figure 9 shows that the model predictions for the penetration length under high rail pressure conditions are longer than those under low rail pressure conditions, which is consistent with the experimental data. Figure 10 shows the initial spray penetration (0.0–0.3 ms); even under the rail pressure conditions in all three cases, these data cannot properly capture the initial ramp for both vapor and liquid penetration. The simulation results show a trend that is consistent with the experimental data. The spray model and other sub-parameters make this model more effective in its predictions, which is consistent with experimental data.

The rail pressure directly affected the air fluctuation in the injected fuel due to the impulse exchange between the gas and the liquid. The spraying behavior under high rail pressure caused the cavitation flow to efficiently accelerate at the nozzle exit. The high rail pressure cavity also increased the spray distribution angle [22]. Figure 11 shows a comparison of the spray angles (left graph) and the spray cone angles (right graph) for different rail pressures, in which the spray distribution angle is measured by the same technique described in the previous section. It was found that the size of the spray angle is not constant, but the spray cone angle from the measured simulation results shows a similar trend to the experiment. This confirms that the spray cone angle size setup in this model is correct and can thus predict the experimental data with acceptable accuracy.

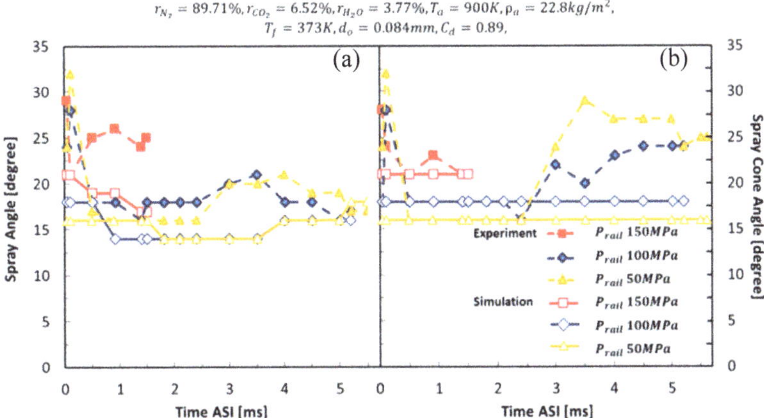

Figure 11. Comparison of the spray distribution angle for different rail pressures. (**a**) Spray angle and (**b**) spray cone angle.

The above results show that different rail pressure conditions affect the spray's infiltration behavior. The generated model shows good predictive performance when there is a change in the working conditions. These results demonstrate that this model can predict the spray's permeability and injection characteristics at an acceptable level. This model will be used for spray behavior predictions to analyze the behavior and mixing processes of diesel spray with different injection rate shapes in the next section.

4. Effect of the Injection Rate Shape

The injection rate shape is an important factor that affects the spray formation. After leaving the injector, the fuel spray atomizes into droplets, vaporizes, and mixes with the air. In this work, models with different injection rate shapes are analyzed for their spray and fuel mixing behavior using the basic conditions from the previous work on ECN case No. 2, as shown in Table 2. The four injection rate shapes shown in Figure 12 were selected for this study. The rectangular injection rate (RECT) shape is the constant injection rate and is a commonly used injection rate shape. The other three shapes consist of the Quick Increase Gradual Decrease injection rate (QIGD), the Gradual Increase Gradual Decrease injection rate (GIGD), and the Gradual Increase Quick Decrease injection rate (GIQD). These shapes were designed to examine the effects of an increase and a decrease of the injection rate over a short injection duration to study the influence of spray and combustion behavioral control factors based on previously reviewed research. The above factors include the peak injection rate, the initial injection rate, and the injection velocity of each injection period, which are useful for understanding the spray and fuel mixing behavior. The four injection rate shapes must have the same injection duration and fuel quantity to be used as a benchmark. The three shapes (QIDG, GIGD, and GIQD) had peak injection rates higher than the experimental conditions in case No. 2, which means that the rail pressure would be higher than 150 MPa. As shown in the simulation model, these injection rate shapes can create a peak rail pressure of approximately 600 MPa. A maximum rail pressure of 600 MPa is too high for current engine technology. In this study, the maximum rail pressure generated by the injection rate was determined from the study conditions according to the constant value of the injection duration with the same amount of fuel as used in the experimental data to ensure that the model calibrated from the experiment data would be accurate. The injection pressure can be increased by changing the injection rate of the fuel pump and adjusting the injector area. If the injection pressure is too high, the ignition delay period will be shorter. This may cause a homogeneous decrease in mixing and reduce the combustion efficiency. In practice, changing the injection rate and injector area is necessary to help reduce these effects.

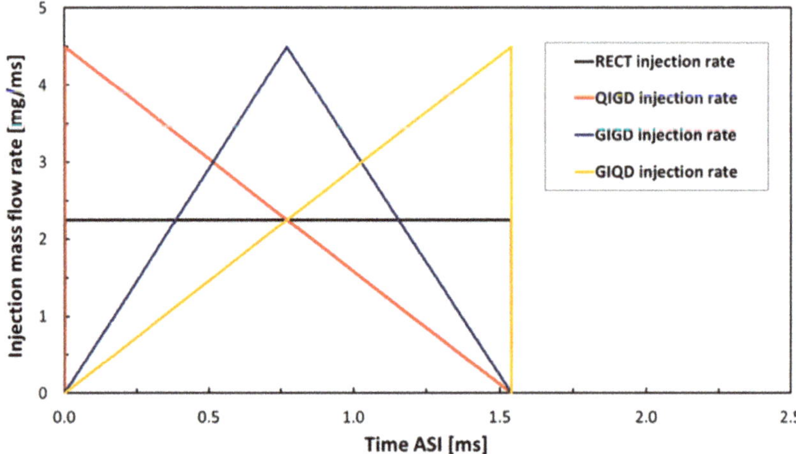

Figure 12. The varying shape of the fuel injection rate with a constant injection duration and fuel quantity. The rectangular injection rate (RECT), the Quick Increase Gradual Decrease injection rate (QIGD), the Gradual Increase Gradual Decrease injection rate (GIGD), and the Gradual Increase Quick Decrease injection rate (GIQD).

As a result that the newly created injection rate shape for this study is determined by a constant injection duration and fuel mass (including under ambient conditions, which are similar to those of the experimental data), under some conditions, this novel injection rate provides a peak injection rate that is higher than the injection rate of the experimental data used to calibrate the model. This means that higher injection rates will also result in higher rail pressure. Although this study explored the maximum rail pressure conditions up to 600 MPa, the results calibrated with the experimental data for rail pressure of 50, 100, and 150 MPa showed that the model can provide reliable simulation results, even under conditions where the pressure is different. In particular, the simulation results show that the study conditions at a maximum rail pressure of 150 MPa can provide better simulation results than under conditions of a lower maximum rail pressure. This means that the newly created model is effective for high rail pressure conditions. In addition, in this paper, all case studies (including the cases of maximum rail pressure conditions of 600 MPa) use ambient conditions. The injection timing and the fuel mass are the same as those of the experimental conditions at a rail pressure of 150 MPa because the simulation results for the calibration under rail pressure conditions of 150 MPa give good results for both quantity and volume. The prediction results for the spray shape are compatible with the photos from the experimental data; this shape reflects the accuracy of the distribution angle, penetration, and evaporation of the spray. The injection rate and injection pressure have a great influence on these physical characteristics. Therefore, the model created can predict the spray behavior very well, especially under high rail pressure conditions. Based on the study of the effects of different rail pressures, the spray shape data from the experiment demonstrate that the size of the spray cone angle is different under the same ambient conditions, as shown in Figure 13. Figure 13 shows the spray cone angle size obtained from the spray picture of the experiment data. The result indicates that the spray cone angle used as the input data in this study depends on the size of the maximum rail pressure. The model calibration demonstrated that using the spray cone angles obtained from the experimental spray images as inputs can provide good predictive results (consistent with the experimental data). Therefore, this study used the prediction equation (Equation (8)) to predict the size of the spray cone angle at different rail pressures:

$$Sray\ cone\ angle = 0.05P_{rail} + 13.333. \tag{8}$$

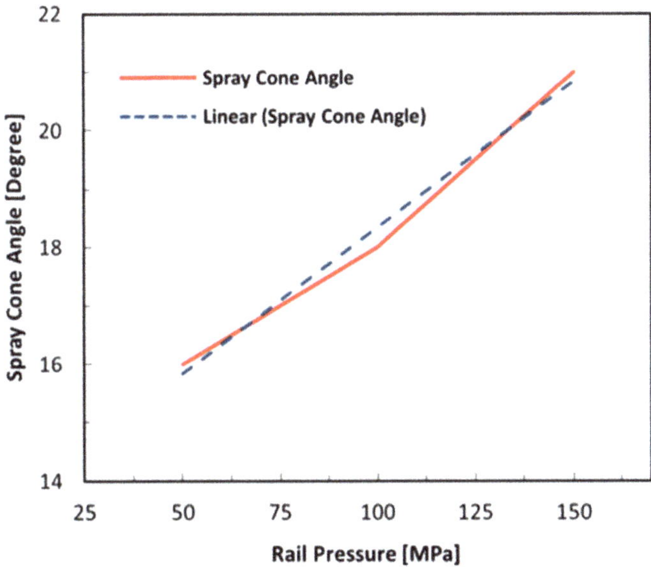

Figure 13. Comparison spray cone angle for different rail pressures.

4.1. Microscopic Spray Characteristics

In spraying simulations, the turbulent distribution has a significant impact on the spray parameters. The spray penetration distance depends on the surrounding conditions, including the injection rate, the injection time, the injected mass quantity, and others. The injection rate design affects the spray penetration and spray distribution angle. This section presents the spray penetration distance consisting of liquid and vapor penetration both during the injection time (0–1.54 ms) and after the EOI, by comparing the results of different injection rate shapes and using these shapes to analyze their effects on the microscopic spray characteristics.

Figure 14 shows the simulation results of the spray penetrations. The black, red, blue, and yellow lines represent the simulation results for the RECT, QIGD, GIGD, and GIQD injection rates, respectively. The results of vapor penetration are shown in the top graph, while the middle and bottom graphs show the liquid penetration and injection mass flow rates, respectively. Figure 14 (top graph) shows the vapor penetration for the four injection rates with different shapes. The QIGD injection rates show the longest vapor penetration (approximately 25 mm), with an injection rate of around 4.5 mg/ms. Vapor penetration lengths with the same peak injection rates will ultimately lead to the same vapor penetrations. This occurs because, during injection, vapor penetration increases with the injection acceleration rate. After the EOI, the vapor penetration distance increases along with the injection rate at the EOI, when the spray penetration distance is affected by the momentum flux ratio. For the GIQD injection rate shape after the EOI, the vapor penetration continues to increase continuously, and a high injection rate results in rapid fuel movement, while the vapor penetration of the QIGD injection rate shape decreases continuously because of the injection rate at the EOI. Figure 14 (middle graph) shows the liquid spray penetration, for which the different injection rates also have a significant impact on liquid penetration. The QIGD injection rate provides the longest initial liquid penetration due to having the highest injection rate, while the GIQD injection rate shows the opposite. Liquid spray penetration increases relative to the injection rate shape and terminates at the end of injection. The liquid spray penetration results show the same trend as the injection rate for all cases. Since the greatest penetration distance is primarily affected by the momentum flux ratio, the liquid penetration distance will increase when the injection rate increases and decrease when the injection rate decreases. This is due to slowdown in the movement of the liquid fuel.

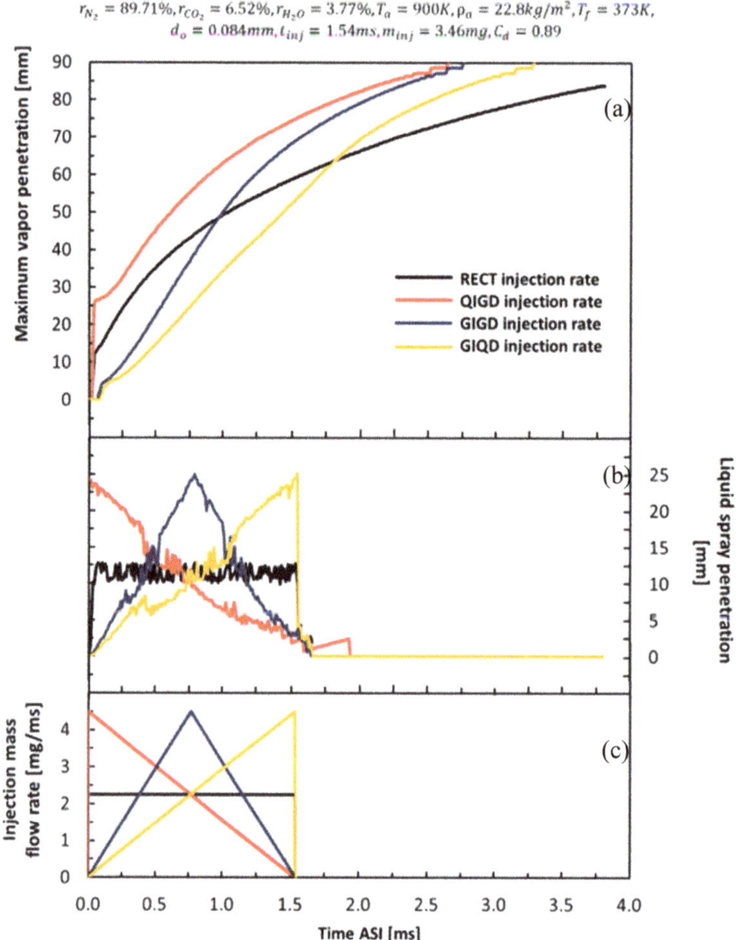

Figure 14. Comparison of the simulated spray penetration with different injection rate shapes. (**a**) Maximum vapor penetration, (**b**) liquid spray penetration and (c) injection mass flow rate.

In addition, because the spray penetration is a phenomenon that occurs in conjunction with the spray distribution, the spray distribution angle is an important parameter that affects the spray mixing process and is important for the analysis of spray performance. Therefore, the spray distribution angle was investigated by measuring the distribution angle via the spray shape obtained from the simulation. Figure 15 shows the spray distribution angles from the simulation results, where the black, red, blue, and yellow dots represent the measurement results of the RECT, QIGD, GIGD, and GIQD injection rates, respectively. As shown in the Figure 14, comparisons were made between the spray angle (left graph) and spray cone angle (right graph) under different injection rate shapes. Similar techniques to those in the previous section were used to measure the spray distribution angles.

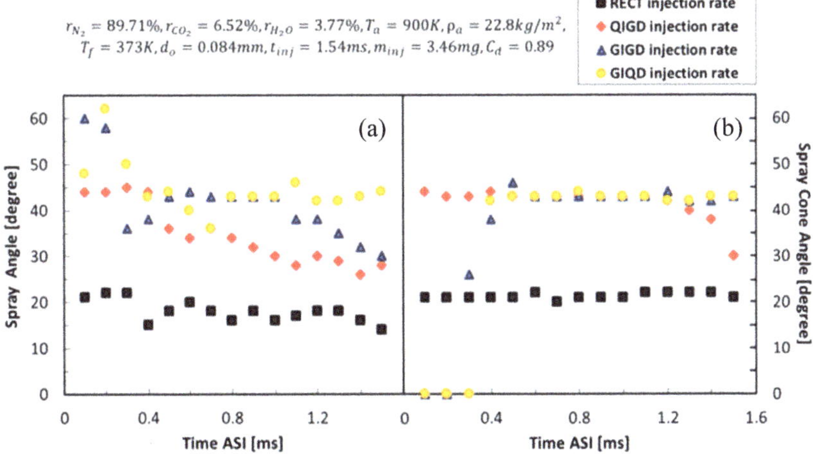

Figure 15. Comparison of the spray distribution angle for different injection rate shapes. (**a**) Spray angle and (**b**) spray cone angle.

The same spray cone angle measurement results are shown in Figure 15 (right graph) for the same peak injection rate (same peak rail pressure) with different injection rate shapes; the results agree with those in the previous section. The shape with a peak rail pressure of 600 MPa and a spray cone angle of 43° will have spray cone angle larger than approximately twice that of the peak rail pressure of 150 MPa (21°).

Figure 15 (left graph) shows a difference in the spray angle event under the same peak rail pressure. The spray angle increases when the acceleration of the injection rate increases. It can be observed that the QIGD injection rate is significantly small but features a continuous decrease in the spray angle compared to the other injection rates. On the other hand, the GIQD injection rate recorded the largest spray angle at the beginning, followed by a slight decrease, and then remained constant (compared to the other injection rates). This phenomenon demonstrates that the increased injection acceleration rate has a significant effect on the fuel distribution ability, which directly affects the spray angle size. The spray angle is higher when the injection rate is higher because the injection rate can increase the rail pressure, thereby resulting in a higher particle force that can cause a higher penetration force leading to better distribution. The spray angle is an important parameter that helps us understand the global characteristics of the spray. The spray angle and quantity evaluation provide useful information about the airflow in the spray [29–31], where the spray angle is an indicator of gas. In general, the greater the spray angle, the higher the increase in gas entrainment, resulting in improved mixing [32]. Another interesting observation is the effect of the injection rate shapes on spray penetration. The simulation results reveal longer vapor penetration at a higher peak injection rate due to the efficient spray distribution, while the lower peak injection rate yielded poor vapor penetration because the low injection rate resulted in poor spray distribution and spray penetration performance. A spray tip that penetrates too long will result in wet combustion chamber walls, causing excessive soot formation and a waste of fuel. On the other hand, if the penetration time is too short, the mixing efficiency and optimum combustion will be compromised. In addition, the simulation results show that the QIGD injection rate with a high initial fuel injection rate quickly causes the initial penetration. This suggests that the injected fuel is very well atomized and has a significant effect on the onset speed of the combustion phenomenon. In the case of the GIQD injection rate, a very low initial injection rate may result in poor fuel atomization and cause an increase in the ignition delay. The evaporation process and mixing behavior will be discussed further in the next section.

4.2. Evaporation Process

We next studied the evaporation process influenced by the fuel injection rate by considering the distribution of droplet sizes. This section presents the simulation results of the evaporation ratio under different injection rate shape conditions, as shown in Figure 16. The simulation results show that different injection rate shapes result in different evaporation ratios. The results of the evaporation ratios are shown in the top graph, and the injection mass flow rates are shown in the bottom graph, where the black, red, blue, and yellow lines represent the simulation results of the RECT, QIGD, GIGD, and GIQD injection rates, respectively. Figure 16 shows that the RECT and the QIGD injection rate shapes undergo more rapid evaporation than the other injection rate shapes. As demonstrated by the RECT and the QIGD injection rate shapes, the evaporation ratio increases to nearly one before approximately 0.01 ms. For the GIGD and GIQD injection rates, the evaporation ratio increases to nearly one at approximately 0.1 ms. These results are due to both injection rate shapes having a quickly increasing initial injection rate, thereby resulting in high rail pressure, which can improve the evaporation rate because higher rail pressure results in a high shear of the fuel particles, which can change the fuel state from liquid to gas very quickly with higher mass flow rates, as well as accelerate the fuel evaporation process.

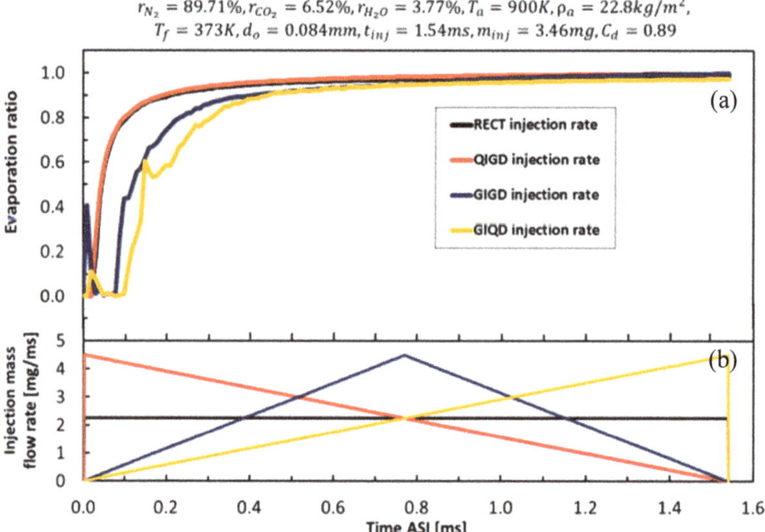

Figure 16. Comparing the evaporation ratios from the simulation results. (**a**) Evaporation ratio and (**b**) injection mass flow rate.

To better understand the spray breakup characteristics, the SMD is an important parameter that should be considered to reflect the spray performance. The size of the SMD is related to droplet breakup, in which a smaller SMD result in better droplet breakup. Due to the lack of experimental data to calibrate the simulated SMD results, only the relationship between the SMD and injection rate have been considered. Figure 17 shows the relationships of the SMD values for different injection rate shapes, where the black, red, blue, and yellow lines represent the simulation results of the RECT, QIGD, GIGD, and GIQD injection rates, respectively. The results show that the SMD decreases rapidly when the injection rate increases sharply, at the same time, the SMD gradually decreases as the injection rate gradually increases. This occurs because the droplet tends to breakup under conditions with higher injection rates. A high initial injection rate can clearly reduce the SMD. As can be considered from the high initial injection rate conditions (RECT, QIGD), the SMD decreases to nearly zero at approximately 0.02 ms; with gradual increases in the injection rate conditions (GIGD, GIQD), the SMD decreases to nearly zero at approximately 0.1 ms. These values are worth noting for the RECT and QIGD injection rates. Although the initial injection rates are not the same, the injection rate is sufficiently higher to result in a rapid decrease in the SMD. These behaviors support the spray breakup phenomenon. The evaporation rate is higher for the droplet under a higher initial injection rate, as shown in Figure 16. From Figure 17, it can be concluded that the initial injection rate is the main factor affecting the size of the droplets, as an increase in the injection rate results in a decrease in the SMD. Therefore, under higher initial injection rates, the droplets will become smaller and lead to faster evaporation.

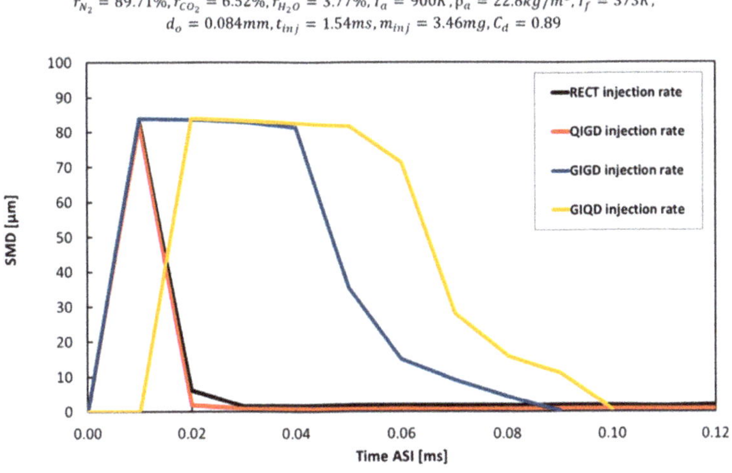

Figure 17. Comparison of the Sauter mean diameter (SMD) for different injection rate shapes.

The fuel evaporation phenomena can be understood by considering the temperature distribution phenomena. Figures 18 and 19 show the simulation results comparing the temperature distribution with different injection rate shapes during injection and after the EOI. The spray images were obtained from the cut-plane in the direction of the spray. The gradient color region represents the temperature data and shows the temperature value as a gradient color bar in the bottom right corner.

Figures 18 and 19 show that the spray penetration area is cooler than the surrounding combustion chamber, as the fuel absorbs heat for vaporization. The RECT and QIGD injection rates show better vapor penetration at the beginning when considering the temperature distribution contours and comparing them with the other injection rate shapes. This phenomenon occurs due to the higher initial injection rate compared to the GIGD and GIQD injection rates. In addition, the area near the nozzle exit showed different vapor penetration values for each case. The very high injection rate that resulted in initial vapor penetration also occurred further away from the nozzle exit. In Figures 18 and 19, under the GIQD injection rate, the vapor penetration at the nozzle exit is farther away from the nozzle exit and clearly farther from the injector exit than other shapes after the EOI. This is because the gradual increase in the injection rate results in fuel breakup capability.

These phenomena occur because faster injection rates can accelerate the evaporation of fuel, with the airflow in the cylinder having a great impact on the evaporation and atomization of the fuel. Therefore, under the conditions of higher injection rates, the droplets will be smaller and evaporate faster, resulting in better acceleration in the formation of the air–fuel mixture. It can be predicted that a high initial injection rate would result in a decrease in the ignition delay period due to better fuel atomization at the beginning of the injection, which may more quickly lead to the start of combustion. The influence of injection rate shapes on the characteristics of the mixture properties will be studied in the next section.

$r_{N_2} = 89.71\%, r_{CO_2} = 6.52\%, r_{H_2O} = 3.77\%, T_a = 900K, \rho_a = 22.8 kg/m^2, T_f = 373K, d_a = 0.084mm, t_{inj} = 1.54ms, m_{inj} = 3.46mg, C_d = 0.89$

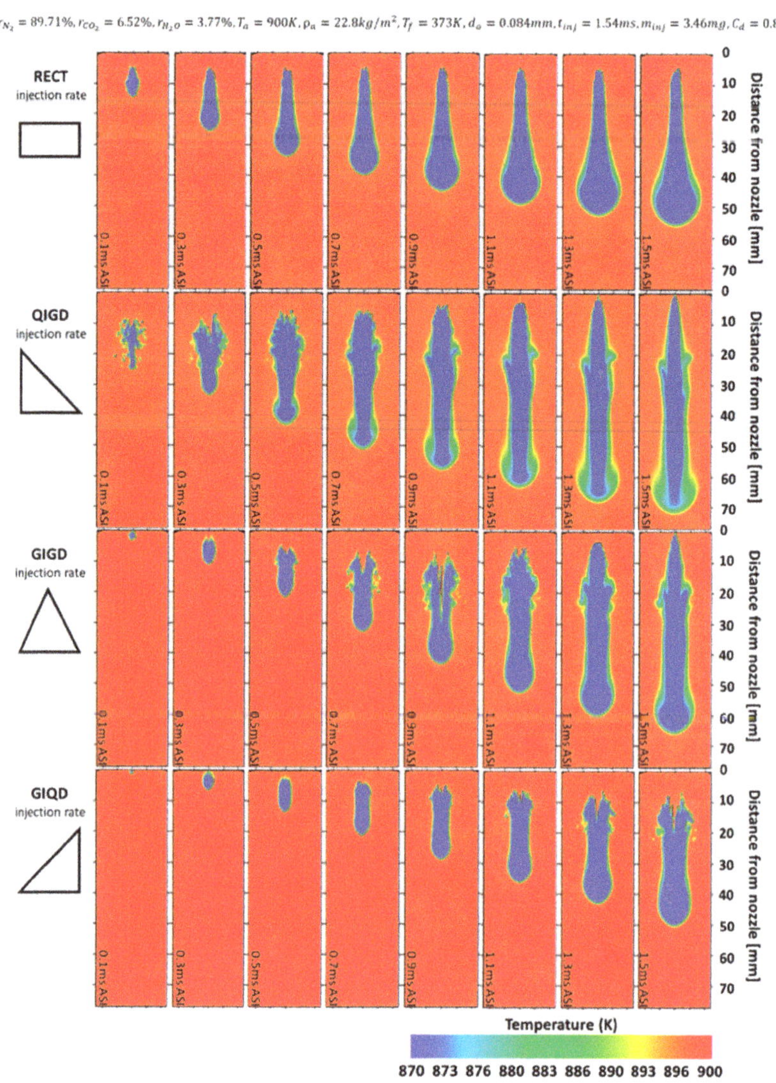

Figure 18. Comparison of the temperature distribution during injection for different injection rate shapes.

$r_{N_2} = 89.71\%, r_{CO_2} = 6.52\%, r_{H_2O} = 3.77\%, T_a = 900K, \rho_a = 22.8 kg/m^2, T_f = 373K, d_o = 0.084mm, t_{inj} = 1.54ms, m_{inj} = 3.46mg, C_d = 0.89$

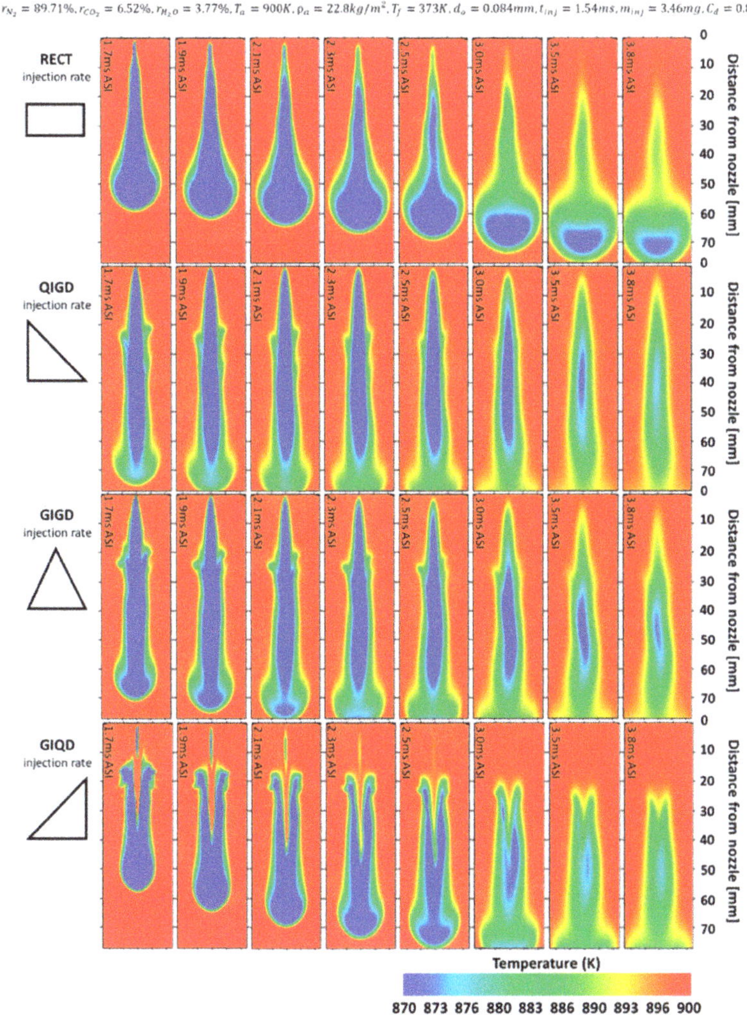

Figure 19. Comparison of temperature distribution after the EOI for different injection rate shapes.

4.3. Mixture Properties

For a deeper understanding of the diesel spray mixing process, the mixture properties are studied in this section by comparing their equivalence ratios and turbulence kinetic energy (TKE) values at different injection rates.

The mixture properties are analyzed by considering the region of the equivalence ratio predicted from the 3D model, which plays an important role in the analysis of the diesel spray mixing process. Therefore, the equivalence ratio is calculated from the basic data of the mass fraction. In this study, the conditions of the case study are the conditions under which the ambient component does not contain oxygen ($r_{O_2} = 0$). We applied the following equation of chemical combustion:

$$Fuel + O_2 \rightarrow CO_2 + H_2O. \tag{9}$$

Here, we assume that by replacing the oxygen ratio with the total gas ratio (including the oxygen ratio), the general air contains 21% oxygen and 79% other gases. Therefore, the mass fraction is given as Fuel/(Fuel + O_2 + other gases), where the equivalence ratio is equal to 1. From the simulation results, the different injection rate shapes show different spray behaviors and equivalence ratio histories, although the injection duration and injection mass quantity are the same. Figures 20 and 21 provide a comparison of the spray behavior and equivalence ratio histories of the simulation results during injection (Figure 20) and after the EOI (Figure 21). The spray images were obtained from the cut-plane in the direction of spray. The gradient color region represents the equivalence ratio data and shows the equivalence ratio values as a gradient color bar in the bottom right corner, while the black particles are the liquid fuel data from the simulation.

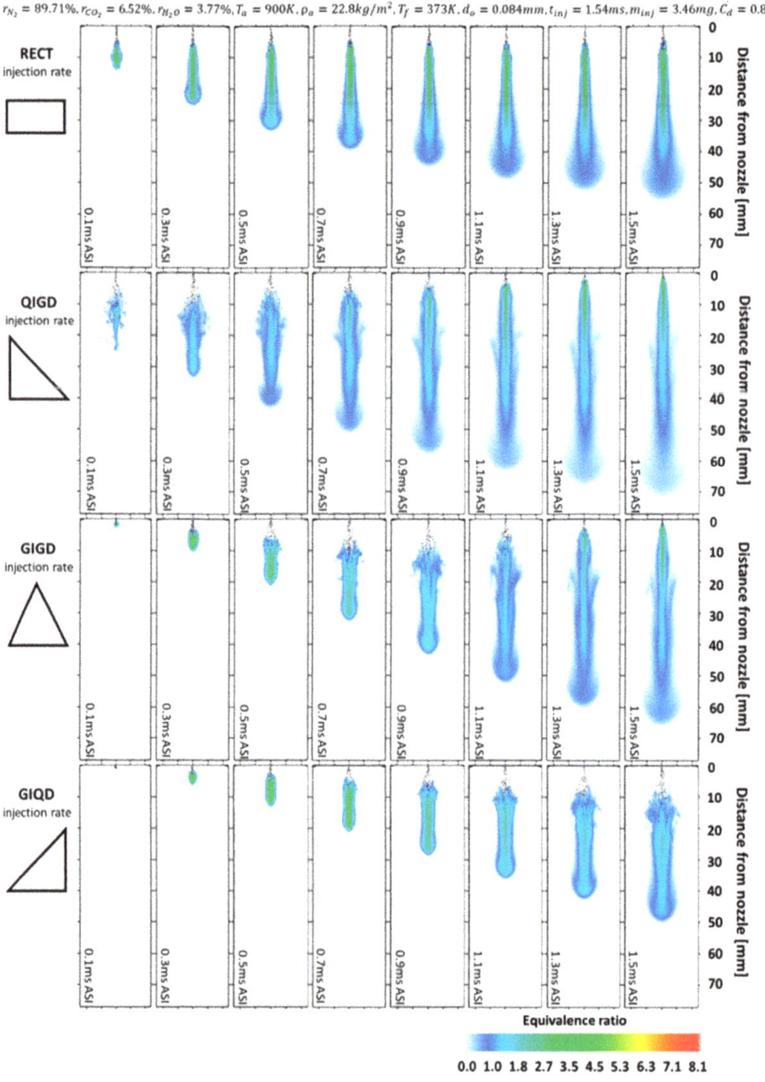

Figure 20. Comparison of the spray behavior and equivalence ratio histories during injection for different injection rate shapes.

$r_{N_2} = 89.71\%, r_{CO_2} = 6.52\%, r_{H_2O} = 3.77\%, T_a = 900K, \rho_a = 22.8kg/m^2, T_f = 373K, d_o = 0.084mm, t_{inj} = 1.54ms, m_{inj} = 3.46mg, C_d = 0.89$

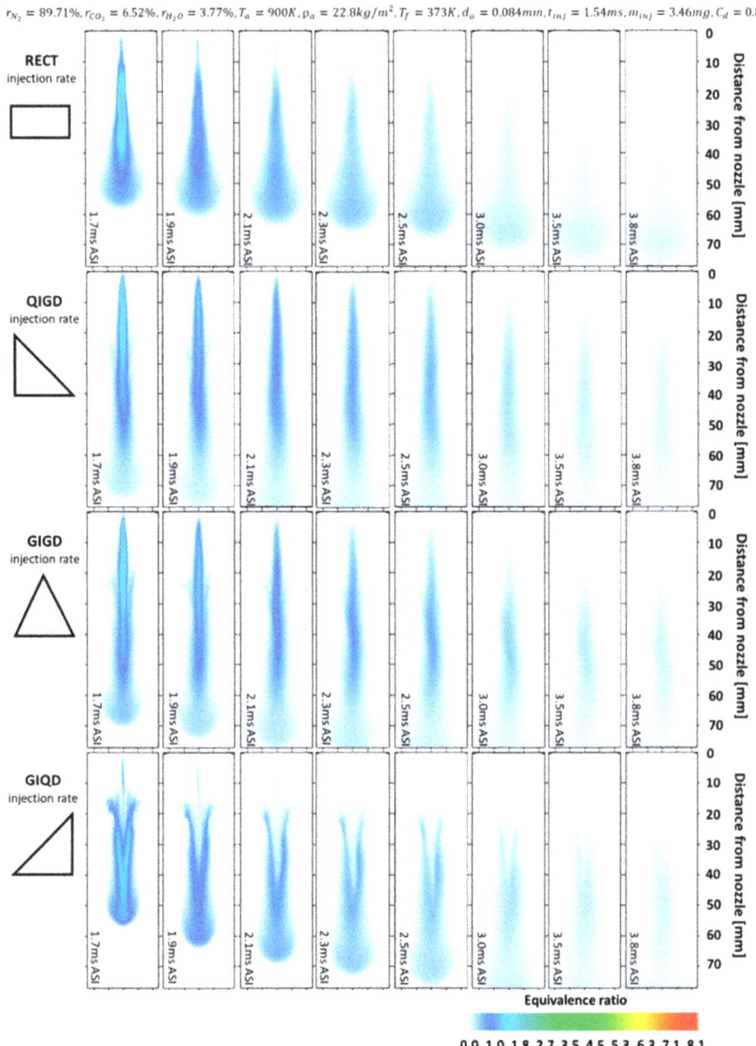

Figure 21. Comparison of the spray behavior and equivalence ratio histories after the EOI for different injection rate shapes.

Figure 20 shows the spray behavior and equivalence ratio histories for the four injection rates. It can be observed that although the injection mass and time for all simulated cases are the same, the spray behavior and equivalence ratio histories for each injection rate shape are very different. There is a high fuel distribution near the nozzle exit of the injection rate shapes with a high peak rail pressure (QIGD, GIGD, and GIQD), which is different from the case with the lower peak rail pressure (RECT). In general, fuel distribution is influenced by changes in the fuel injection rates. These phenomena are clearly reflected in Figure 21. Figure 21 clearly shows that the GIQD injection rate with the highest injection rate at the EOI provides the leanest equivalence ratio near the nozzle exit. This is due to the influence of the injection rate, which affects the fuel breakup and the surrounding air crossflow. The momentum arising from increasing the injection rates yields a complementary momentum. The initial injection rate is related to the droplet breakup. For example, for high initial injection rate conditions (RECT and QIGD), the black particles represent the liquid fuel intensely near the nozzle exit and descend quickly when the injection time has passed. This is an example of quick SMD reductions, as shown in Figure 17.

For a better understanding of the spray behavior in diesel spray mixing, TKE is an important influencing factor that can explain the phenomenon of the equivalence ratio. TKE can reflect the intensity of the turbulent movement in the cylinder. Figure 22 shows a comparison of TKE with different injection rates. The TKE is displayed as the average value of the TKE in the control volume model, where the black, red, blue, and yellow lines represent the simulation results of the RECT, QIGD, GIGD, and GIQD injection rates, respectively. The top graph shows the TKE and the bottom graph shows the injection mass flow rate.

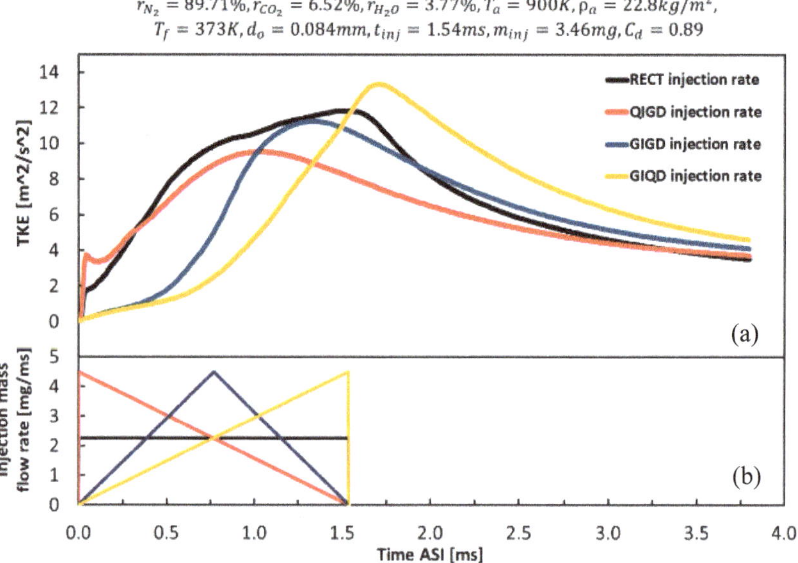

Figure 22. Comparison of the turbulence kinetic energy (TKE) for different injection rate shapes. (**a**) TKE and (**b**) injection mass flow rate.

The injection rate has a significant relationship with the TKE. At low mass flow rates, the field flow will be smooth without a recirculation zone. On the other hand, when the mass flow rate increases, the vortex will increase. This intense recirculation will increase heat transfer compared to the smooth channels. Figure 22 shows the strongest TKE during the initial injection under the QIGD injection rate, while the RECT injection rate shows the strongest TKE during the injection. Subsequently, after the EOI, the GIQD injection rate shows the strongest TKE. These phenomena can be explained by the influence of TKE production, which is related to the effect of the spray structure. The simulation results of TKE distribution with different injection rates, both during injection and after the EOI, are shown in Figures 23 and 24, respectively. The turbulence phenomenon is considered in the boundary of the spray region, where the model provides sufficient mesh density. The spray images were obtained from the cut-plane of the direction of the spray. The gradient color region represents the TKE distribution, which presents the TKE data as a gradient color bar in the bottom right corner.

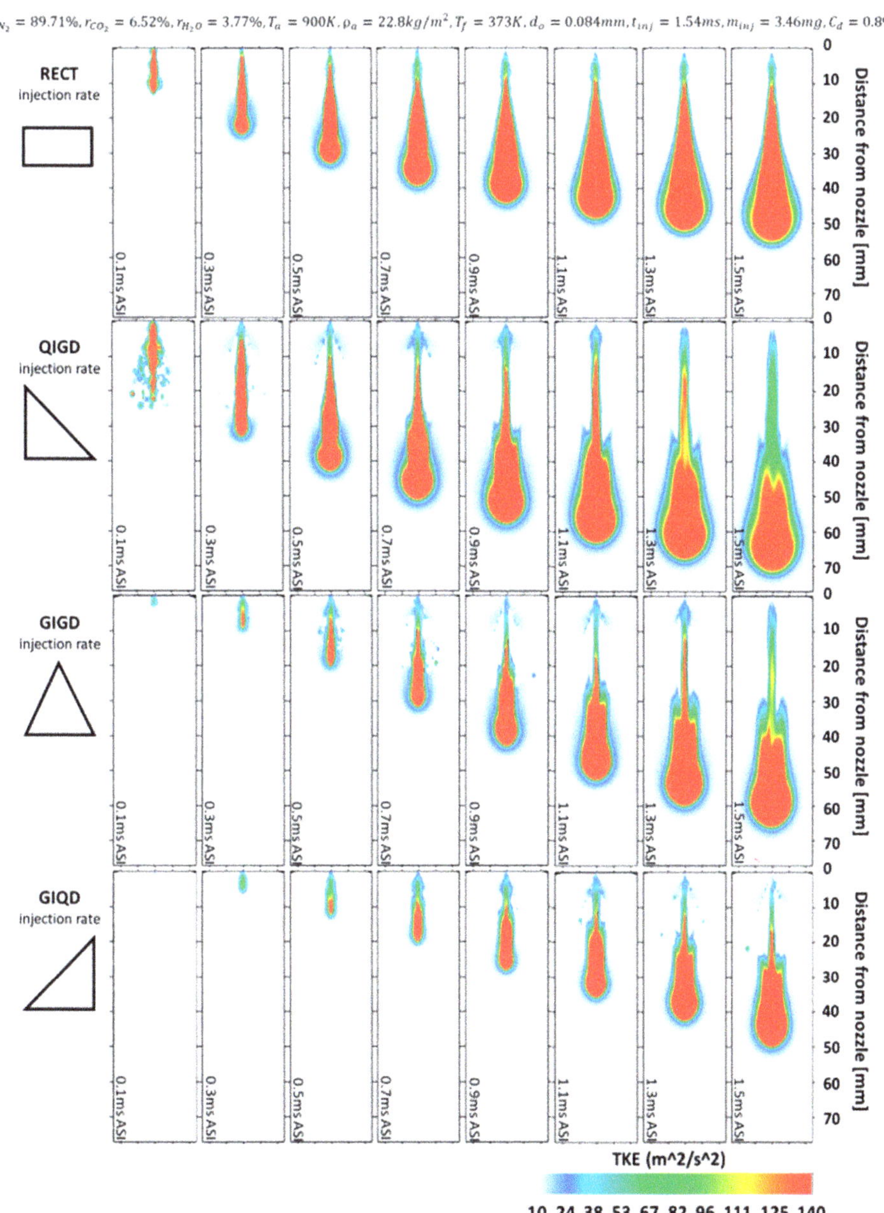

Figure 23. Comparison of the TKE distribution during injection for different injection rate shapes.

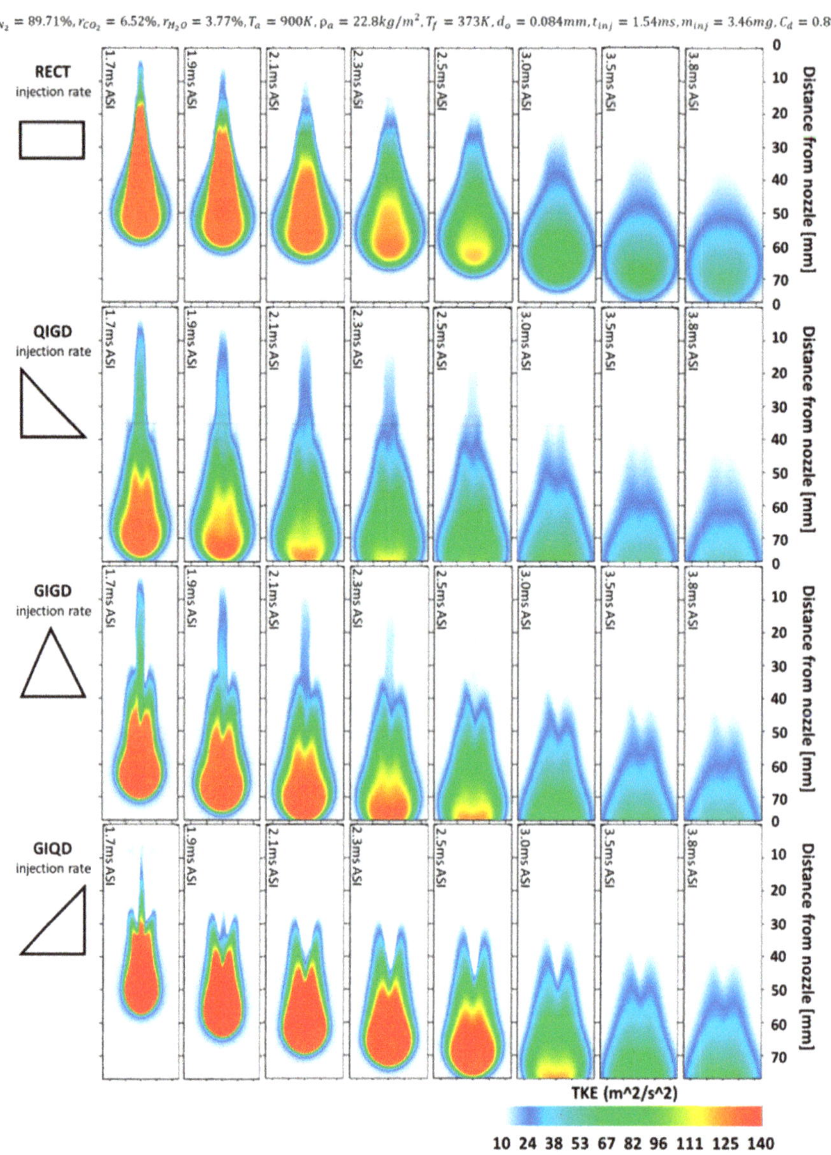

Figure 24. Comparison of the TKE distribution after the EOI for different injection rate shapes.

Figures 23 and 24 show the contour of the TKE with different injection rate shapes. The state of turbulence distribution affects the proportions of the length and width of the spray shape. For example, Figures 23 and 24 show the strong TKE of the QIGD injection rate case that moved to the area along the length of the spray, which resulted in long spray penetration and a lean equivalence ratio in the spray tip area, as can be seen in Figures 20 and 21. This is because a high injection rate directly affects the crossflow, resulting in significant disintegration and displacement. Figures 23 and 24 show that the TKE production of the RECT injection rate is strong in the axial area of the spray, which indicates that the TKE distribution occurs along the width of the spray. Although the TKE of the RECT injection rate is the strongest during the injection period (see Figure 22), the TKE produced cannot offer penetration longer than the QIGD injection rate (Figures 23 and 24). The QIGD injection rate initially shows the strongest TKE; then, the TKE distribution moves along the length of the spray faster than the other injection rates and has the highest TKE at the spray tip area. This is due to this spray's high speed and acceleration, causing more increased turbulence levels in the downstream areas than in other injection rate shapes. This gives the QIGD injection rate the highest efficiency of spray turbulence, providing the leanest equivalence ratio at the spray tip area.

The phenomenon of TKE distribution can be well illustrated by the temperature distribution in Figures 18 and 19, which show that the QIGD injection rate has a higher temperature in the spray tip area than other shapes. When considering the temperature distribution behavior after the EOI in Figure 19, the RECT injection rate shape retains the lowest temperature at the spray tip area compared to the other injection rate shapes. The highest temperature at the spray tip area of the QIGD injection rate and the lowest temperature at the spray tip area of the RECT injection rate can predict the spray turbulence behavior. A high temperature means that the area may produce high turbulence results with a lean equivalence ratio but show opposite spray behavior at low temperatures. These results are due to the high temperature influencing the droplet size and evaporation. These phenomena occur when the increased fuel injection rate causing the TKE grows, thereby leading to faster fuel spread and resulting in faster fuel and air mixing. The TKE can reflect the intensity of the turbulent motion in the cylinder, with a large TKE indicating that the agitation is more intense in the cylinder. Examination of the turbulence demonstrates that a high TKE flow provides a clear driving force for mixing and evolution. In addition, strong turbulent mixing leads to the highest chance of saturation, which results in better mixing opportunities. Therefore, the higher the initial injection rate, the greater the variability of the cutting force field in the spray area. This phenomenon is related to the fact that the shape of the injection rate has a significant impact on the potential of spray behavior and TKE production. The mixing efficiency is very potent when the TKE distribution has high potential due to a sufficiently large increase in rail pressure.

The results of this simulation study show that high initial injections will produce high turbulence energy. The GIQD injection rate is highly efficient in atomization and creating good fuel–air mixing, thus resulting in good combustion. The duration of the combustion at each injection rate can be arranged, from long to short, as follows: GIQD, GIGD, RECT, and QIGD. Since the QIGD injection rate has the highest initial injection rate, combustion may start earlier, while the GIQD injection rate has the lowest initial injection rate, which will result in less atomization and poorer combustion.

In addition, when analyzing the mixing behavior of sprays (which can affect emissions), CO emissions increase when the initial injection rate is low, which will shorten the flame lift-off length [6]. This means that the fuel-injection does not have enough time for good air entrainment before the start of combustion, resulting in higher CO emissions levels. A high initial injection will result in high NO because a high initial injection will lead to an earlier start of combustion and a higher peak heat-release. We expect that the QIGD injection rate will have the highest amount of heat released because the temperature contours and the QIGD injection rate case will provide the fastest mixing. This mixing will result in the shortest ignition delay duration with the highest observed heat-release. The strategy for creating high initial injection rates often results in NO_x emissions and a higher engine noise level. In our numerical simulation on the influence of injection rates on spray behavior affecting the mixing and combustion processes, we chose the QIGD injection rate as the optimal rate for the injection strategy, which requires the injection time and combustion to be short, because this injection rate's mixing efficiency is greater than that of the other injection rate shapes. The RECT injection rate was used for reducing NO_x and engine noise due to lower pressure in the combustion chamber. Future studies will be developed on the efficiency of injection strategies that can simultaneously reduce NO_x emissions, engine noise levels, and soot emissions.

5. Conclusions

In this study, diesel spray was tested with a 3D-CFD numerical simulation. First, the 3D-CFD numbers were checked for their accuracy from the experimental data. Then, we studied the effects of various injection rate shapes on the properties of the spray and analyzed them using the 3D-CFD simulation results. The preliminary conclusions can be summarized as follows:

(1) The 3D-CFD model can predict the spray behavior's agreement with the experimental data both during the injection and after the end of the injection, covering different temperatures and injection pressure situations.

(2) The injection rate shapes affect the penetration of both liquid and vapor, which in turn influence spray evaporation and spray mixing. The spray behavior can change even when the injection duration and injection mass are the same for all simulation cases.

(3) Injection rates with high peak injection rates (QIGD, GIGD, and GIQD) can provide higher vapor penetration capabilities compared to injection rates with a low peak injection rate (RECT). Furthermore, injection rates with the same peak injection rates will ultimately yield the same vapor penetration.

(4) Injection rates with a high initial fuel injection, such as the RECT and QIGD injection rates, will make the evaporation more rapid. When the injection rate is rapid enough, resulting in a high temperature, the high TKE and SMD decrease rapidly. For the QIGD injection rate, the TKE grows and moves to an area according to the length of the spray, which results in a longer spray penetration and a leaner equivalence ratio in the spray tip area than the other injection rates. For the RECT injection rate, the TKE grows and moves to an area according to the width of the spray due to the constant injection rate. This means that the initial injection rate shape has a significant impact on the evaporation efficiency and TKE production.

(5) Injection rate shapes with good evaporation efficiency, such as the QIGD injection rates, will likely result in a shorter ignition delay, including a more rapid onset of combustion and a shorter duration of combustion. The QIGD injection rate has high rail pressure and the strongest turbulent mixing at initial injection, resulting in faster mixing. For a low initial injection rate, the GIQD injection rate shows the opposite, and CO emissions increase because the fuel injection does not have enough time to develop good air entrainment before the start of combustion. Strategies for creating high pressure in the cylinder often result in NO_x emissions and high engine noise levels. This means that the initial injection rate and rail pressure have a significant impact on combustion time, emissions, and engine noise levels.

Finally, the analysis of the numerical simulation of the effects of different fuel injection rate shapes on the diesel spray mixing process shows that the variation in the structure of the air and fuel mixture in diesel engines depends on the shape of the fuel injection rate. The injection rate influences the diesel spray mixing process both during injection and after the EOI.

Author Contributions: Conceptualization, L.L.; methodology, I.N.; simulation, I.N.; validation, L.L., K.N. and D.L.; formal analysis, L.L. and I.N.; data curation, I.N.; writing—original draft preparation, I.N.; writing—review and editing, L.L.; project administration, X.M. All authors have read and agreed to the published version of the manuscript.

Funding: This research was funded by the National Natural Science Foundation of China (grant number 51509051), and the International Exchange Program of Harbin Engineering University for Innovation-oriented Talent Cultivation.

Acknowledgments: This paper was funded by the International Exchange Program of Harbin Engineering University for Innovation-oriented Talent Cultivation. The experimental data were taken from the Engine Combustion Network website that is maintained and operated by the Engine Combustion Department of Sandia National Laboratories.

Conflicts of Interest: The authors declare no conflict of interest.

Abbreviations

ECN	Engine Combustion Network	r_{O_2}	Oxygen Ratio [%]
CMT	CMT-Motores Térmicos	t_{inj}	Injection Time [ms]
CFD	Computational Fluid Dynamics	m_{inj}	Injection Mass [mg]
3D	Three-Dimensional	P_{rail}	Rail Pressure [MPa]
EOI	End of Injection	$\wedge$	Wavelength
ASI	After Start of Injection	B_0	KH Model Breakup Size Constant
SMD	Sauter Mean Diameter	We	Weber Number
TKE	Turbulence Kinetic Energy	C_d	Steady Flow Discharge Coefficient
RECT	Rectangular Injection Rate Shape	T_a	Ambient Temperature [K]
QIGD	Quick Increase Gradual Decrease Injection Rate Shape	P_a	Ambient Pressure [MPa]
GIGD	Gradual Increase Gradual Decrease Injection Rate Shape	ρ_a	Ambient Density [kg/m^2]
GIQD	Gradual Increase Quick Decrease Injection Rate Shape	T_f	Fuel Temperature [K]
RANS	Reynolds Averaged Navier–Stokes	r	Droplet Radius
RNG	Re-Normalization Group Theory	τ	Breakup Time
NTC	No Time Counter Model	B_1	KH Model Breakup Time Constant
KH	Kelvin-Helmholtz Instability Model	L_b	Breakup Length
RT	Rayleigh-Taylor Instability model	C_{bl}	Breakup Length Constant
NO$_x$	Nitrogen Oxides	ρ_l	Fuel Density
CO	Carbon Monoxide	ρ_g	Ambient Gas Density
UHC	Unburnt Hydrocarbons	d_0	Orifice Diameter
$C_{12}H_{26}$	*n*-Dodecane	C_τ	RT Model Breakup Time Constant
N_2	Nitrogen	Ω	Maximum Growth Rate of Varicose Waves
CO_2	Carbon Dioxide	C_{RT}	RT Model Size Constant
H_2O	Water	K_{RT}	RT Model Weave Number
O_2	Oxygen		

References

1. Guan, B.; Zhan, R.; Lin, H.; Huang, Z. Review of the state-of-the-art of exhaust particulate filter technology in internal combustion engines. *J. Environ. Manag.* **2015**, *154*, 225–258. [CrossRef] [PubMed]
2. Imtenan, S.; Varman, M.; Masjuki, H.; Kalam, M.; Sajjad, H.; Arbab, M.; Fattah, I.R. Impact of low temperature combustion attaining strategies on diesel engine emissions for diesel and biodiesels: A review. *Energy Convers. Manag.* **2014**, *80*, 329–356. [CrossRef]
3. Samojeden, B.; Motak, M.; Grzybek, T. The influence of the modification of carbonaceous materials on their catalytic properties in SCR-NH 3. A short review. *C. R. Chim.* **2015**, *18*, 1049–1073. [CrossRef]
4. Komninos, N.; Rakopoulos, C. Modeling HCCI combustion of biofuels: A review. *Renew. Sustain. Energy Rev.* **2012**, *16*, 1588–1610. [CrossRef]
5. Reitz, R.D.; Duraisamy, G. Review of high efficiency and clean reactivity controlled compression ignition (RCCI) combustion in internal combustion engines. *Prog. Energy Combust. Sci.* **2015**, *46*, 12–71. [CrossRef]
6. Juneja, H.; Ra, Y.; Reitz, R.D. *Optimization of Injection Rate Shape Using Active Control of Fuel Injection*; SAE Technical Paper Series, 2004–01-0530; SAE: Warrendale, PA, USA, 2004. [CrossRef]
7. Long, L.; Bo, L.; Changfu, H.; Magagnato, F.A. Study on Mixing Process of Diesel Jets with Short Injection Duration. *ISAIF-S-0092* **2017**, *13*, 1–5.
8. Ivan, A.; Rocco, D.L.; Cesare, P.; Matteo, D.C. Enhanced combustion model with fuel-wall impingement oriented to injection pattern tuning in automotive Diesel engines. *IFAC-PaperOnLine* **2016**, *49*, 476–483.
9. Tay, K.L.; Yang, W.; Zhao, F.; Yu, W.; Mohan, B. Effects of triangular and ramp injection rate-shapes on the performance and emissions of a kerosene-diesel fueled direct injection compression ignition engine: A numerical study. *Appl. Therm. Eng.* **2017**, *110*, 1401–1410. [CrossRef]
10. Mohan, B.; Yang, W.; Yu, W.; Tay, K.L.; Chou, S.K. Numerical investigation on the effects of injection rate shaping on combustion and emission characteristics of biodiesel fueled CI engine. *Appl. Energy* **2015**, *160*, 737–745. [CrossRef]
11. Tay, K.L.; Yang, W.; Zhao, F.; Yu, W.; Mohan, B. A numerical study on the effects of boot injection rate-shapes on the combustion and emissions of a kerosene-diesel fueled direct injection compression ignition engine. *Fuel* **2017**, *203*, 430–444. [CrossRef]

12. Zhou, D.; Tay, K.L.; Tu, Y.; Li, J.; Yang, W.; Zhao, D. A numerical investigation on the injection timing of boot injection rate-shapes in a kerosene-diesel engine with a clustered dynamic adaptive chemistry method. *Appl. Energy* **2018**, *220*, 117–126. [CrossRef]

13. Morgan, R.; Banks, A.; Auld, A.; Heikal, M.; Ienartowicz, C. *The Benefits of High Injection Pressure on Future Heavy Duty Engine Performance*; SAE Technical Paper Series, 2015-24-2441; SAE: Warrendale, PA, USA, 2015. [CrossRef]

14. Stanton, D.W. Systematic development of highly efficient and clean engines to meet future commercial vehicle greenhouse gas regulations. *SAE Int. J. Engines* **2013**, *6*, 1395–1480. [CrossRef]

15. Huang, W.; Wu, Z.; Gao, Y.; Zhang, L. Effect of shock waves on the evolution of high-pressure fuel jets. *Appl. Energy* **2015**, *159*, 442–448. [CrossRef]

16. Agarwal, A.K.; Dhar, A.; Srivastava, D.K.; Maurya, R.K.; Singh, A.P. Effect of fuel injection pressure on diesel particulate size and number distribution in a CRDI single cylinder research engine. *Fuel* **2013**, *107*, 84–89. [CrossRef]

17. Shuai, S.; Abani, N.; Yoshikawa, T.; Reitz, R.D.; Park, S.W. Evaluation of the effects of injection timing and rate-shape on diesel low temperature combustion using advanced CFD modeling. *Fuel* **2009**, *88*, 1235–1244. [CrossRef]

18. Xu, L.; Bai, X.-S.; Jia, M.; Qian, Y.; Qiao, X.; Lu, X. Experimental and modeling study of liquid fuel injection and combustion in diesel engines with a common rail injection system. *Appl. Energy* **2018**, *230*, 287–304. [CrossRef]

19. Naruemon, I.; Long, L.; Qihao, M.; Xiuzhen, M. Investigation on an injection strategy optimization for diesel engines using a one-dimensional spray model. *Energies* **2019**, *12*, 4221. [CrossRef]

20. Tetrault, P.; Plamondon, E.; Breuze, M.; Hespel, C.; Mounaïm-Rousselle, C.; Seers, P. *Fuel Spray Tip Penetration Model for Double Injection Strategy*; SAE Technical Paper Series, 2015-01-0934; SAE: Warrendale, PA, USA, 2015. [CrossRef]

21. Sazhin, S.; Feng, G.; Heikal, M.R. A model for fuel spray penetration. *Fuel* **2001**, *80*, 2171–2180. [CrossRef]

22. Yin, B.; Yu, S.; Jia, H.; Yu, J. Numerical research of diesel spray and atomization coupled cavitation by Large Eddy Simulation (LES) under high injection pressure. *Int. J. Heat Fluid Flow* **2016**, *59*, 1–9. [CrossRef]

23. Pickett, L.M. Engine Combustion Network 2012. Available online: http://www.sandia.gov/ecn (accessed on 15 August 2019).

24. Richards, K.J.; Senecal, P.K.; Pomraning, E. *CONVERGETM (Version 2.2.0) Theory Manual*; Convergent Science Inc.: Madison, WI, USA, 2014.

25. Kook, S.; Pickett, L.M. Liquid length, and vapor penetration of conventional, Fischer–Tropsch, coal-derived, and surrogate fuel sprays at high temperature and high-pressure ambient conditions. *Fuel* **2012**, *93*, 539–548. [CrossRef]

26. Som, S.; Aggarwal, S.K. Effects of primary breakup modeling on spray and combustion characteristics of compression ignition engines. *Combust. Flame* **2010**, *157*, 1179–1193. [CrossRef]

27. Reitz, R.D. Modeling atomization processes in high-pressure vaporizing sprays. *At. Sprays Technol* **1987**, *3*, 309–337.

28. Hwang, S.S.; Liu, Z.; Reitz, R.D. Breakup Mechanisms and Drag Coefficients of High-Speed Vaporizing Liquid Drops. *At. Sprays* **1996**, *6*, 353–376.

29. Naber, J.; Siebers, D.L. *Effects of Gas Density and Vaporization on Penetration and Dispersion of Diesel Sprays*; SAE Technical Paper Series, 960034; SAE: Warrendale, PA, USA, 1996. [CrossRef]

30. Siebers, D.L. *Scaling Liquid-Phase Fuel Penetration in Diesel Sprays Based on Mixing-Limited Vaporization*; SAE Technical Paper Series, 0528; SAE: Warrendale, PA, USA, 1999. [CrossRef]

31. Arai, M. Diesel spray behaviour and air entrainment. *JNNA* **2018**, *2*, 1–17.

32. Gimeno, J.; Bracho, G.; Martí-Aldaraví, P.; Peraza, J.E. Experimental study of the injection conditions influence over n-dodecane and diesel sprays with two ECN single-hole nozzles. Part I: Inert Atmosphere. *Energy Convers. Manag.* **2016**, *126*, 1146–1156. [CrossRef]

applied sciences

Article

Investigations for a Trajectory Variation to Improve the Energy Conversion for a Four-Stroke Free-Piston Engine

Robin Tempelhagen [1,*], Andreas Gerlach [2], Sebastian Benecke [2], Kevin Klepatz [1], Roberto Leidhold [2] and Hermann Rottengruber [1]

[1] Department of Mobile Systems, Otto-von-Guericke University, 39016 Magdeburg, Germany; kevin.klepatz@ovgu.de (K.K.); hermann.rottengruber@ovgu.de (H.R.)
[2] Department of Electric Power Systems, Otto-von-Guericke University, 39016 Magdeburg, Germany; andreas.gerlach@ovgu.de (A.G.); sebastian.benecke@ovgu.de (S.B.); roberto.leidhold@ovgu.de (R.L.)
* Correspondence: robin.tempelhagen@ovgu.de; Tel.: +49-391-67-52307

Featured Application: This work deals with the influence of the trajectory variation on the combustion process. The stroke trajectory is varied for a constant period and the losses that occur in the internal combustion engine and the electrical machine are examined. The developed methods were used for a four-stroke free-piston engine.

Abstract: Internal combustion engines with a crankshaft have been successfully developed for many years. They are lacking in the fact that the piston trajectory, i.e., position as a function of time, is limited by the crankshaft motion law. Position-controlled electric linear machines directly coupled to the piston allow to realize free-piston engines. Unlike the crankshaft-based engines, they allow for a higher degree of freedom in shaping the piston trajectory, including adaptive compression ratios, which enables optimal operation with alternative fuels. The possibility of adapting the stroke course results in new degrees of freedom with which the combustion process can be optimized. In this work, four-stroke trajectories with different amplitudes and piston dynamics have been proposed and analyzed regarding efficiency. A simulation model was created based on experimental measurements for testing the proposed trajectories. It could be proved that the variation of the trajectory resulted in an improvement of the overall efficiency. The trajectories were described analytically so that they can be used for a prototype in a future work.

Keywords: free-piston engine; linear permanent magnet synchronous machine; four-stroke; energy-conversion; efficiency improvement

Citation: Tempelhagen, R.; Gerlach, A.; Benecke, S.; Klepatz, K.; Leidhold, R.; Rottengruber, H. Investigations for a Trajectory Variation to Improve the Energy Conversion for a Four-Stroke Free-Piston Engine. *Appl. Sci.* **2021**, *11*, 5981. https://doi.org/10.3390/app11135981

Academic Editors: Cinzia Tornatore and Luca Marchitto

Received: 31 May 2021
Accepted: 24 June 2021
Published: 27 June 2021

Publisher's Note: MDPI stays neutral with regard to jurisdictional claims in published maps and institutional affiliations.

1. Introduction

The internal combustion engine is one of the most common engines. With this motor, chemically stored energy is converted into mechanical energy. The energy conversion unfortunately produces exhaust gases and losses, which is the focus of research of many scientific elaborations [1]. To improve efficiency [2,3], various components were optimized [4], control strategies were adapted, valve timing was varied [5], alternative fuels [6] were investigated and much more.

Due to the shortage of fossil resources and the environmental impact of the combustion engine, alternative concepts for optimal energy conversion are being investigated. One possible future-oriented concept here is the free-piston engine [7–9]. The free-piston engine consists of an internal combustion engine, where the piston is connected to a linear electrical machine [10–12]. This means that no crankshaft is required, which makes it possible to vary the stroke trajectory, and thus, also the compression ratio. Various prototypes have already been developed [13,14]. With these prototypes, it could be shown that the operation of a two- or four-stroke free-piston engine is possible [15–17].

The free piston engine offers different research focuses. One major focus is the optimization of the control strategy [16,18–21]. Another focus is the analyzation of the combustion process [22–24]. As a third major research focus, the modeling investigation of the modeling of a free-piston engine can be defined [25,26]. The fourth research topic is the optimal design of electric machines [27–29].

In addition to varying the compression ratio, it is also possible to adjust the stroke trajectory if the linear machine is adequately controlled [8]. The influence of the piston trajectory on the combustion is also a very interesting research topic [30,31]. In the present work, different stroke trajectories are proposed and analyzed with respect to the losses in the combustion process and conversion of mechanical energy to electric energy with an electric machine. The novelty of this work is the analyzation of the system efficiency of a free-piston engine with respect to the piston trajectory. There was particular interest in how the losses can be varied with a variation in the stroke amplitude, the acceleration of the piston and a time shift based on the TDCF. It could be proved that the stroke course has a major influence on the efficiency of the internal combustion engine. In Section 2, it is explained how the chemically stored energy is converted into electrical energy. Section 3 explains how the experimental prototype is constructed. For the prototype, a thermodynamic model and an electromechanical model were designed, which are described in Sections 4 and 5, respectively. With these models, it is possible to determine the energy conversion losses. The same piston stroke curves were specified for the models. The generation of trajectories is explained in Section 6. The results are presented in Section 7 and summarized in Section 8.

2. Consideration of the Energy Losses for Converting Chemically Stored Energy into Electrical Energy

To justify the pros and cons for different piston trajectories regarding the thermodynamic and electric efficiency of the free-piston engine, a loss analysis is used. Depending on the piston trajectory, different losses may result in terms of combustion characteristics, heat losses, gas exchange, compression and expansion characteristics.

The chemical fuel energy E_{fuel}, which enters the engine via the injector, represents the maximal supplied energy for the engine and is the basis of the loss analysis. This energy only depends on the injected mass m_{fuel} and the lower heat value LHV according to:

$$E_{fuel} = m_{fuel} \cdot LHV \, . \tag{1}$$

This chemical energy is transformed to thermal energy during the combustion process. Due to the fact that an ideal combustion is not realizable, there is always an amount of unburned fuel m_{unb}, which leads to an energy loss during the transformation from chemical to thermal energy E_{th}. Besides the imperfect combustion, incomplete combustion could occur when injected fuel is not completely thermally converted and oxidized. When fuel reaches the cylinder wall or the upper face of the piston during combustion, the flame extinguishes and is, thus, not available for further thermal conversion. Out of the difference of the incomplete converted fuel and the injected fuel, incomplete combustion efficiency can be derived. The parameter η_{comb} describes the efficiency of the combustion based on the converted fuel mass, whereby the thermal energy yields to:

$$E_{th} = \left(m_{fuel} - m_{unb} \right) \cdot LHV = \eta_{comb} \cdot E_{fuel} \, . \tag{2}$$

This thermal energy is the basis for the volume change work. Due to the fact that an energy loss analysis is used for the comparison, the volume change work is called here the volume change energy E_{vc}. The combustion process creates high temperature gradients, which lead to an energy loss through the walls E_{wal}. This is described by the heat flux through the walls of the combustion chamber, which is defined by.

$$E_{wal} = \alpha \cdot A \cdot \Delta T \cdot \Delta t \, . \tag{3}$$

where α is the heat transfer coefficient (HTC), A is the area of the combustion chamber, ΔT is the temperature gradient between the surrounding solid material and the gas inside the combustion chamber and Δt is the time in which the process is performed. All parameters are space- and time-dependent. The parameters in Equation (3) are mean values over one work cycle to simplify the formulas for a better understanding [32]. The time-dependent temperature at the wall, which is causing ΔT, is influenced by all three ways of heat transfer. Due to the very low thermal conductivity coefficient of air, the influence of the conduction within the combustion chamber is low as well. The influence of the radiation to the wall temperature depends on the used ignition strategy and the corresponding injection system. Diesel engines with direct injection are more influenced as gasoline engines in general and especially with port fuel injection. The last influence is the convection. Due to the fast motion of the gas mixture inside the combustion chamber, it has strongest influence.

Not all of the thermal energy can be transformed into volume change energy within the given time for the expansion stroke. Therefore, thermal energy E_{exh} is lost within the exhaust gas when leaving the combustion chamber over the exhaust valve, as defined as:

$$E_{exh} = E_{int\ out} - E_{int\ in} = (T_{out} \cdot c_{p_{out}} - T_{in} \cdot c_{p_{in}}) \cdot m_{gas} . \tag{4}$$

where $E_{int\ out}$ is the internal energy within the exhaust gas, $E_{int\ in}$ is the internal energy within the supplied air-fuel-mixture, T is the mean temperature, c_p is the mean specific heat capacity and m_{gas} is the gas mass which is supplied and exhausted. The values for T and c_p of the supplied and exhaust gas are averaged over the time during which the specific valve is open.

The volume change energy E_{vc} is the balance of the thermal energy, the wall heat losses and the thermal energy losses through the exhaust. The losses can be described by the thermal efficiency η_{th}. The volume change energy, thus, results in:

$$E_{vc} = E_{th} - E_{wal} - E_{exh} = \eta_{th} \cdot E_{th} . \tag{5}$$

The volume change energy leads to a motion of the piston, which in turns is always subject to friction. The friction losses arise mainly in our prototype in the piston rings, bearings and the valve train. The mechanical energy E_M is the subtraction of the friction energy from the volume change energy and can be described by the friction efficiency loss η_{Fr} as follows:

$$E_M = E_{vc} - E_{Fr} = \eta_{Fr} \cdot E_{vc} \tag{6}$$

As described in Equation (6), friction occurs, which can also be described as:

$$P_{Fr} = v^2 \cdot u = F_{Fr} \cdot v , \tag{7}$$

where μ the coefficient of friction and v is the speed of the translator.

The mechanical power P_M for the electrical machine can be expressed as:

$$P_M = F_{EM} \cdot v, \tag{8}$$

where F_{EM} is the force of the electric machine. Due to the oscillation of the translator, the system requires acceleration power P_{acc} in many operating points. The acceleration power can be expressed with:

$$P_{acc} = m \cdot \frac{dv}{dt} \cdot v , \tag{9}$$

where m is the accelerated mass of the system and $\frac{dv}{dt}$ the time derivative of the speed. The power output from the internal combustion engine P_{ICE} can be described with:

$$P_{ICE} = F_p \cdot v = A_{cyl} \cdot p \cdot v, \tag{10}$$

where p is the cylinder pressure, A_{cyl} is the area of the piston and F_p is the gas force. In addition, in the electrical machine the reluctance power P_{Rel} arises, which can be described as a function of position:

$$P_{Rel} = f(x) \cdot v. \tag{11}$$

The power equilibrium on the mechanical transmission element results as follows:

$$P_{ICE} + P_M + P_{Fr} + P_{Rel} = P_{acc}. \tag{12}$$

In the case of a directly driven free-piston engine [17], the trajectory can be varied highly dynamically with the electric machine using the field oriented control. This is possible with the electrical machine force F_{EM}, which is proportional to the current i_q via the force constant k_F. The force of the electric machine can be expressed with:

$$F_{EM} = \frac{3}{2} k_F i_q. \tag{13}$$

The mechanical energy can be converted into electrical energy with the help of the electrical machine.

The special feature of the free-piston engine is that it does not have a crankshaft. The required oscillation movement of the piston can be realized with the help of a linear electric machine. The electrical machines can be constructed in different ways [27,33,34]. Depending on the machine type, different power losses can arise. The power loss occurs when the iron is re-magnetized, which is referred to as hysteresis power loss P_{Hys}. Eddy current power dissipation P_{Ft} can also arise. The sum of the hysteresis power loss and the eddy current power loss is referred to as the iron power loss P_{Fe}:

$$P_{Fe} - P_{Ft} + P_{Hys}. \tag{14}$$

In addition to the iron power loss, there is also a power loss due to the current i_q in a coil winding with ohmic resistance R. This power loss is referred to as the copper power loss. The copper power losses are calculated in the d/q-reference frame as follows:

$$P_{Cu} = \frac{3}{2} (i_q^2 + i_d^2) \cdot R. \tag{15}$$

Due to the iron and copper power losses, the entire mechanical power on the electrical machine cannot completely be converted into electrical power. This can be expressed stationary as follows:

$$P_M + P_{Cu} + P_{Fe} = P_{el}. \tag{16}$$

The electrical power P_{el} can be determined with the voltages $u_{d,q}$ and currents $i_{d,q}$.

$$P_{el} = \frac{3}{2} (u_d \cdot i_d + u_q \cdot i_q) \tag{17}$$

By integrating the calculated powers over a cycle, it is possible to calculate the energies.

3. Experimental Prototype

The experimental measurements for the validation of the simulation model were obtained with the help of the prototype in Figure 1 and is shown schematically in Figure 2.

Figure 1. Experimental Prototype of the free-piston engine.

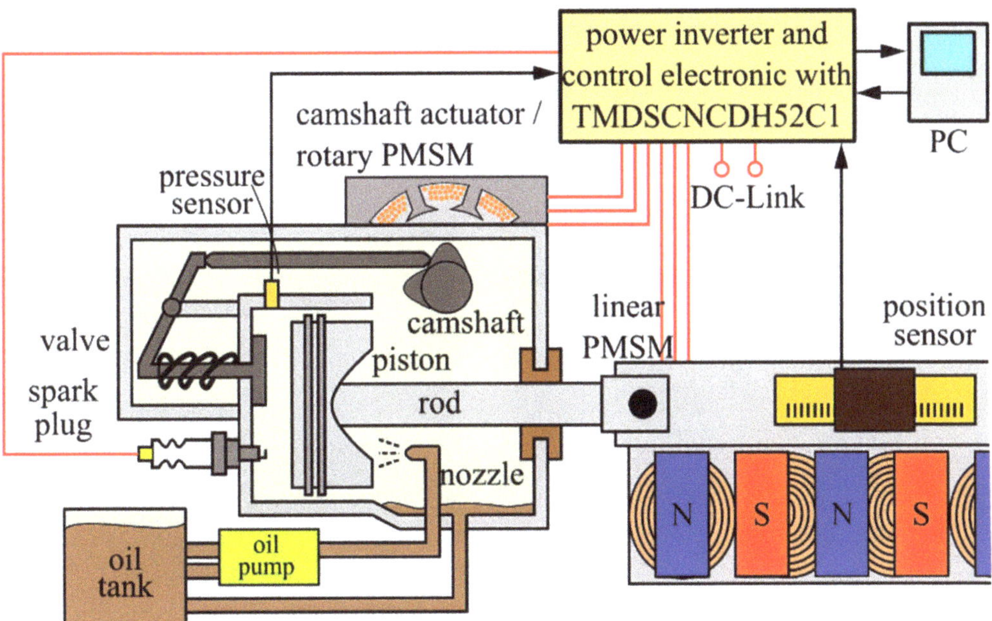

Figure 2. Schematic representation of the free-piston engine.

The internal combustion engine was originally used in a lawn mower. The crankshaft was removed and replaced with a piston and rod. This rod was passed out of the crankcase and connected to a linear electrical machine. The electrical machine is designed by the manufacturer Tecnotion (2xUXX6) as a permanent magnet excited synchronous machine (PMSM).

The mover of the electrical machine is ironless, where the primary part is moved. In this way, no cogging forces are present and the mass to be moved is low. The position of the electric machine was controlled using an inverter, a linear sensor and a microcontroller from Texas Instruments F28M35H52C. This microcontroller is able to calculate different trajectories, evaluate sensors and calculate manipulated variables (such as ignition sparks, injection times and transistor switch-on times). The data were then sent to a PC, saved and evaluated. The camshaft is connected to a highly dynamic PMSM. With the help of a valve drive, it is possible to operate the valves independently of the piston position. The camshaft and piston are synchronized digitally. To lubricate the cylinder surfaces, a nozzle was inserted which sprays a thin layer of oil under the piston. The most important parameters of the free-piston engine are shown in Table 1.

Table 1. Characteristic data of the test stand.

Linear PMSM		Combustion Engine	
Rated Power	3.24 kW	Engine Type	Four-Stroke
Rated Speed	6.6 m/s	Engine Displacement	99.25 cm^3
Rated Current	5.6 A$_{rms}$	Rated Power	1.9 kW
Peak Current	28 A$_{rms}$	Bore Diameter	0.0554 m
Rated Force	491 N	Number of Cylinders	1
Peak Force	2455 N	Conventional Stroke	0.0402 m
Mover Mass	3.271 kg	Maximal Stroke	0.0566 m
Mover Length	0.505 m	Rated Speed	4.27 m/s
Magnet Yoke length	0.627 m	Rated Force	444.96 N
Maximum stroke	0.122 m	Moving Mass	0.286 kg
Resistance per phase	1.29 Ω		

The problem with the electrical machine was that high copper losses occur when the four-stroke internal combustion engine is operated. It will be replaced with a self-developed electric machine [29]. The primary part of the electrical machine is not moved. It has a flat geometry, a short stator and is double-sided. The essential characteristic data is illustrated in Figure 3 and shown in Table 2.

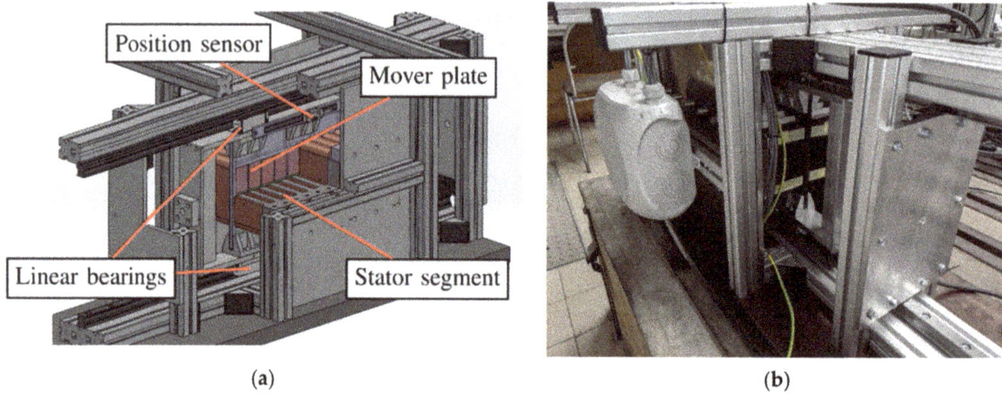

Figure 3. (**a**) CAD illustration of the developed electrical machine. (**b**) Photo of the developed electric machine.

The electrical machine was designed based on a measured force curve through the first prototype. A special feature of the electrical machine is that the primary part has the same length as the secondary part. This creates reluctance forces when the secondary part moves out of the stator. This force acts like a mechanical spring and can minimize copper losses. The reluctance force for this electrical machine is shown in Figure 4a. However, iron losses occur due to the magnetic structure. The iron power loss P_{Fe} was measured as a function of the electrical frequency f_{el} at various operating points and a function for this was approximated [35]. The function here corresponds to a Steinmetz model:

$$P_{Fe} = C_{hy} \cdot \hat{B}^{1,6} \cdot f_{el} + C_{wb} \cdot \hat{B}^2 \cdot f_{el}^2, \tag{18}$$

where $\hat{B}$ is the magnetic flux density and C_{hy} and C_{wb} are two coefficients. The iron power loss is shown in Figure 4b.

Table 2. Characteristic data of the developed electrical machine.

	Linear PMSM
Rated Power	7.35 kW
Rated Speed	2.26 m/s
Rated Current	16.55 A$_{rms}$
Peak Current	21.21 A$_{rms}$
Rated Force	3251 N
Peak Force	3689 N
Mover Mass	7.84 kg
Mover Length	0.357 m
Magnet Yoke length	0.357 m
Maximum stroke	0.17 m
Resistance per phase	0.441 Ω

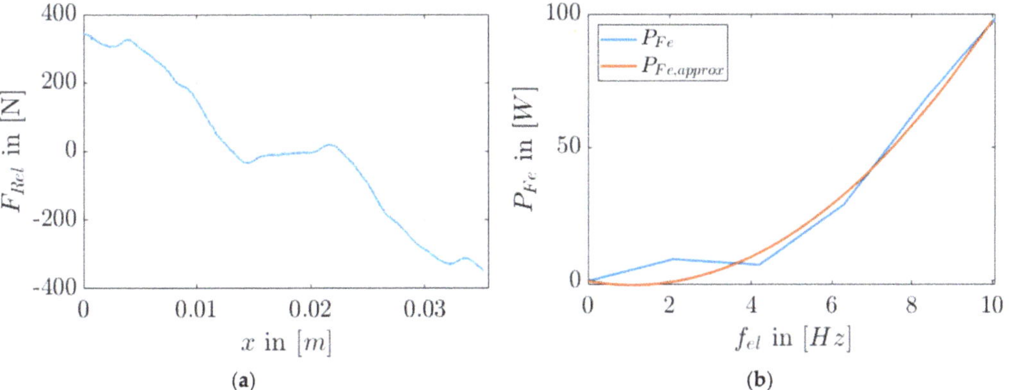

Figure 4. (a) Reluctance force F_{Rel} depending on the position x. (b) Iron power losses P_{Fe} depending on frequency f_{el}.

These special properties were inserted into a simulation model in order to determine the losses in the respective trajectories. For the simulations that were carried out, the parameters of the newly developed machine were used.

Since the secondary part moves out of the stator and the proportion of covered magnets to the windings is reduced, the magnetic flux is also reduced. This leads to a reduction in the induced voltage as well as to a reduction in the force per ampere ratio. Therefore, the force of the electrical machine depends on the position and current and is shown in Figure 5. It can be seen in Figure 5 that the force decreases at the stroke limits.

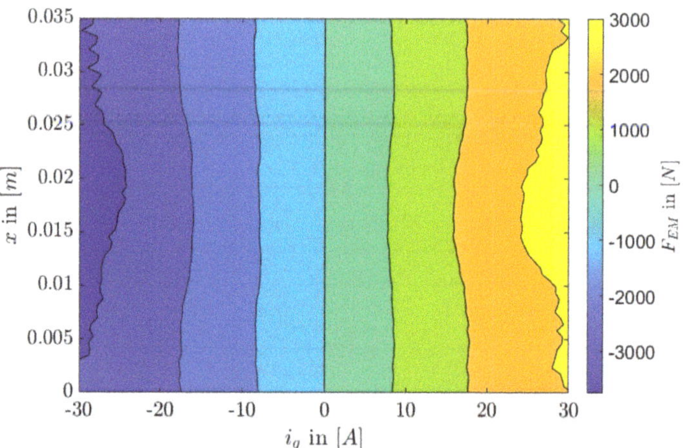

Figure 5. Force F_{EM} depending on position x and current i_q.

4. Thermodynamic Model of the Free-Piston Engine

Based on the data of Table 1, a one-dimensional thermodynamic model for the free-piston engine is built up. The combustion engine is working with an external mixture formation via a gasoline injector with a constant injected fuel mass in the intake manifold. The fuel mass is chosen this way so that the equivalence ratio $\phi = 1$. This model of the free-piston engine is quite similar to an engine with a conventional crank train. To model the combustion process and the corresponding heat release, the predictive combustion model, the so-called Spark-Ignition Turbulent Flame Model (SITurb) from GT-Suite, was used. This model can be used for several operating points after it has been parameterized once. Additionally, it is normally used for gasoline engines, which is the test bench engine. The heat release is calculated by the burned fuel mass m_b:

$$\frac{dm_b}{dt} = \frac{m_e - m_b}{\tau}$$

(19)

where m_e is the entrained mass and τ is the time constant.

The entrained mass is calculated by:

$$\frac{dm_e}{dt} = \rho_u \cdot A_e \cdot (S_T + S_L)$$

(20)

where ρ_u is the unburned density, A_e is the surface area at flame front, S_T is the turbulent flame speed and S_L is the laminar flame speed.

The turbulent flame speed is calculated by:

$$S_T = C_{TFS} \cdot u' \cdot \left(1 - \frac{1}{1 + C_{FKG} \cdot \left(\frac{R_f}{L_i} \right)^2} \right)$$

(21)

where C_{TFS} is the turbulent flame speed multiplier, u' is the turbulent intensity, C_{FKG} is the flame kernel growth multiplier, R_f is the flame radius and L_i is the integral length scale.

The laminar flame speed is calculated by:

$$S_L = \left(B_m + B_\phi \cdot [\phi - \phi_m]^2 \right) \cdot \left(\frac{T_u}{T_{ref}} \right)^{\alpha_e} \cdot \left(\frac{p}{p_{ref}} \right)^{\beta} \cdot f(Dilution)$$

(22)

where B_m is the maximum laminar speed, B_ϕ is the laminar speed roll-off value, ϕ is the equivalence ratio, ϕ_m is the equivalence ratio at maximum speed, T_u is the unburned gas temperature, T_{ref} is defined as 298 K, α_e is the temperature exponent, p is the pressure, p_{ref} is defined as 101,325 Pa and β is the pressure exponent.

The heat transfer is defined by the WoschniClassic model from GT-Suite. This model is used for engines without swirl flows, which correspond to the test bench engine. First, a convective heat transfer is calculated; then, it is calibrated to the measurement data by a multiplier to take the influences of conduction and radiation into account. The heat transfer coefficient is calculated by:

$$\alpha = \frac{K_1 \cdot p^{0.8} \cdot w^{0.8}}{B^{0.2} \cdot T^{K_2}} \tag{23}$$

where K_1 and K_2 are model constants, p is the cylinder pressure, w is the average cylinder gas velocity, B is the cylinder bore and T is the cylinder temperature. All multipliers are used to calibrate the model to the measurement data.

This heat transfer coefficient only takes the convection into account. To consider the influence of the radiation and conduction, an additional multiplier is used to calculate α for Equation (3).

The major difference of the model and the conventional engine models is the completely free definition of the piston motion. Specifically, the determination of the piston movement in the model is done via a table with free choice of the piston position as a function of time. In contrast to that, the piston motion normally is defined by the geometrical parameters of the crank train and the corresponding relation between crank angle and piston position. The thermodynamic model is illustrated in Figure 6. This model included the combustion and heat transfer process represented by the "cylinder" symbol and the mechanical components, which are represented by the "crank train" symbol. The boundary conditions of the incoming air and the outcoming exhaust gas are symbolized by the green symbols "inlet" and "outlet." The intake system consists of the inlet manifold, the inlet pipe, the throttle valve, the inlet channel, the injector and the inlet valve, as illustrated by the symbols on the left side of Figure 6. Thus, the engine works with an external mixture formation and a quantity control via the throttle valve. The symbols on the right side of Figure 6 represent the exhaust valve, the exhaust channel and the exhaust pipe.

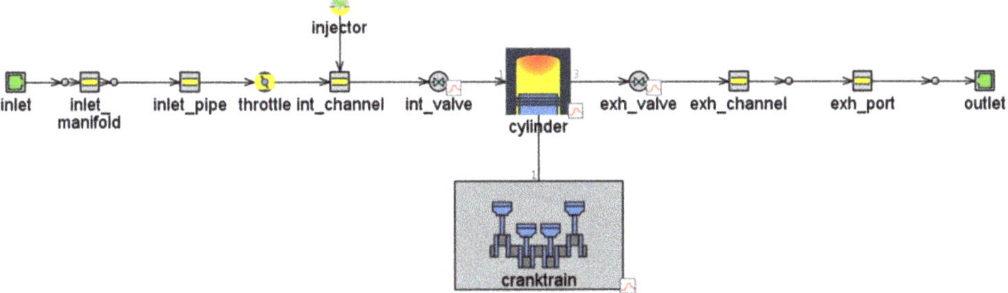

Figure 6. Thermodynamic model of a free-piston engine.

To analyze the influence of different trajectories on the thermodynamic losses, a stationary operating point is chosen and validated with test bench data. In the high-pressure area, which represents the combustion in this operation point, the measurement and simulation data fit quite well (see Figure 7).

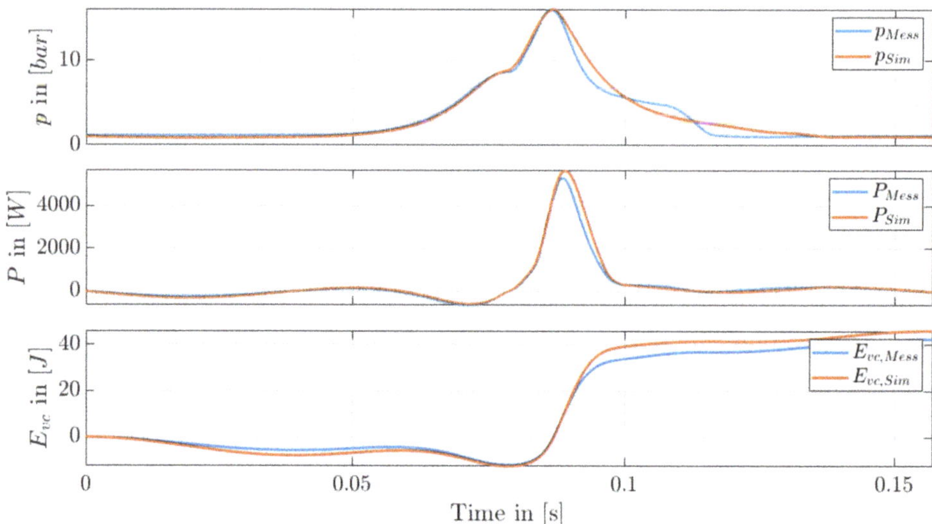

Figure 7. Comparison of the measured and simulated cylinder pressure, the engine power and the volume change energy.

This point, which is shown in Table 3, is located at an engine speed of 764 min^{-1} and an IMEP of 4.7 bar. The stroke is at an initial value of 0.03 m for the reverence trajectory, which leads to a compression ratio of 5.368. As fuel, the reference fuel indolene is used to describe automotive gasoline according to [36]. To model the combustion process, the already described predictive approach called "SI-Turb" is used, which describe the combustion by defining the turbulence intensity within the combustion chamber. The parameters, or rather the multipliers and their values which are used for this approach, are summarized in Table 3. The turbulent flame speed multiplier has influence on the turbulence level of the flame front during the combustion. The flame kernel growth multiplier describes how fast the flame kernel spreads during the combustion. With a higher multiplier of the Taylor length scale, the amount of time the air–fuel mixture is entrained into the flame front is decreased, which leads to short burning durations [37].

Table 3. Boundary conditions of the thermodynamic analysis.

Combustion Engine		Combustion Model: SI-Turb	
Engine Speed	764 min^{-1}	Turbulent Flame Speed Multiplier	0.025
IMEP	4.7 bar	Flame Kernel Growth Multiplier	0.01
Injected fuel mass	4.32 mg	Taylor Length Scale Multiplier	0.001
Injected fuel	indolene		
LHV of the fuel	43.95 MJ/kg		
Initial Stroke	0.03 m		
Effective compression ratio	5.368		

5. Simulation Model of the Linear Permanent Magnet Synchronous Machine

The mentioned properties of the electrical machine were implemented in a simulation model. The simulation model for this is shown schematically in Figure 8.

The acceleration force F_{ACC} is calculated with the acceleration a and the mass m. The gas force F_{ICE} is calculated from the product of cylinder pressure p and area A_{cyl}. It is possible to determine the reluctance force F_{Rel} from the position x and the course from Figure 4a. The frictional force was neglected. With the help of these forces, the force of the electrical machine is calculated. The map from Figure 5 is then used to calculate the target current from the force F_{EM} and the position x. The copper power loss P_{Cu} can be calculated from the current i_q and ohmic resistance R. The mechanical power of the electrical machine P_M can be determined from the force F_{EM} and the speed v. The iron losses P_{Fe} can be calculated from the electrical frequency f_{el} and the map from Figure 4. The sum of all powers results in the electrical power.

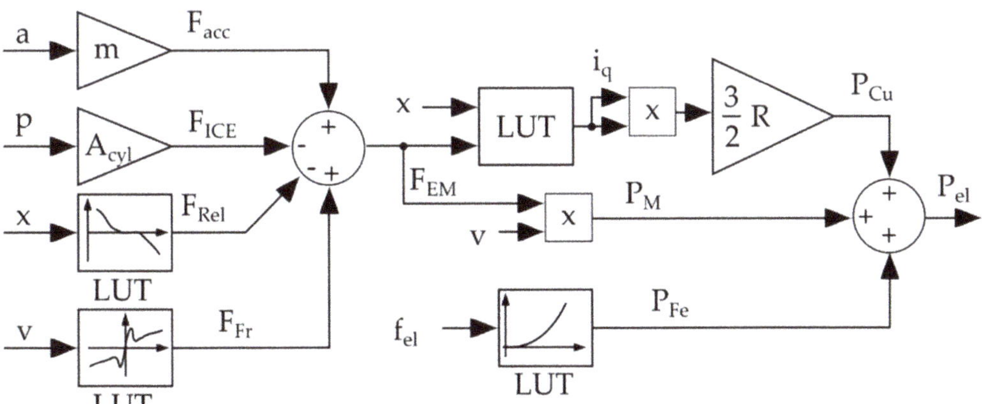

Figure 8. Simplified simulation model of the LMPSM.

6. Trajectory Variation

6.1. Specificity of the Free-Piston Engine

The free-piston engine allows the piston stroke to be adjusted in a highly dynamic manner so that different trajectories can be followed. For practical implementation, care must be taken that there are no discontinuities in the acceleration, as this on the one hand cannot be implemented and on the other hand represents a high load on the drive train. Furthermore, it is advantageous for the trajectory variation if it can be described analytically. Here, various parameters can be set that can specifically adapt the shape of the trajectory. In the following sections, various functions will be presented which allow the piston stroke course to be adjusted.

6.2. Reference Stroke

A sinusoidal curve is used to compare the trajectories. The trajectory can be described analytically as follows:

$$x = -\frac{x_{peak}}{2} \cdot \cos(\varphi) + \frac{x_{peak}}{2} \tag{24}$$

$$v = \frac{x_{peak} \cdot \frac{d\varphi}{dt}}{2} \cdot \sin(\varphi) \tag{25}$$

$$a = \frac{x_{peak} \cdot \frac{d\varphi}{dt}^2}{2} \cdot \cos(\varphi) + \frac{x_{peak}}{2} \cdot \frac{d^2\varphi}{dt^2} \cdot \sin(\omega t) \tag{26}$$

The first and second derivatives of the stroke with respect to time result in the velocity and acceleration. For the four strokes, it is assumed that the stroke amplitude x_{peak} and the virtual angular velocity ω do not change (Assumption: $\omega = const.$). The used reference trajectory is shown in Figure 9.

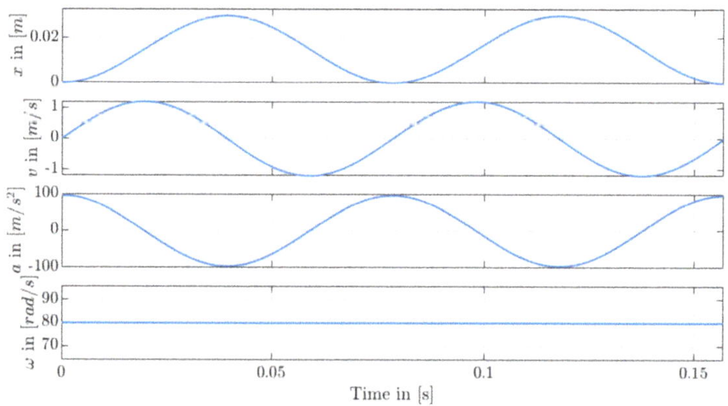

Figure 9. Sinusoidal trajectory for the four strokes with the parameters $x_{peak} = 0.03$ m and $\omega = 80\,\frac{rad}{s}$ depending on time.

As can be seen here, the time intervals for the TDC, BDC and the stroke amplitude are constant and the change in the piston position is very symmetrical. This trajectory is used in the following as a reference trajectory in order to show the changes in the piston position.

6.3. Time Displacement of the Dead Center (Trajectory A)

First, a function has to be determined with which the duration of one stroke can be changed without changing the period duration of the four strokes. The following individual stroke duration (T_i) and their setting parameters ($\Delta T_{i,rel}$) are defined depending on the period time T_P:

$$T_1 = \Delta T_{1,rel} \cdot T_P \tag{27}$$

$$T_2 = \Delta T_{2,rel} \cdot T_P \tag{28}$$

$$T_3 = \Delta T_{3,rel} \cdot T_P \tag{29}$$

$$T_4 = T_P - T_1 - T_2 - T_3 \tag{30}$$

To change the duration of a stroke, the average angular speed of one stroke must be adjusted. In addition, the transition between the cycles must take place without discontinuities in the acceleration. This is achieved in that the initial angular speed ω_{Begin} is identical to the final speed ω_{End} and the average speed ω_P over the four cycles:

$$\omega_{Begin} = \omega_{End} = \omega_P = \frac{4\pi}{T_P}. \tag{31}$$

These conditions can be met with various functions. A sinusoidal change was chosen to implement the time displacement. The mean value of a half sine wave can be calculated as follows:

$$\overline{\omega}_i = \frac{1}{\pi} \int_0^\pi \hat{\omega}_i \cdot \sin(\varphi)d\varphi = \frac{2}{\pi}\,\hat{\omega}_i = \frac{\pi}{T_i}. \tag{32}$$

In order to meet the boundary conditions, the sine function has to be shifted by ω_P:

$$\omega = \frac{\pi}{2} \cdot (\overline{\omega}_i - \omega_P) \cdot \sin(t\,\overline{\omega}_i) + \omega_P. \tag{33}$$

The integration of Equation (33) results in the angle φ, which can be used in Equations (24) and (25) in order to obtain an analytical description of the speed and acceleration. The trajectory for the dead center shift is shown in Figure 10.

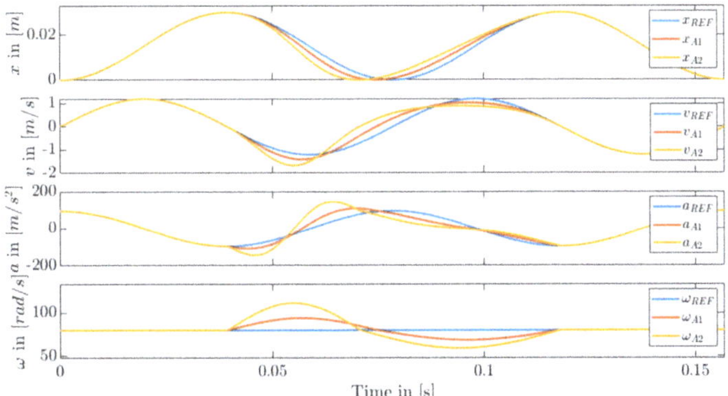

Figure 10. Comparison between the reference curve and dead center shift in the operating point $\omega = 80\,\frac{rad}{s}$, $x_{peak} = 0.03$ m, $T_1 = 0.25 \cdot T_p$, $T_2 = 0.225 \cdot T_p$, $T_3 = 0.275 \cdot T_p$ and $\omega = 80\,\frac{rad}{s}$, $x_{peak} = 0.03$ m, $T_1 = 0.25 \cdot T_p$, $T_2 = 0.2 \cdot T_p$, $T_3 = 0.3 \cdot T_p$.

With the advanced features, it is possible to vary the period of three cycles. The period of the fourth cycle is calculated from the specified period for the entire cycle.

6.4. Compression and Extension of the Stroke (Trajectory B)

Furthermore, it was investigated how the variation of the piston holding time affects the combustion process. For this, it was necessary to develop a function with which it should be possible to compress or stretch the stroke curve without changing the temporal position of the dead centers. The piston stroke trajectory can be varied by changing the virtual angular velocity during a cycle. The position of the dead centers can be kept constant if the mean angular velocity remains the same. A sinusoidal modulation was also chosen here:

$$\omega = \Delta\omega_{Ti} \cdot \sin(2\varphi) + \omega_P \tag{34}$$

where $\Delta\omega_{Ti}$ is an adjustable parameter which corresponds to the amplitude of the superimposed sinusoidal oscillation from the i-th stroke. The frequency is twice as high so that the mean value is not changed within one stroke. A combination of the method for shifting the dead centers is also possible. All you have to do is replace ω_P with $\overline{\omega}_i$. The trajectory for a stroke compression and extension is shown in Figure 11.

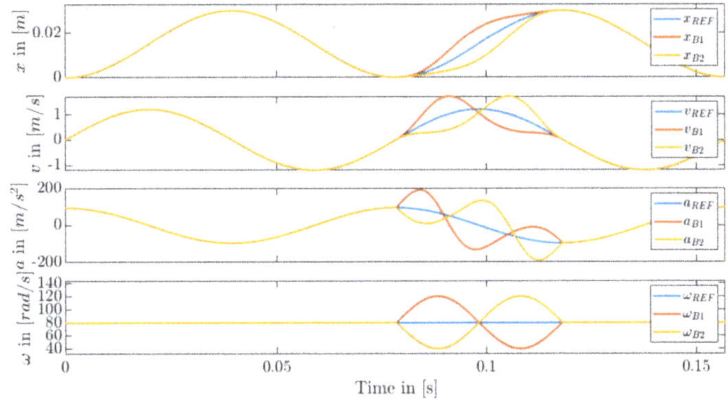

Figure 11. Comparison between the reference curve and compression and extension in the operating point $\omega = 80\,\frac{rad}{s}$, $x_{peak} = 0.03$ m, $\Delta\omega_{T1} = 0\,\frac{rad}{s}$, $\Delta\omega_{T2} = 0\,\frac{rad}{s}$, $\Delta\omega_{T3} = 40\,\frac{rad}{s}$, $\Delta\omega_{T4} = 0\,\frac{rad}{s}$ and $\omega = 80\,\frac{rad}{s}$, $x_{peak} = 0.03$ m, $\Delta\omega_{T1} = 0\,\frac{rad}{s}$, $\Delta\omega_{T2} = 0\,\frac{rad}{s}$, $\Delta\omega_{T3} = -40\,\frac{rad}{s}$, $\Delta\omega_{T4} = 0\,\frac{rad}{s}$.

It can be seen in Figure 11 that the piston stroke trajectory can be changed significantly without changing the position of the dead centers. It can also be seen here that the acceleration has no points of discontinuity.

6.5. Variation of the Stroke Amplitude (Trajectory C)

The adjustment of the stroke amplitude is another interesting degree of freedom for modulating the trajectory. The position of the top dead center was left constant and the bottom dead center was shifted. The piston stroke has already been described analytically with Equation (24). If x_{peak} is changed here, it is possible to vary the target trajectory. This change in stroke should also take place in such a way that there are no discontinuities in the acceleration. Various functions can be used for this. A sinusoidal change in the stroke amplitude was decided:

$$x_{peak} = -\frac{\hat{x}_1 - \hat{x}_2}{2} \cdot \sin\left(\frac{\varphi}{2}\right) + \frac{\hat{x}_1 - \hat{x}_2}{2}. \tag{35}$$

where $\hat{x}_2$ corresponds to the stroke amplitude at the bottom dead center between the intake and compression stroke and $\hat{x}_1$ is the stroke amplitude at the bottom dead center between the expansion and exhaust stroke. The trajectory is shown in Figure 12.

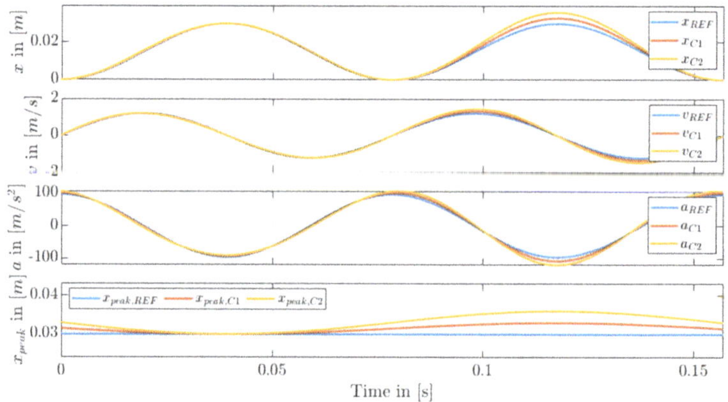

Figure 12. Comparison between the reference curve and stroke amplitude variation in the operating point ω = 80 rad/s, $\hat{x}_1 = 0.03$ m, $\hat{x}_2 = 0.033$ m and ω = 80 rad/s, $\hat{x}_1 = 0.03$ m, $\hat{x}_2 = 0.036$ m.

As can be seen here, the different stroke amplitudes can be set separately from one another. In addition, no steps in acceleration can be seen here either, so practical implementation is possible.

6.6. Combination of Proposed Trajectories (Trajectory D)

As the last type of variation, it is shown that it is possible to combine the mentioned trajectory generation methods. It was decided to adjust the position of the TDC and to increase the acceleration of the piston after the TDC. The trajectories can be calculated as a combination of the methods from Sections 6.3 and 6.4. The trajectory is shown in Figure 13.

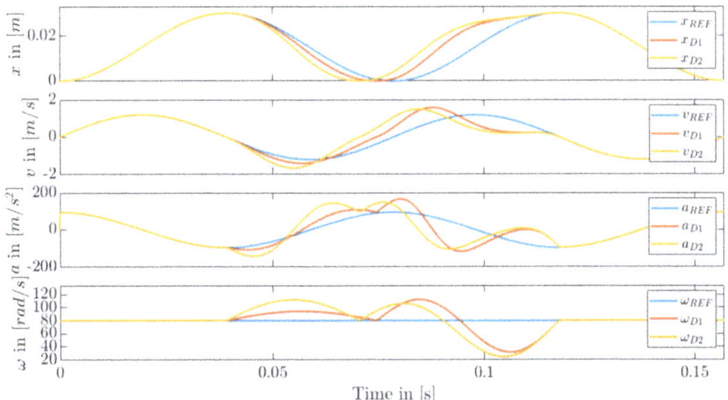

Figure 13. Comparison between the reference curve and stroke amplitude variation in the operating point $\omega = 80$ rad/s, $x_{peak} = 0.03$ m, $T_2 = 0.225 \cdot T_p$, $T_3 = 0.275 \cdot T_p$, $\Delta\omega_{T3} = 40\ \frac{rad}{s}$ and $\omega = 80$ rad/s, $x_{peak} = 0.03$ m, $T_2 = 0.2 \cdot T_p$, $T_3 = 0.3 \cdot T_p$, $\Delta\omega_{T3} = 40\ \frac{rad}{s}$.

7. Results

This chapter shows the results of the efficiency loss analysis for the trajectory variation. The losses of the internal combustion engine and the electrical machine were shown together in a diagram in order to evaluate the influence of the trajectory. For the evaluation of the energy conversion from chemical to mechanical energy the courses for the piston motion, the corresponding pressure and temperature as well as a PV diagram of the compared trajectories will be shown and discussed. For the energy conversion from mechanical to electrical energy, the position and speed of the piston, the electrical machine force and the resulting power are shown and discussed.

The loss analysis for the reference is shown in Figure 14. The summation of all energy shares is the chemical energy from the fuel. This energy is divided into the losses for unburned fuel, the wall heat losses, the losses inside the exhaust gas, the friction losses, the iron losses and the copper losses. The last share is the electrical energy, which is issued by the free-piston engine.

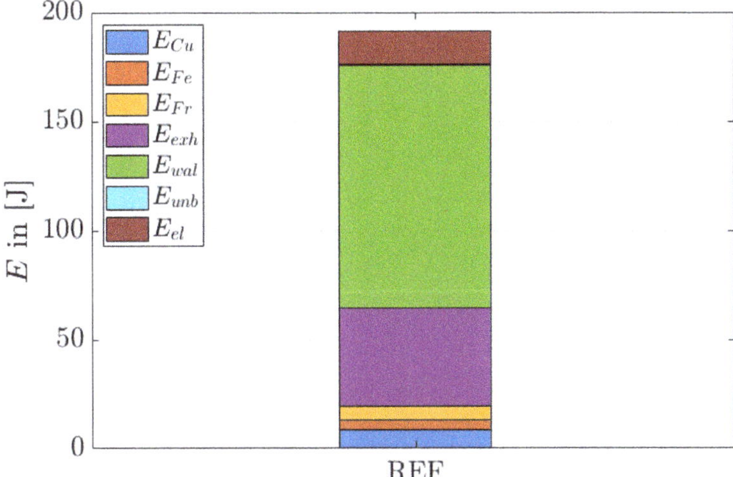

Figure 14. Energy losses for trajectory REF.

Corresponding to the trajectory variation described in Section 6, the courses for the piston motion, the pressure and temperature are shown in Figure 15. The PV diagram for variant A1 and A2 in comparison to the reference is shown in Figure 16. The moved position in time of top dead center firing (TDCF) affects the pressure and temperature courses identically. The moved TDCF of variant A1 and A2 leads to earlier increasing pressure and temperature in comparison to the reference due to the fact that the combustion is initialized early. The decrease of the pressure and temperature is shifted in parallel. At the end of the expansion stroke all courses approach until the exhaust valve opens (EVO). After that, the courses are similar.

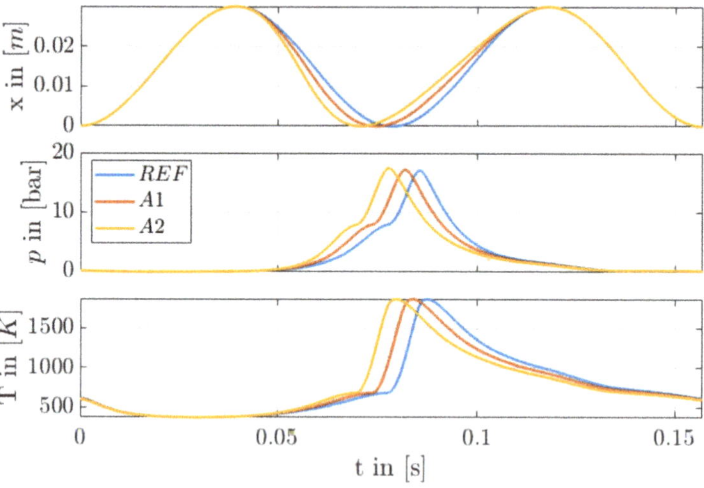

Figure 15. Comparison of trajectories REF, A1 and A2 related to the energy conversion of the internal combustion engine: illustration of the stroke, cylinder pressure and temperature depending on time.

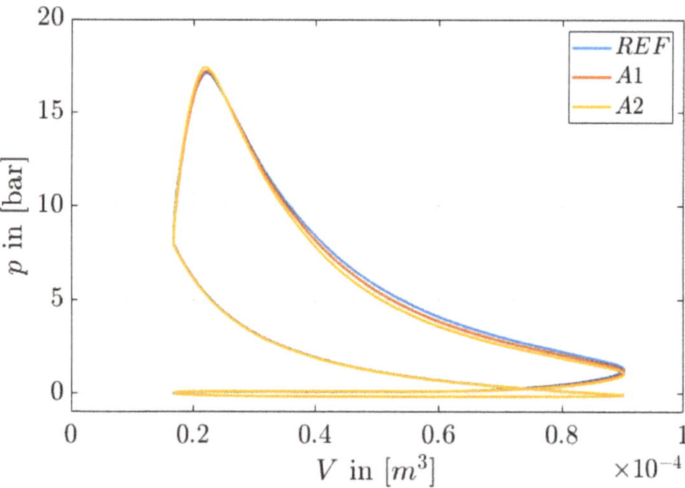

Figure 16. Comparison of trajectories REF, A1 and A2 related to the energy conversion of the internal combustion engine: cylinder pressure depending on volume.

For the evaluation of the electrical energy conversion of the trajectory REF, A1, A2, the stroke, speed and force of the electrical machine are shown in Figure 17 and the volume change, copper, friction and electrical power in Figure 18.

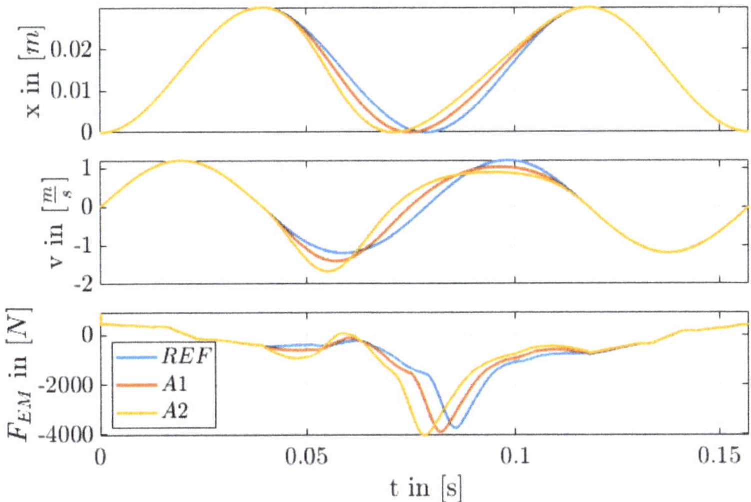

Figure 17. Comparison of trajectories REF, A1 and A2 related to the energy conversion of the electrical machine: illustration of the stroke, velocity and electric machine force depending on time.

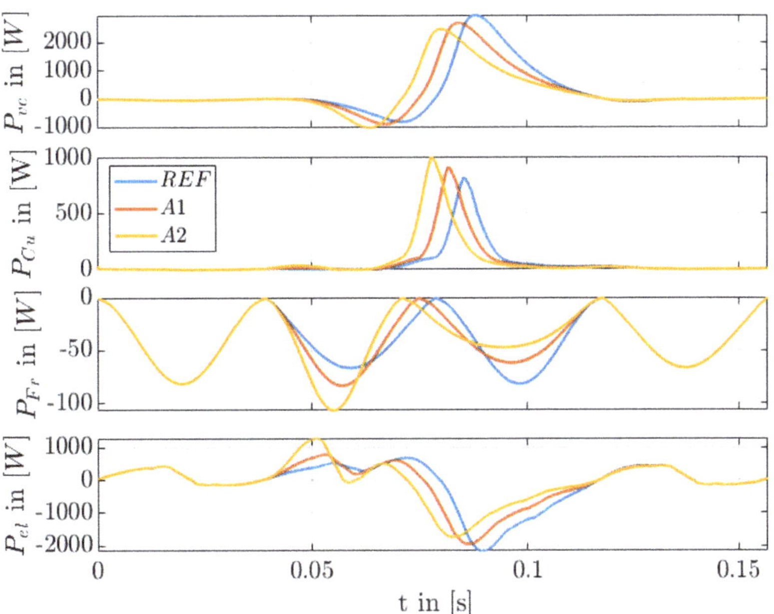

Figure 18. Comparison of trajectories REF, A1 and A2 related to the energy conversion of the electrical machine: volume change power, copper power losses, friction power losses and electric power depending on time.

The shift of the trajectory from REF to A2 is realized by further accelerating the drive train. The required trajectory can be implemented using the force F_{EM} of the electrical machine. Due to the increased force F_{EM}, the required current and, thus, the copper losses P_{Cu} increase. Since the speed is increased, the friction losses also increase. In addition to the volume change work, it can be seen that the electrical machine requires energy in the intake and exhaust cycle so that the four-cycle process can be implemented.

The longer expansion time with a higher temperature and similar pressure leads to the higher wall heat losses and reduced exhaust losses shown in Figure 19. This phenomenon can be seen when moving the TDCF to an earlier point in time. Due to the fact that the increase of the wall heat losses combined with the copper losses is higher than the decrease of the exhaust losses, both variants show a lower effective efficiency than the reference.

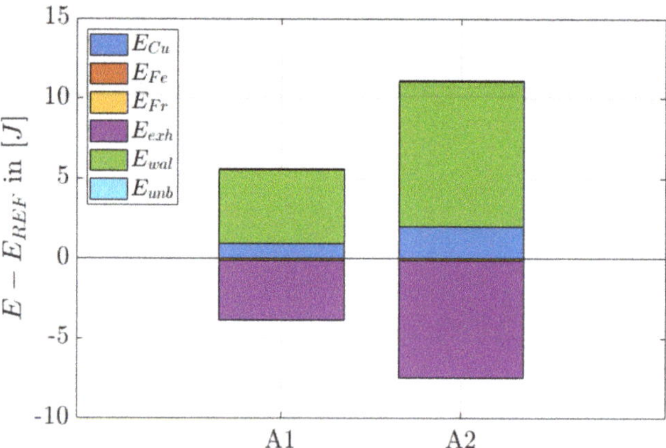

Figure 19. Energy losses for trajectory A1 and A2 with respect to REF.

The courses for the piston motion, the pressure and temperature are shown in Figure 20. The PV diagram for variant B1 and B2 in comparison to the reference is shown in Figure 21.

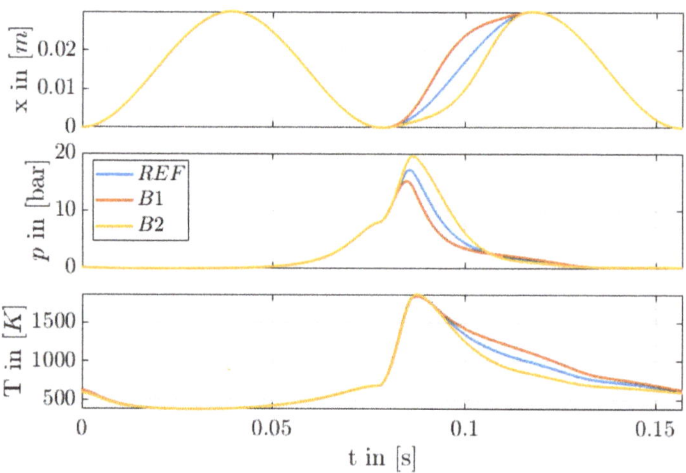

Figure 20. Comparison of trajectories REF, B1 and B2 related to the energy conversion of the internal combustion engine: illustration of the stroke, cylinder pressure and temperature depending on time.

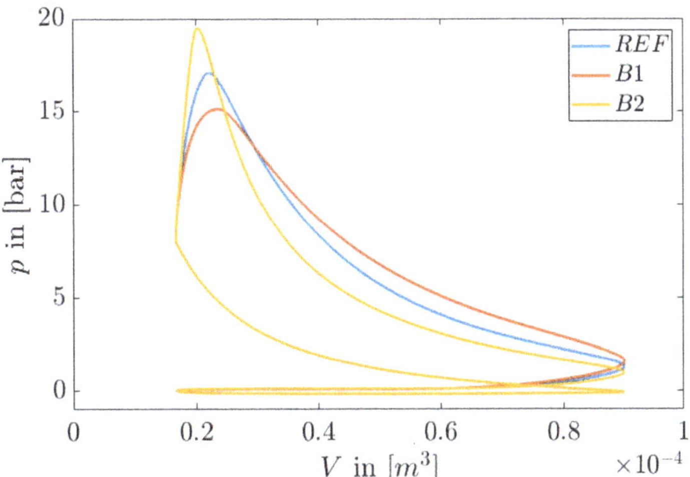

Figure 21. Comparison of trajectories REF, B1 and B2 related to the energy conversion of the internal combustion engine: cylinder pressure depending on volume.

The variant B1 has an increased acceleration of the piston during the expansion close to the TDCF and a reduced acceleration near the BDC. Variant B2 is exactly flipped. This leads to a decrease of the peak pressure near TDCF for B1 and an increased peak pressure for B2. Near BDC, the opposite behavior can be observed. The maximum temperature is not influenced by the different variation of the piston acceleration. The decrease of the temperature from TCDF to BDC is slower for variant B1 and faster for variant B2 in comparison to the reference.

For the evaluation of the electrical energy conversion of the trajectory REF, B1, B2, the stroke, speed and force of the electrical machine are shown in Figure 22 and the volume change, copper, friction and electrical power are shown in Figure 23.

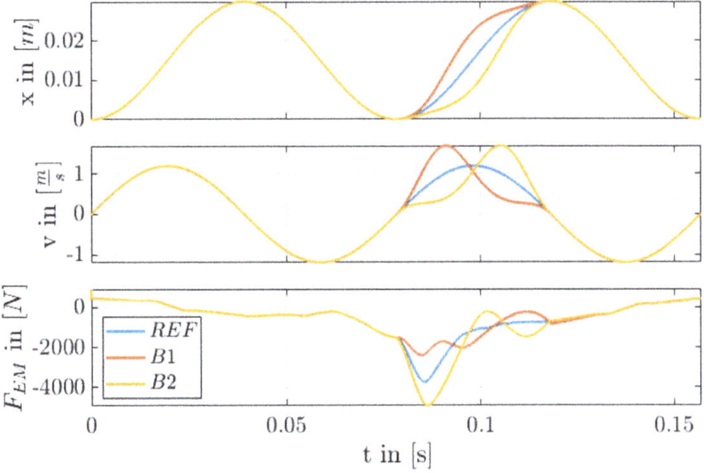

Figure 22. Comparison of trajectories REF, B1 and B2 related to the energy conversion of the electrical machine: illustration of the stroke, velocity and electric machine force depending on time.

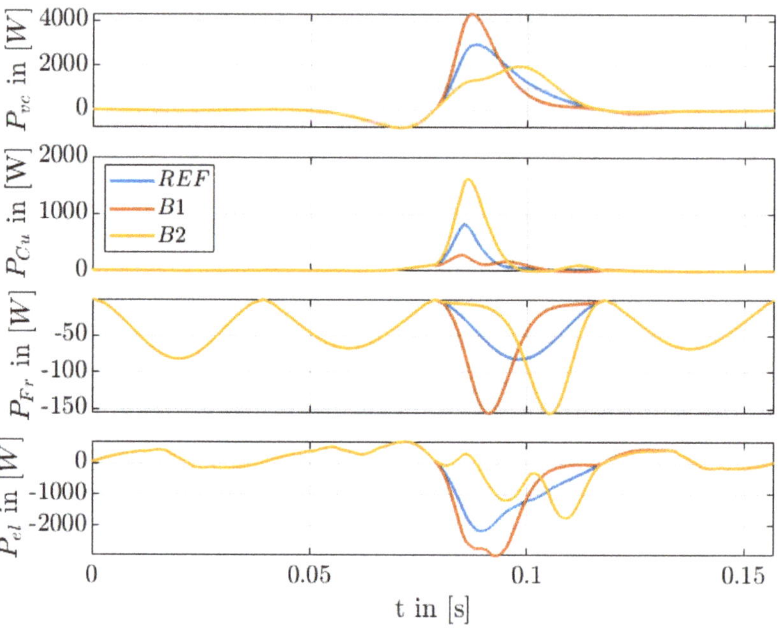

Figure 23. Comparison of trajectories REF, B1 and B2 related to the energy conversion of the electrical machine: volume change power, copper power losses, friction power losses and electric power depending on time.

Due to the slow change in the piston stroke in the trajectory B2, the electric machine must compensate for the force of the internal combustion engine so that a slow movement is possible. In contrast, in the case of trajectory B1, the piston is accelerated almost freely. As a result, the copper losses are high in case B2 and low in case B1. The friction losses are also approximately the same because the speed curve of B1 and B2 can be mirrored. Because of the reduced copper losses in case B1, the electrical energy conversion is good.

The loss analysis for B1 and B2 with respect to the reference is shown in Figure 24. Due to the fact that the maximum pressure has a strong influence on the fluid properties, and therefore, on the HTC, the wall heat losses for B1 are lower and for B2 higher in comparison to the reference. For the exhaust losses, this behavior is the other way around. Due to the higher temperature course for B1, especially when the exhaust valve opens, the exhaust losses increase in comparison to the reference. For B2, the temperature is lower when the exhaust valve opens. This leads to lower exhaust losses.

Figure 25 shows the courses for the piston motion, the pressure and temperature. The PV diagram for variant C1 and C2 in comparison to the reference is shown in Figure 26. The extended maximum stroke leads to no significant variance in the pressure and temperature courses. However, based on the different acceleration of the piston from TDCF to BDC, because of the different maximum strokes, which has to be reached in the same time, the PV diagram shows an extension of the included area to the end of the expansion stroke as well as a small decrease of the peak pressure.

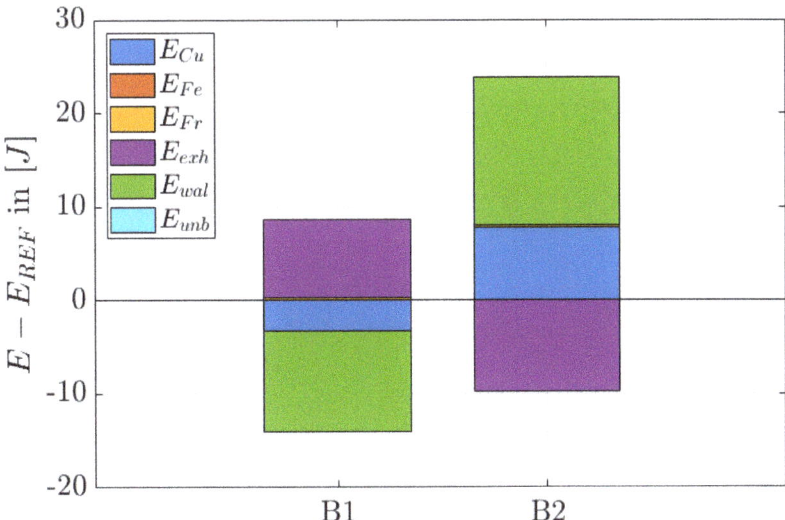

Figure 24. Energy losses for trajectory B1 and B2 with respect to REF.

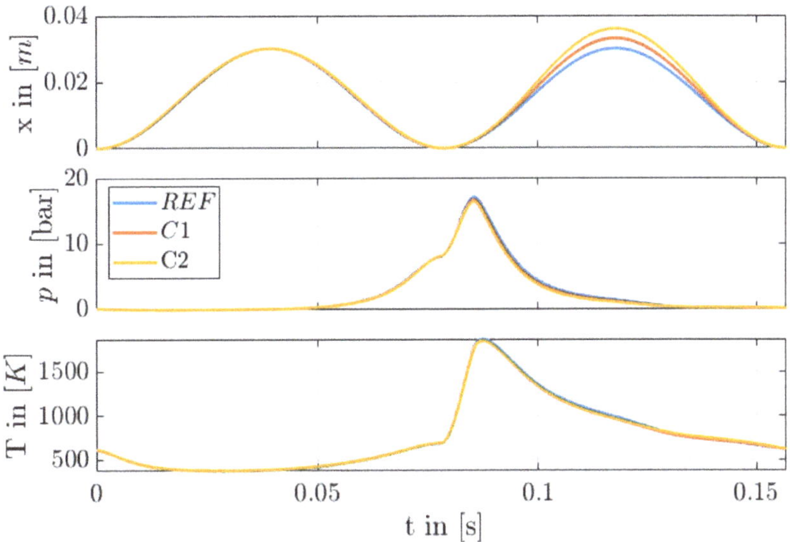

Figure 25. Comparison of trajectories REF, C1 and C2 related to the energy conversion of the internal combustion engine: illustration of the stroke, cylinder pressure and temperature depending on time.

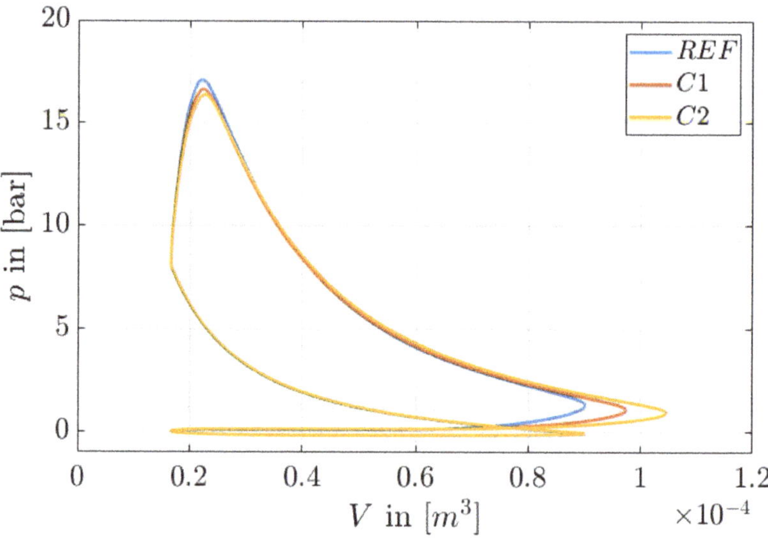

Figure 26. Comparison of trajectories REF, C1 and C2 related to the energy conversion of the internal combustion engine: cylinder pressure depending on volume.

For the evaluation of the electrical energy conversion of the trajectory REF, C1, C2, the stroke, speed and force of the electrical machine are shown in Figure 27 and the volume change, copper, friction and electrical power are shown in Figure 28.

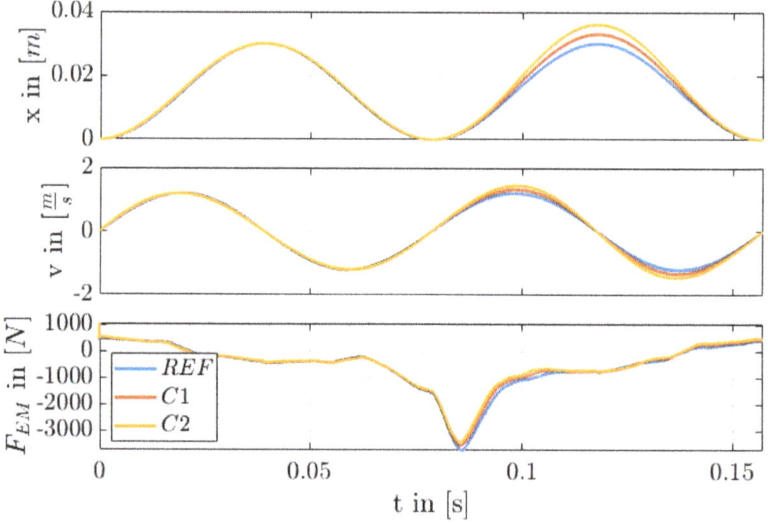

Figure 27. Comparison of trajectories REF, C1 and C2 related to the energy conversion of the electrical machine: illustration of the stroke, velocity and electric machine force depending on time.

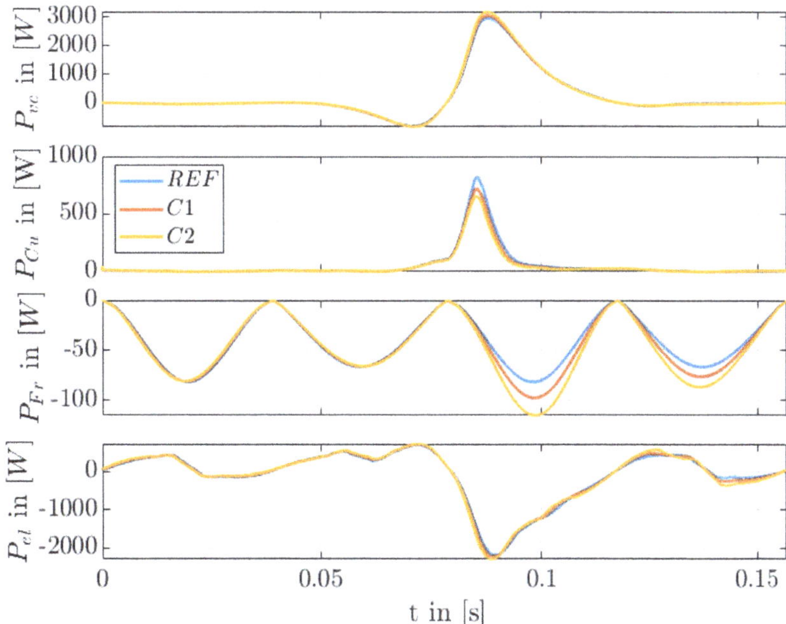

Figure 28. Comparison of trajectories REF, C1 and C2 related to the energy conversion of the electrical machine: volume change power, copper power losses, friction power losses and electric power depending on time.

Due to the stroke amplitude variation, the required speed increases in the expansion and exhaust cycle. The increase in speed leads to higher friction losses. However, the required acceleration in the expansion stroke is also increased. Since the combustion engine releases energy in this cycle, the electric machine does not have to brake the piston as hard. An increase in the stroke amplitude at this operating point has resulted in the copper losses being reduced. The iron losses have also increased, as the stroke variation leads to an increase in the electrical frequency.

The higher acceleration of the piston from TDCF to BDC leads to a small decrease of the peak pressure, which has a positive (reducing) influence on the wall heat losses. Against that, the higher maximum stroke leads to a higher surface area of the combustion chamber. This has a negative (increasing) influence on the wall heat losses. For the variant C1, the second phenomenon has a greater influence. This changes for the variant C2; thus, the first phenomenon has a greater influence. Therefore, the wall heat losses for C1 are higher and the wall heat losses for C2 are lower in comparison to the reference shown in Figure 29. The exhaust losses for C1 and C2 are lower than the exhaust losses from the reference. With higher maximum stroke, the benefit from the reduced exhaust losses decreases.

Figure 30 shows the courses for the piston motion, pressure and temperature. The PV diagram for variant D1 and D2 in comparison to the reference is shown in Figure 31. Because of the combination of moving the TDCF to an earlier point in time and the increased acceleration of the piston during the expansion near the TDCF, the courses of D1 and D2 are the superimposition of the variation of A and B1. This leads to an earlier increase of the pressure similar to the TDCF shift to an earlier time. Contemporaneous with the peak pressure, decreases similar to the increased piston acceleration in comparison to the reference. The temperature course is parallel shifted similar to the TDCF shift to earlier. However, the decrease of the temperature during the expansion is slower for the increased piston acceleration. This leads to higher temperatures when the exhaust valve opens.

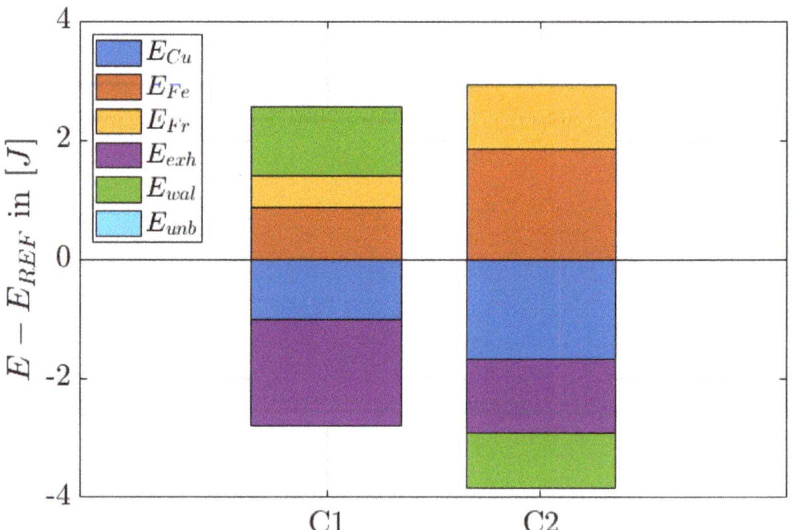

Figure 29. Energy losses for trajectory C1 and C2 with respect to REF.

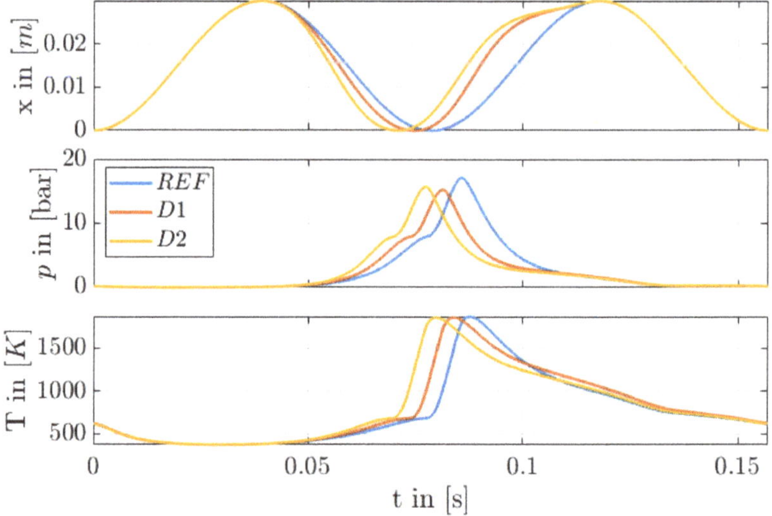

Figure 30. Comparison of trajectories REF, D1 and D2 related to the energy conversion of the internal combustion engine: illustration of the stroke, cylinder pressure and temperature depending on time.

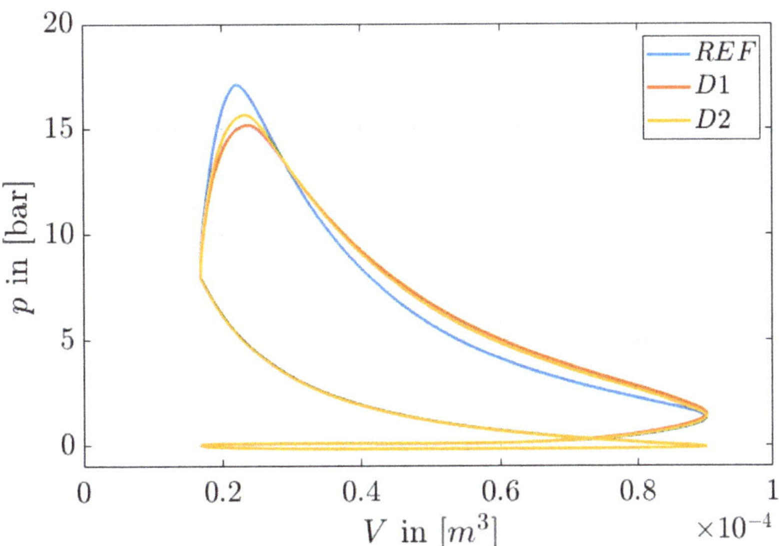

Figure 31. Comparison of trajectories REF, D1 and D2 related to the energy conversion of the internal combustion engine: cylinder pressure depending on volume.

For the evaluation of the electrical energy conversion of the trajectory REF, D1, D2, the stroke, speed and force of the electrical machine are shown in Figure 32 and the volume change, copper, friction and electrical power are shown in Figure 33.

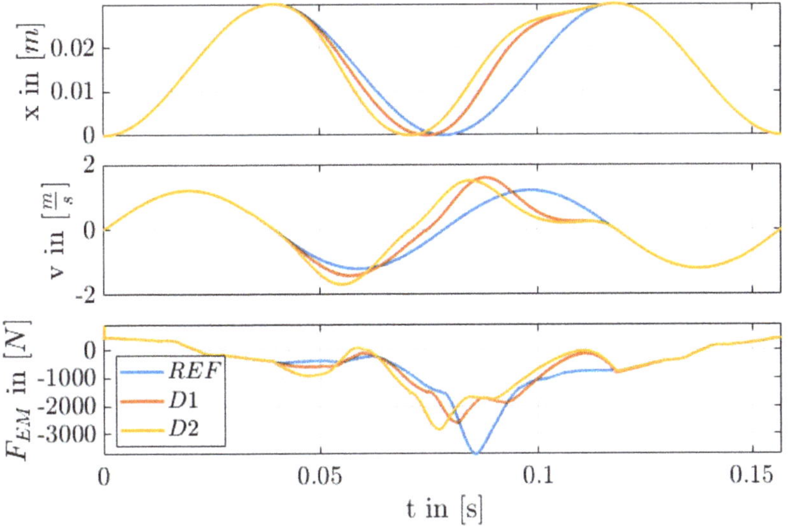

Figure 32. Comparison of trajectories REF, D1 and D2 related to the energy conversion of the electrical machine: illustration of the stroke, velocity and electric machine force depending on time.

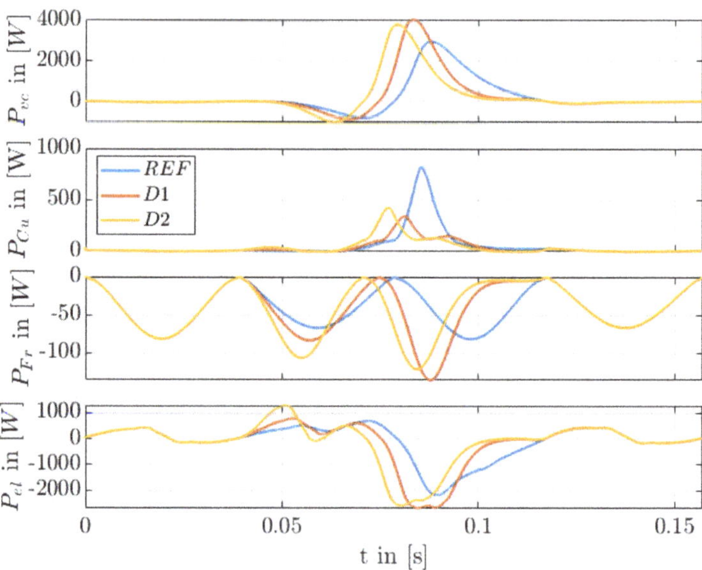

Figure 33. Comparison of trajectories REF, D1 and D2 related to the energy conversion of the electrical machine: volume change power, copper power losses, friction power losses and electric power depending on time.

The combination of variants A and B1 shows that the electrical machine must provide an adapted force curve. In this case, the copper losses in D1 are smaller than in D2, but the friction losses have increased.

The associated loss analysis for D1 and D2 with respect to the reference is shown in Figure 34. Due to the higher temperature when the exhaust valve opens, D1 has higher exhaust losses than D2, which has higher exhaust losses than the reference. Because of the lower pressure, D1 and D2 have lower wall heat losses than the reference. D1 has the lowest peak pressure. Therefore, the decrease of the wall heat losses for D1 is higher than for D2.

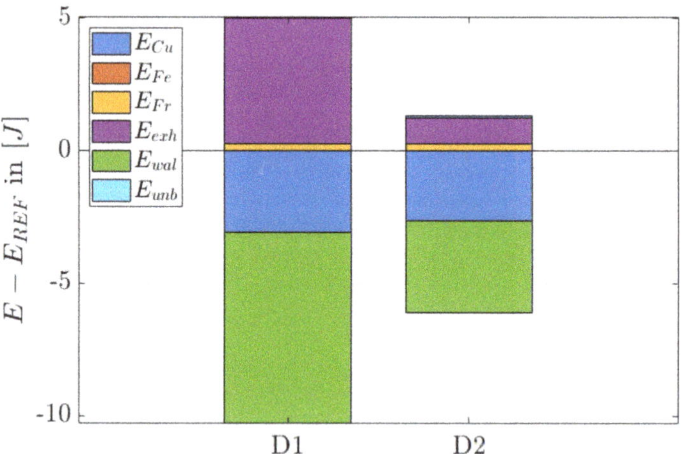

Figure 34. Energy losses for trajectory D1 and D2 with respect to REF.

Figure 35 shows the electrical energy for all variants in comparison to the reference. The shifting of the TDCF to an earlier point in time shows a decrease of the effective efficiency. The lower exhaust losses are compensated by the increasing wall heat and copper losses. The increased piston acceleration shows the best results and the decreased acceleration the worst results. The decreased wall heat and copper losses are way higher than the increased exhaust losses for B1. This is the other way around for B2. The maximum piston stroke extension shows higher efficiency with increasing maximum piston stroke. This is based on the stronger decreasing wall heat losses in comparison to the increasing exhaust losses. The combination from A and B1 (D1 and D2) shows a decreasing efficiency with a higher shift of the TDCF identical to A. Due to the higher acceleration, the efficiency is on a higher level comparable to B1.

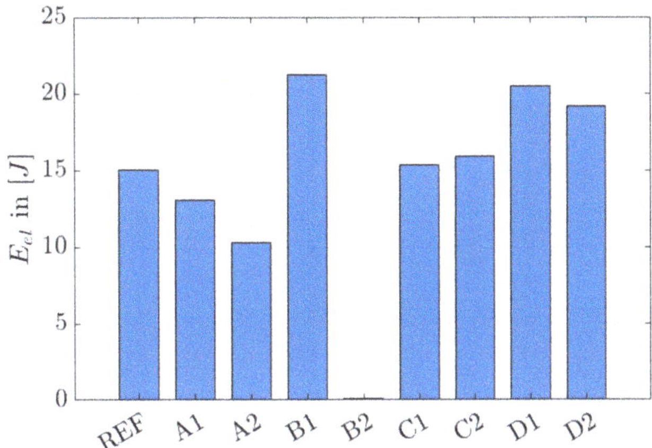

Figure 35. Comparison of the electrical energy conversion for the different trajectories.

Variant B1 shows the highest effective efficiency. The efficiency for the conversion from chemical energy within the fuel to electrical energy could be increased by 41%.

8. Conclusions

The use of a harmonic sinus function for the piston motion leads to an imperfect conversion from chemical energy into electrical energy by using a free-piston engine. The optimization of the energy conversion is a complex challenge, because of several influences by the thermodynamic and electrical conversion. The results of the trajectory variation described in Section 7 have shown that an increased acceleration of the piston from TDCF to BDC leads to a strong decrease of the peak pressure, which results in a strong decrease of the wall heat losses. The analyzed trajectory B1 could show an increase of the efficiency by 41% for the energy conversion. In addition to the efficiency increase, it is possible to increase the inlet pressure to reach the same peak pressure as the reference. This would lead to an increased power density of the free-piston engine. Despite the optimization, the absolute electrical energy share is still low for the test engine. This is based on the fact that the test engine is very small and has a low thermal efficiency. Therefore, the results of this investigation have to be verified for bigger engines. Due to differing surface-to-volume ratios, it is possible that another trajectory shows better results than B1. The further steps for the free-piston engine investigations are the described verification for engines with higher bore diameter as well as an automated trajectory optimization. When optimizing the geometric parameters of the internal combustion engine, it must be considered that this leads to different requirements for the electrical machine. Combustion engines and electrical machines should, therefore, be optimally designed for each other.

Author Contributions: Conceptualization, R.T. and A.G.; methodology, K.K.; software, R.T. and A.G.; validation, R.T. and A.G.; formal analysis, K.K.; investigation, R.T., A.G. and K.K.; resources, A.G. and S.B.; data curation, R.T. and A.G.; writing—original draft preparation, R.T., K.K. and A.G.; writing—review and editing, S.B., R.L. and H.R.; visualization, R.T. and A.G.; supervision, H.R. and R.L.; project administration, R.T. and A.G.; funding acquisition, R.T. All authors have read and agreed to the published version of the manuscript.

Funding: This research received no external funding.

Institutional Review Board Statement: Not applicable.

Informed Consent Statement: Not applicable.

Data Availability Statement: The data presented in this study are available on request from the corresponding author. The authors would like to know who re-uses their research materials and research data, which is why they would only share them on personal request.

Conflicts of Interest: The authors declare no conflict of interest.

Nomenclature and Abbreviations

LHV	Lower heat value		E_i	Energy [J]
HTC	Heat transfer coefficient		m_i	Mass [kg]
PC	Personal computer		η_i	Efficiency [-]
PMSM	Permanent magnet excited synchronous machine		A_i	Area [m^2]
IMEP	Indicated mean effective pressure		T	Temperature [K]
SI-Turb	Spark-Ignition Turbulent Flame Model		T_{in}	Inlet temperature [K]
TDCF	Top dead center firing			
REF	Reference case		T_{out}	Outlet temperature [K]
A1	Low TDC shift		t	Time [s]
A2	High TDC shift		α	Heat transfer coefficient [W/(m^2 K)]
B1	Early piston acceleration		c_p	Mean specific heat capacity [J/(kg K)]
B2	Late piston acceleration		P	Power [W]
C1	Low stroke amplitude change		v	Velocity [m/s]
C2	High stroke amplitude change		F	Force [N]
D1	Combination of early piston acceleration and low TDC shift		μ	Friction coefficient [-]
D2	Combination of early piston acceleration and high TDC shift		p	Cylinder pressure [Pa]
x	Stroke [m]		i_i	Current [A]
T_i	Duration of the cycle [s]		ω	Virtual angular velocity
ΔT_i	Relative duration change [-]		φ	Virtual angle
k_F	Force constant [N/A]		f	Frequency [Hz]
R	Ohmic resistance [Ω]		B	Magnetic flux density [T]
u_i	Voltage [V]		ϕ	Equivalence ratio [-]
ρ_u	Unburned density [kg/m^3]		a	Acceleration [m/s^2]
S_T	Turbulent flame speed [m/s]		S_L	Laminar flame speed [m/s]
C_{TFS}	Turbulent flame speed multiplier [-]		C_{FKG}	Flame kernel growth multiplier [-]
R_f	Flame radius [m]		L_i	Integral length scale [-]
B_m	Maximum laminar speed [m/s]		B_ϕ	Laminar speed roll-off value [m/s]
ϕ_m	Equivalence ratio at maximum speed [m/s]		T_u	Unburned gas temperature [K]
α_e	Temperature exponent [-]		β	Pressure exponent [-]
w	Average cylinder gas velocity [m/s]		K_i	Model constant [-]

References

1. Rahman, S.M.A.; Rizwanul Fattah, I.M.; Ong, H.C.; Zamri, M.F.M.A. State-of-the-Art of Strategies to Reduce Exhaust Emissions from Diesel Engine Vehicles. *Energies* **2021**, *14*, 1766. [CrossRef]
2. Burnete, N.V.; Mariasiu, F.; Moldovanu, D.; Depcik, C. Simulink Model of a Thermoelectric Generator for Vehicle Waste Heat Recovery. *Appl. Sci.* **2021**, *11*, 1340. [CrossRef]
3. Woo, S.; Lee, J.; Han, Y.; Yoon, Y. Experimental Study of the Combustion Efficiency in Multi-Element Gas-Centered Swirl Coaxial Injectors. *Energies* **2020**, *13*, 6055. [CrossRef]
4. Chen, P.-T.; Yang, C.-J.; Huang, K.D. Dynamic Simulation and Control of a New Parallel Hybrid Power System. *Appl. Sci.* **2020**, *10*, 5467. [CrossRef]
5. Gerlach, A.; Fritsch, M.; Benecke, S.; Rottengruber, H.; Leidhold, R. Variable Valve Timing With Only One Camshaft Actuator for a Single-Cylinder Engine. *IEEE/ASME Trans. Mechatron.* **2019**, *24*, 1839–1850. [CrossRef]
6. Wrobel, R.; Sierzputowski, G.; Sroka, Z.; Dimitrov, R. Comparison of Diesel Engine Vibroacoustic Properties Powered by Bio and Standard Fuel. *Energies* **2021**, *14*, 1478. [CrossRef]
7. Sato, M.; Goto, T.; Zheng, J.; Irie, S. Resonant Combustion Start Considering Potential Energy of Free-Piston Engine Generator. *Energies* **2020**, *13*, 5754. [CrossRef]
8. Zhang, Z.; Feng, H.; Zuo, Z. Numerical Investigation of a Free-Piston Hydrogen-Gasoline Engine Linear Generator. *Energies* **2020**, *13*, 4685. [CrossRef]
9. Zhang, Q.; Xu, Z.; Liu, S.; Liu, L. Effects of Injector Spray Angle on Performance of an Opposed-Piston Free-Piston Engine. *Energies* **2020**, *13*, 3735. [CrossRef]
10. Hu, Y.; Xu, Z.; Sun, Y.; Liu, L. Electromagnetic Characteristics Analysis of a Tubular Moving Magnet Linear Generator System. *Appl. Sci.* **2020**, *10*, 3713. [CrossRef]
11. Hu, Y.; Xu, Z.; Yang, L.; Liu, L. Electromagnetic Loss Analysis of a Linear Motor System Designed for a Free-Piston Engine Generator. *Electronics* **2020**, *9*, 621. [CrossRef]
12. Graef, M.; Treffinger, P.; Pohl, S.-E.; Rinderknecht, F. Investigation of a high efficient Free Piston Linear Generator with variable Stroke and variable Compression Ratio A new Approach for Free Piston Engines. *World Electr. Veh. J.* **2007**, *1*, 116–120. [CrossRef]
13. Kock, F.; Haag, J.; Friedrich, H.E. The Free Piston Linear Generator-Development of an Innovative, Compact, Highly Efficient Range-Extender Module. In *SAE 2013 World Congress & Exhibition, APR. 16*; SAE International400 Commonwealth Drive: Warrendale, PA, USA, 2013.
14. Kosaka, H.; Akita, T.; Moriya, K.; Goto, S.; Hotta, Y.; Umeno, T.; Nakakita, K. Development of Free Piston Engine Linear Generator System Part 1-Investigation of Fundamental Characteristics. In *SAE 2014 World Congress & Exhibition, APR. 08*; SAE International400 Commonwealth Drive: Warrendale, PA, USA, 2014.
15. Virsik, R.; Heron, A. Free piston linear generator in comparison to other range-extender technologies. *World Electr. Veh. J.* **2013**, *6*, 426–432. [CrossRef]
16. Zhang, C.; Chen, F.; Li, L.; Xu, Z.; Liu, L.; Yang, G.; Lian, H.; Tian, Y. A Free-Piston Linear Generator Control Strategy for Improving Output Power. *Energies* **2018**, *11*, 135. [CrossRef]
17. Gerlach, A.; Rottengruber, H.; Leidhold, R. Control of a Directly Driven Four-Stroke Free Piston Engine. In Proceedings of the IECON 2018-44th Annual Conference of the IEEE Industrial Electronics Society, Washington, DC, USA, 21–23 October 2018; pp. 4525–4530, ISBN 978-1-5090-6684-1.
18. Li, K.; Zhang, C.; Sun, Z. Precise piston trajectory control for a free piston engine. *Control Eng. Pract.* **2015**, *34*, 30–38. [CrossRef]
19. Vu, D.N.; Lim, O. Piston motion control for a dual free piston linear generator: Predictive-fuzzy logic control approach. *J. Mech. Sci. Technol.* **2020**, *34*, 4785–4795. [CrossRef]
20. Bashir, A.; Zulkifli, S.A.; Abidin, E.Z.Z.; Aziz, A.R.A. Control of Piston Trajectory in a Free-Piston Linear Electric Generator by Load Current Modulation. In Proceedings of the 2020 IEEE Student Conference on Research and Development (SCOReD), Batu Pahat, Malaysia, 27–29 September 2020; pp. 465–470, ISBN 978-1-7281-9317-5.
21. Zhang, C.; Li, K.; Sun, Z. Modeling of piston trajectory-based HCCI combustion enabled by a free piston engine. *Appl. Energy* **2015**, *139*, 313–326. [CrossRef]
22. Kim, J.; Bae, C.; Kim, G. Simulation on the effect of the combustion parameters on the piston dynamics and engine performance using the Wiebe function in a free piston engine. *Appl. Energy* **2013**, *107*, 446–455. [CrossRef]
23. Feng, H.; Guo, C.; Yuan, C.; Guo, Y.; Zuo, Z.; Roskilly, A.P.; Jia, B. Research on combustion process of a free piston diesel linear generator. *Appl. Energy* **2016**, *161*, 395–403. [CrossRef]
24. Yuan, C.; Feng, H.; He, Y.; Xu, J. Combustion characteristics analysis of a free-piston engine generator coupling with dynamic and scavenging. *Energy* **2016**, *102*, 637–649. [CrossRef]
25. Chiang, C.-J.; Yang, J.-L.; Lan, S.-Y.; Shei, T.-W.; Chiang, W.-S.; Chen, B.-L. Dynamic modeling of a SI/HCCI free-piston engine generator with electric mechanical valves. *Appl. Energy* **2013**, *102*, 336–346. [CrossRef]
26. Jia, B.; Smallbone, A.; Feng, H.; Tian, G.; Zuo, Z.; Roskilly, A.P. A fast response free-piston engine generator numerical model for control applications. *Appl. Energy* **2016**, *162*, 321–329. [CrossRef]
27. Arshad, W.M.; Thelin, P.; Backstrom, T.; Sadarangani, C. Use of Transverse-Flux Machines in a Free-Piston Generator. *IEEE Trans. Ind. Applicat.* **2004**, *40*, 1092–1100. [CrossRef]

28. Xu, Z.; Chang, S. Improved Moving Coil Electric Machine for Internal Combustion Linear Generator. *IEEE Trans. Energy Convers.* **2010**, *25*, 281–286. [CrossRef]
29. Benecke, S.; Gerlach, A.; Leidhold, R. Design Principle for Linear Electrical Machines to Minimize Power Loss in Periodic Motions. *IEEE Trans. Ind. Applicat.* **2020**, *56*, 4820–4828. [CrossRef]
30. Zhang, C.; Sun, Z. Trajectory-based combustion control for renewable fuels in free piston engines. *Appl. Energy* **2017**, *187*, 72–83. [CrossRef]
31. Xu, J.; Yuan, C.; He, Y.; Wang, R. An optimization of free-piston engine generator combustion using variable piston motion. *Adv. Mech. Eng.* **2017**, *9*, 168781401772087. [CrossRef]
32. Tipler, P.A.; Mosca, G. *Physics for Scientists and Engineers: With Modern Physics*, 6th ed.; Extended Version; Chapter 1–41; Freeman: New York, NY, USA, 2008; ISBN 1429202653.
33. Rinderknecht, F. A highly efficient energy converter for a hybrid vehicle concept-focused on the linear generator of the next generation. In *Proceedings of the 8th International Conference and Exhibition on Ecological Vehicles and Renewable Energies (EVER), Grimaldi Forum, Monte Carlo, Monaco, 27–30 March 2013*; IEEE: Piscataway, NJ, USA, 2013; ISBN 9781467352710.
34. Li, Q.-F.; Xiao, J.; Huang, Z. Flat-type permanent magnet linear alternator: A suitable device for a free piston linear alternator. *J. Zhejiang Univ. Sci. A* **2009**, *10*, 345–352. [CrossRef]
35. Zhang, B. *Modellierung und Hocheffiziente Berechnung der Lastabhängigen Eisenverluste in Permanentmagneterregten Synchronmaschinen*; KIT Scientific Publishing: Karlsruhe, Germany, 2019.
36. Rhodes, D.B.; Keck, J.C. Laminar Burning Speed Measurements of Indolene-Air-Diluent Mixtures at High Pressures and Temperatures. In *SAE International Congress and Exposition, FEB. 25*; SAE International: Warrendale, PA, USA, 1985.
37. Gamma Technologies LLC. *GT-SUITE-Engine Performance-Application Manual*; Gamma Technologies LLC: Westmont, IL, USA, 2019.

Article

Influence of Pre-Turbine Small-Sized Oxidation Catalyst on Engine Performance and Emissions under Driving Conditions

José Ramón Serrano, Pedro Piqueras *, Joaquín De la Morena and María José Ruiz

CMT-Motores Térmicos, Universitat Politècnica de València, Camino de Vera s/n, 46022 Valencia, Spain; jrserran@mot.upv.es (J.R.S.); joadela@mot.upv.es (J.D.l.M.); maruilu@mot.upv.es (M.J.R.)
* Correspondence: pedpicab@mot.upv.es; Tel.: +34-963-877-650

Received: 29 September 2020; Accepted: 29 October 2020; Published: 31 October 2020

Abstract: The earlier activation of the catalytic converters in internal combustion engines is becoming highly challenging due to the reduction in exhaust gas temperature caused by the application of CO_2 reduction technologies. In this context, the use of pre-turbine catalysts arises as a potential way to increase the conversion efficiency of the exhaust aftertreatment system. In this work, a small-sized oxidation catalyst consisting of a honeycomb thin-wall metallic substrate was placed upstream of the turbine to benefit from the higher temperature and pressure prior to the turbine expansion. The change in engine performance and emissions in comparison to the baseline configuration are analyzed under driving conditions. As an individual element, the pre-turbine catalyst contributed positively with a relevant increase in the overall CO and HC conversion efficiency. However, its placement produced secondary effects on the engine and baseline aftertreatment response. Although small-sized monoliths are advantageous to minimize the thermal inertia impact on the turbocharger lag, the catalyst cross-section is in trade-off with the additional pressure drop that the monolith causes. As a result, the higher exhaust manifold pressure in pre-turbine pre-catalyst configuration caused a fuel consumption increase higher than 3% while the engine-out CO and HC emissions did around 50%. These increments were not completely offset despite the high pre-turbine pre-catalyst conversion efficiency (>40%) because the partial abatement of the emissions in this device conditioned the performance of the close-coupled oxidation catalyst.

Keywords: internal combustion engine; emissions; fuel consumption; efficiency; aftertreatment; pre-turbine

---

1. Introduction

New emission regulations applied to the ground transport in major countries are focused on the reduction of both greenhouse gases and pollutant emissions [1]. In the particular case of Europe, an agreement was reached to further reduce the exhaust CO_2 targets beyond 2020. Thus, the New European Driving Cycle (NEDC)-based CO_2 limit in 2020 (95 g/km for passenger cars) will be reduced by a 15% in 2025 and 37.5% in 2030 [2]. The new regulation also defines zero- and low-emission vehicles (ZLEV) as those with CO_2 emission below 50 g/km [3]. This ZLEV category includes battery full electric vehicles (BEV) and plug-in hybrid electric vehicles (PHEV). This strategy opens the door to original engine manufacturers to meet CO_2 fleet requirements offering PHEV powered by internal combustion engines (ICEs), without the need to resort to BEV extensively [4].

In this context of hybridization, several advanced technologies for ICEs are available to reach near-zero CO_2 emissions while allowing for the sustenance of the automotive industry and market [5]. ICE improvements promote benefits in CO_2 [6] and pollutant emissions [7], being the exhaust aftertreatment systems (ATS) key to meet the limits on pollutant emissions in a context of highly dynamic operating conditions [8]. However, the early activation of the catalytic converters is becoming a critical challenge. On the one hand, the optimization of the engine thermal efficiency has resulted in the reduction of the exhaust gas temperature [9]. On the other hand, the electric propulsion periods make the engine to face long switch-off phases during which the ATS is cooled down [10]. Specific warm-up strategies must be applied to reach early catalyst light-off, what comes at the expense of fuel consumption and subsequently (CO_2) penalty. These strategies comprise active techniques, such as fuel post-injection [11] and electric heating [12], or passive techniques, as those related to thermal insulation of the combustion chamber [13] and the exhaust line [14] or management of the exhaust valve timing [15,16].

The availability on demand of electric power is leading the focus to the use of pre-turbine ATS [4]. This layout of the exhaust line has been studied in the past and is presently revisited as a passive way to increase the conversion efficiency of the catalytic converters. The most evident advantage of the pre-turbine ATS location is the increase of the catalyst temperature. However, besides the earlier thermal catalyst activation, the potential to reduce the fuel consumption exists when the pre-turbine ATS substitutes some of the post-turbine elements. If any ATS device is moved from downstream to upstream of the turbine, its pressure drop is reduced due to the higher gas density (only from certain boosting pressure so that the turbine inlet pressure offsets the higher temperature) and the fact that it is not multiplied by the turbine expansion ratio to set the engine backpressure (exhaust manifold pressure) [17]. This is particularly relevant for wall-flow monoliths, whose baseline pressure drop is high and increases as the soot is collected (while lower soot accumulation would occur in pre-turbine placement due to higher passive oxidation). As a drawback, a loss of enthalpy is found at the turbine inlet, mainly during accelerations, caused by the thermal inertia and heat losses of the ATS prior to the turbine [18]. Consequently, the turbocharger lag is increased damaging the drivability. A balanced solution consists of the pre-turbine ATS downsizing looking for an equilibrium between fuel consumption benefit while decreasing the thermal inertia (both decreased as the ATS size is smaller) [19].

As a result, the downsized ATS is able to keep high conversion efficiency of gaseous pollutants [20] and filtration efficiency in wall-flow particulate filters [21], but the turbocharger lag cannot be fully recovery to the traditional post-turbine ATS performance without additional measures. In addition, the difficulty of the ATS to deal with current emission limits does not enable the placement of all the ATS upstream of the turbine. With these boundaries, hybrid pre- and post-turbine layouts are currently taking the attention [22]. Lindemann et al. [23] combined pre-turbine diesel oxidation catalyst (DOC), selective catalytic reduction (SCR) and SCR filter (SCRF) with an underfloor SCR in a diesel engine with 48V mild hybridization. This layout provided a reduction in CO_2 ranging from 6 to 19% with respect to the baseline post-turbo ATS layout. Heavy-duty applications have also found synergies between mild hybridization and pre-turbine ATS as a complement to the baseline ATS, as discussed by Amar and Li [24] concerning the benefits of the high pressure in pre-turbine SCR systems because of the increased reactants partial pressure [25].

As an alternative to full-size pre-turbine ATS and hybridization to deal with turbocharger lag, this work explores the impact of installing a small-sized pre-turbine diesel oxidation catalyst (pre-DOC) on the engine fuel consumption as well as engine-out and tailpipe CO and HC emissions. The pre-DOC consists of a metallic substrate with triangular cells added to the baseline post-turbine ATS, which is composed of a DOC and a combination of selective catalytic reduction and particulate filter in a single monolith (SCRF). The proposed ATS configuration is evaluated in cold and warm engine operation under driving conditions represented by the Worldwide harmonized Light vehicles Test Cycle (WLTC). Despite its small size, which was selected to reduce the thermal inertia and, hence, the turbocharger lag, high CO and HC conversion efficiency was found in the pre-DOC. However, such a small size altered the flow path, deteriorating the engine performance in terms of fuel consumption and engine-out emissions. These effects and their root causes are analyzed in detail along with the change in the response of the close-coupled DOC.

2. Setup and Methods

For this study, a 4-cylinder 1.6 L diesel engine, whose main characteristics are summarized in Table 1, was employed. The fuel injection system included a high-pressure pump capable of delivering up to 200 MPa, a common-rail and Denso G4.5s solenoid fuel injectors with 8-holes, 155° included angle and a flow number of 340 cc/s. As sketched in Figure 1, the engine was equipped with a variable swirl actuator and two cooled exhaust gas recirculation (EGR) systems, i.e., high and low pressure. The high-pressure exhaust gas recirculation (HP-EGR) line extracted the gases from the exhaust manifold through the cylinder head. This HP-EGR route was composed of a valve, a cooler with a bypass route to avoid potential fouling issues at low load conditions, and a mixer introducing these exhaust gases on one side of the intake manifold. Instead, the low-pressure EGR (LP-EGR) line took the gases at the outlet of the close-coupled aftertreatment system and integrated them upstream of the compressor. The amount of recirculated gases in the low-pressure system was controlled by a three way valve, which throttled the intake flow before the compressor when the pressure difference in the LP-EGR line was low.

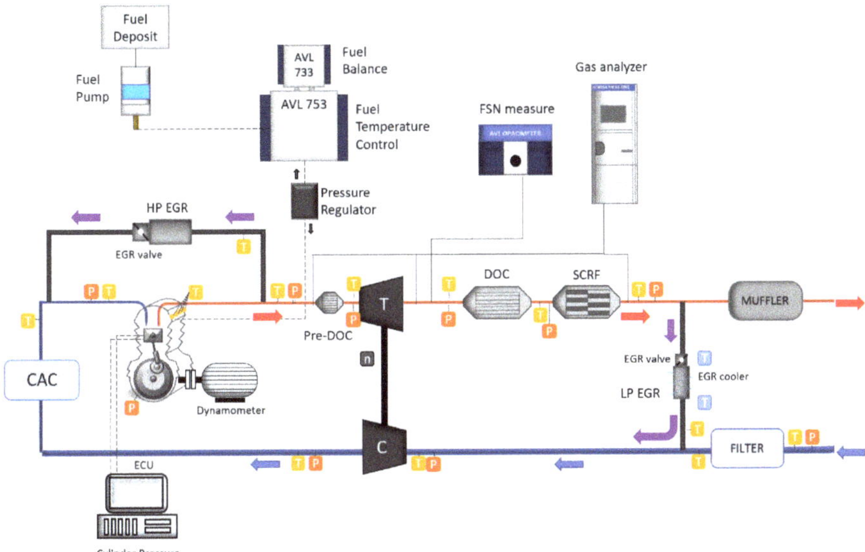

Figure 1. Scheme of the engine (with pre-turbine pre-DOC) and experimental setup.

Table 2 lists the main instrumentation used in this work. The engine was coupled to an asynchronous dynamometer controlled by the HORIBA SPARC automation system accessed through the STARS user interface. The dynamometer can be used for engine dynamic tests being able to reproduce driving cycles as the WLTC tested in this work. An AVL 733S fuel balance measured the fuel consumption. ETAS INCA v7.1 was used to register the main actuations performed by the electronic control unit (ECU).

The emissions were monitored by an AVL 439 opacimeter placed at the turbine outlet and up to three samples of HORIBA MEXA-7100 DEGR to analyze the composition of the exhaust gases in different points of the exhaust line depending on the ATS layout. Figure 1 shows the case of installed pre-turbine pre-DOC. The same systems measured the CO_2 mole fraction at the compressor outlet, for the determination of the low-pressure EGR rate, as well as at the intake manifold, to obtain the combination of both EGR systems when used simultaneously. The air path was monitored with an ABB hot-wire anemometer placed upstream of the engine airbox. Pressure and temperature probes were installed at the most significant sections of the intake and exhaust lines. Finally, a Picoturn sensor was placed in the compressor housing to measure the instantaneous rotational speed of the turbocharger.

Table 1. Main characteristics of the engine.

Engine type	HSDI diesel
Emission standards	Euro 6
Displacement	1606 cm^3
Bore	80.1 mm
Stroke	79.7 mm
Number of cylinders	4 in-line
Compression ratio	16:1
Rated power @ speed	100 kW @ 3500–4000 rpm
Rated torque @ speed	320 Nm @ 2000–2500 rpm
Fuel injection	Common-rail direct fuel injection
Turbocharger	VGT
EGR	Cooled high and low pressure

Table 2. Main characteristics of the instrumentation.

Magnitude	Instrument	Range	Accuracy
Speed control	SIEMENS dynamometer	6000 rpm	±2 rpm
Torque control	SIEMENS dynamometer	±450 Nm	±0.5 Nm
Air mass flow	Sensiflow DN80	20 to 720 kg/h	±2%
Fuel mass flow	AVL 733S Fuel meter	0 to 150 kg/h	±0.2%
Temperature	K-type thermocouples	−200 to 1250 °C	±1.5 °C
Mean pressure	Kistler piezo-resistive sensor	0–10 bar	linearity 0.2%
CO_2	NDIR	0–20 (% Vol)	±1% (full scale)
CO_L	NDIR	0–5000 (ppm)	±1% (full scale)
CO_H	NDIR	0–12 (% Vol)	±1% (full scale)
THC	HFID	0–5000 (ppm)	±1% (full scale)
NO&NOx	CLD	0–10,000 (ppm)	±1% (full scale)

The experimental campaign was driven to the analysis of the impact of adding a small-sized DOC at the exhaust manifold outlet upstream of the variable geometry turbine (VGT) in order to improve the removal of HC and CO emissions. The nominal ATS was composed of a closed-coupled DOC, coated with Pt and a zeolite layer for HC adsorption, followed by a SCRF. In this work, the urea injection was canceled to focus the analysis on the engine-out NOx emission and CO and HC oxidation in the pre-DOC and DOC. The geometry of these catalysts is detailed in Table 3. The pre-DOC substrate was metallic with Pt coating but without HC adsorption capability (no zeolite layer). Metallic substrates are interesting for its use in pre-turbine location because of their thin walls, which provide high geometric surface area and low pressure drop in comparison to ceramic counterparts. In addition, these substrates have good mechanical durability and resistance to thermal shock at the same time ease the adaptation as electric catalysts to compensate the thermal inertia effects (not considered in this work). The pre-DOC was obtained from a full-size post-turbo metallic substrate cutting it with the diameter of the exhaust manifold outlet (turbine inlet) to reduce the added canning volumes. The original length (85 mm) was kept providing a volume of 60 cm^3. These dimensions balanced the effects on thermal inertia (small volume to minimize the turbocharger lag) and conversion efficiency (maximum residence time under the diameter constraint). The turbocharger was relocated due to the pre-DOC presence being necessary to adapt the oil supply. However, the small pre-DOC size avoided any interference of the turbocharger and close-coupled ATS with the engine block nor other components. Therefore, the original exhaust line was only modified by the installation of the pre-DOC.

The performance of the ATS was analyzed with the baseline ATS and with pre-turbine pre-DOC by means of driving cycles defined by the WLTC. For each configuration, both cold (starting with engine coolant at room temperature) and warm (starting with coolant at 90 °C) were performed consecutively. Before each WLTC, the DPF was regenerated to avoid the effects of different initial DPF soot loading on the engine backpressure. Finally, the procedure was done twice to ensure the repeatability of the results.

Table 3. Geometrical parameters of the oxidation catalysts.

	DOC	Pre-DOC
Substrate	Cordierite	Metallic
Channel cross-section	Square	Triangular
Diameter [mm]	125	30
Length [mm]	70	85
Volume [cm^3]	860	60
Cell density [cpsi]	400	300
Cell size [mm]	1.16	2.1
Wall thickness [mm]	0.11	0.08
Catalytic area [m^2]	2.22	0.175

3. Discussion of the Results

Figure 2a shows the vehicle speed during the WLTC superimposed over the engine torque provided by the engine to reach it. The driving cycle is composed of four speed phases, from low to extra-high speed, which will be taken as reference to conduct the analysis of the results along with the total cumulative result of mass-based magnitudes. With this approach, Figure 2b,c represent the cumulative fuel consumption in every speed phase and the total amount for each ATS configuration in cold and warm WLTCs respectively. The fuel consumption was lower in the warm cycles but the difference between baseline ATS and pre-turbine pre-DOC configurations was similar in all cases. An increase of fuel consumption was brought by the installation of the pre-turbine pre-DOC. The cumulative fuel consumption at the end of the cycle increased in 3.22% (46.7 g) and 3.89% (53.7 g) in cold and warm WLTCs respectively. On this regard, the variation in fuel consumption brought by the pre-turbine pre-DOC configuration in every WTLC phase is detailed in Figure 2b,c. In cold conditions, the pre-turbine pre-DOC configuration improved the fuel consumption during the low-speed phase (−3.05% (−6.5 g)). However, a progressive deterioration of the fuel consumption was found from the medium speed phase on. In warm cycles, the trend was similar in these medium- to extra-high-speed phases but the low-speed phase also resulted in a fuel consumption penalty for the pre-turbine pre-DOC configuration (2.96% (5.3 g)).

The fuel consumption penalty has its origin in two linked pressure drop phenomena. In this study, the addition of a new element (pre-turbine pre-DOC) necessarily involves an increase of the exhaust line pressure drop, so that some fuel consumption damage was expected. Figure 3 shows the pressure drop of the ATS elements in baseline and pre-turbine pre-DOC configurations for each WLTC. The pre-turbine location contributes to decrease the pressure drop of a given element because of the higher gas density provided that the pressure at the turbine inlet is high enough to offset the higher temperature with respect to post-turbine location [26]. This condition is only fulfilled from a certain level of boost pressure. However, the pre-DOC shows high pressure drop in comparison to the DOC+SCRF during all WLTC phases. The pre-DOC pressure drop ranged 25–30% of the DOC-SCRF one in pre-turbine pre-DOC configuration. It is interesting to note that the ATS pressure drop, and in particular the pre-DOC one, was higher in the low-speed phase of the warm WTLC than in the cold one. This result contributes to explain the higher fuel penalty damage for the former during such WLTC phase. This was due to the lower gas temperature and SCRF soot loading in the cold WLTC (cold WLTC was run from clean wall-flow filter while warm WLTC did it from the soot loading corresponding to cold WLTC). The magnitude of the pre-DOC pressure drop was caused by the reduced effective cross-section of the monolith. Although it had the same diameter than the turbine inlet, the open frontal area (OFA) distorted the flow path due to the decrease of the effective section by 18% (OFA = 0.82).

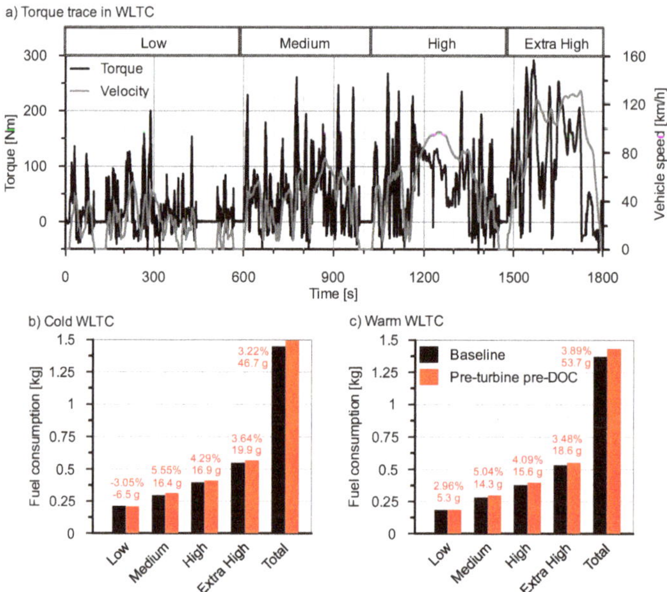

Figure 2. (a) Vehicle speed trace during WLTC and cumulative fuel consumption in every WLTC phase in (b) cold and (c) warm tests with baseline ATS and pre-turbine pre-DOC configurations.

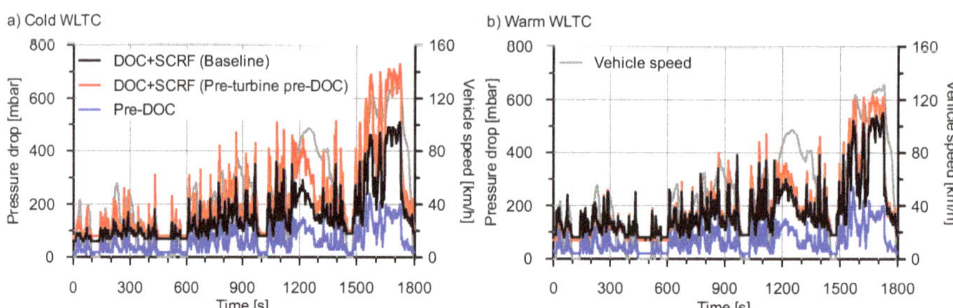

Figure 3. Pressure drop in DOC+SCRF and pre-turbine pre-DOC in (a) cold and (b) warm tests with baseline ATS and pre-turbine pre-DOC configurations.

The addition of this pressure drop to the exhaust path penalizes the engine backpressure by itself, but also induces a negative effect on the boost control, as described next. On the one hand, the pre-turbine location directly affects the amplitude of the instantaneous pressure pulses at the turbine inlet. As seen in previous works, this location produces a change from pulsating to constant pressure turbocharging that can have a positive impact on the turbine efficiency [27]. At the same time, the rarefaction wave affects the gas exchange process increasing the internal EGR. On the other hand, these changes in the gas exchange induce an increase in the soot emission, as shown in Figure 4a,b by the instantaneous engine-out opacity and found in previous studies regarding pre-turbine ATS [18]. Consequently, higher SCRF soot loading was obtained, what explains the difference in DOC+SCRF pressure drop between pre-turbine pre-DOC configuration and the baseline ATS observed in Figure 3. Due to the change in the flow pattern at the VGT inlet and, mainly, the increase of the tailpipe backpressure, the engine control varied the VGT position (closing it) to reach the target boost pressure (the HP- and LP-EGR control strategy was unaltered, so no differences in EGR rate were produced). Figure 5 shows in plots (a) and (b) how the VGT was closer (higher VGT position value) until reaching its maximum closure (90%) in pre-turbine pre-DOC than in the baseline configuration for both cold and warm WLTCs. The closer VGT position and the trends in pressure drop resulted in a relevant increase of the exhaust manifold pressure with pre-DOC, which was more evident in sudden accelerations and as the engine load increased (Figure 5c,d). As for fuel consumption, the penalty of pre-turbine pre-DOC configuration on exhaust manifold pressure was higher in the low-speed phase for warm conditions than cold ones. In fact, this pressure is even lower (more open VGT) along some periods of the low-speed phase with pre-turbine pre-DOC configuration in the cold WLTC. This is because of the benefits of constant pressure turbocharging (higher turbine efficiency) govern on the pressure drop damage for this initial stage in the cold WLTC.

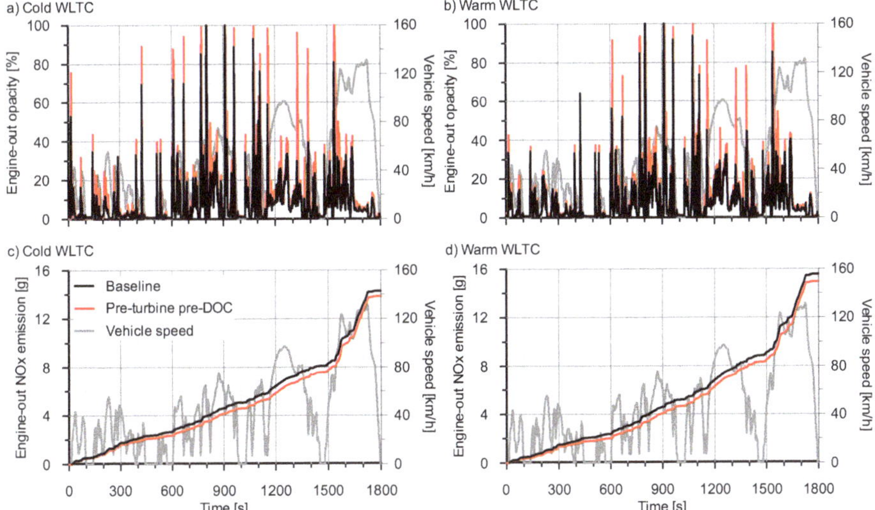

Figure 4. Engine-out opacity and cumulative NOx emission in cold and warm WLTCs with baseline ATS and pre-turbine pre-DOC configurations. (**a**) Instantaneous engine-out opacity in cold WLTC, (**b**) instantaneous engine-out opacity in warm WLTC, (**c**) cumulative engine-out NOx emission in cold WLTC, (**d**) cumulative engine-out NOx emission in warm WLTC.

Figure 4a,b show that the higher opacity in pre-turbine pre-DOC configuration with respect to the baseline ATS case is present throughout the cold and warm cycles. The penalty is similar between them during medium-, high- and extra-high-speed phases. In contrast, the pre-turbine pre-DOC case increased more the opacity in cold operation than in warm driving during the low-speed phase. Figure 4c,d depict the cumulative engine-out NOx emission. According to the increase of the internal EGR because of the exhaust manifold pressure increase, the engine-out NOx emissions decreased by the installation of the pre-turbine pre-DOC. At the end of the cycle, the NOx emission reduction reached 0.43 g (3%) in the cold WLCT and 0.60 g (3.9%) in the warm case.

The most harmful effect of pre-turbine ATS on the engine performance is the marked turbocharger lag during sudden accelerations caused by the monolith thermal inertia [18]. As this effect is critical in nominal-sized ATS movement upstream of the turbine, it is progressively avoided as the monolith size is decreased. Despite the potential for ATS downsizing of the pre-turbine location [28], the complete removal of the turbocharger lag requires very small monoliths complemented with post-turbo counterparts for full pollutant abatement or an electric-assisted turbocharger [29]. In this work, the tested pre-DOC produced a very slight turbocharger lag, as observed in intake manifold pressure and turbocharger speed represented in Figure 6 for the more demanding high- and extra-high-speed phases. Accordingly, the effect on the turbine inlet temperature, which is represented in top charts of Figure 7 for cold and warm WLTCs as a function of the ATS configuration, was also negligible. As observed, the pre-DOC inlet temperature is higher than the one in the exhaust manifold in baseline

ATS configuration because of the higher exhaust backpressure effects. However, the temperature dropped across the pre-DOC because of the heat losses and thermal inertia, which were not offset by the heat released due to the HC and CO oxidation. Nevertheless, the change in DOC inlet temperature is positive and much more relevant, as illustrated in Figure 7c,d. The turbocharger operation with quasi-steady flow led to lower temperature drop in the expansion, favoring a faster warm-up of the close-coupled DOC.

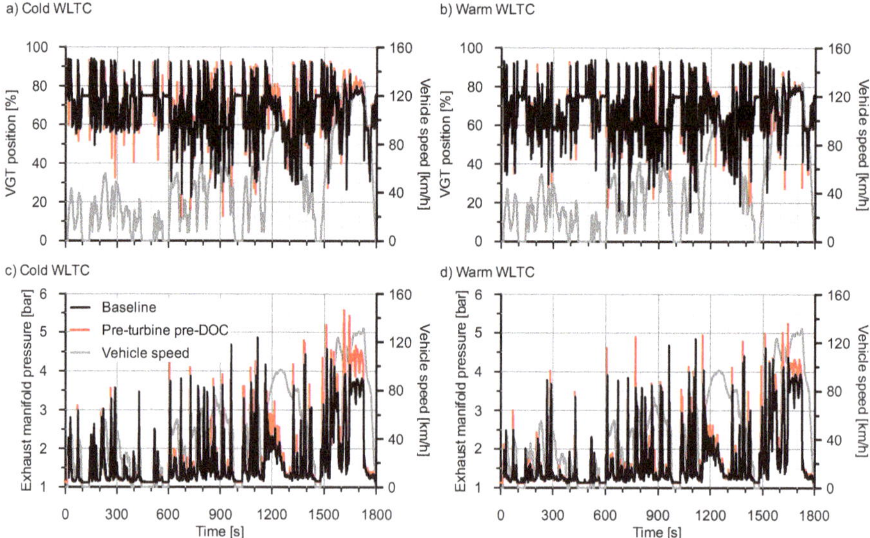

Figure 5. Exhaust manifold pressure and VGT position in cold and warm WLTCs with baseline ATS and pre-turbine pre-DOC configurations. (**a**) VGT position in cold WLTC, (**b**) VGT position in warm WLTC, (**c**) exhaust manifold pressure in cold WLTC, (**d**) exhaust manifold pressure in warm WLTC.

With these fluid-dynamic boundaries, Figure 8 shows the cumulative engine-out and tailpipe emissions of CO and HC in cold and warm WLTCs with baseline ATS and pre-turbine pre-DOC configurations. Complementary, Table 4 summarizes the cumulative engine-out CO and HC emissions at the end of each WLTC as a function of the ATS configuration. The higher exhaust manifold pressure deteriorated the combustion process with pre-turbine pre-DOC configuration due to the increase of the internal EGR. Consequently, a slight decrease in air mass flow was found, with an increase in fuel consumption to reach the vehicle speed target. Therefore, an increase of the equivalence ratio took place in pre-turbine pre-DOC configuration. This engine response made the CO and HC emission increase around 50% both in cold and warm WLTCs with respect to the baseline ATS case. As observed in Figure 8, this response avoided decreasing the tailpipe emissions with respect to the baseline configuration, despite the earlier activation of the pre-DOC, which highly contributed to the emission abatement. These results are next discussed in detail to bring out the involved phenomena.

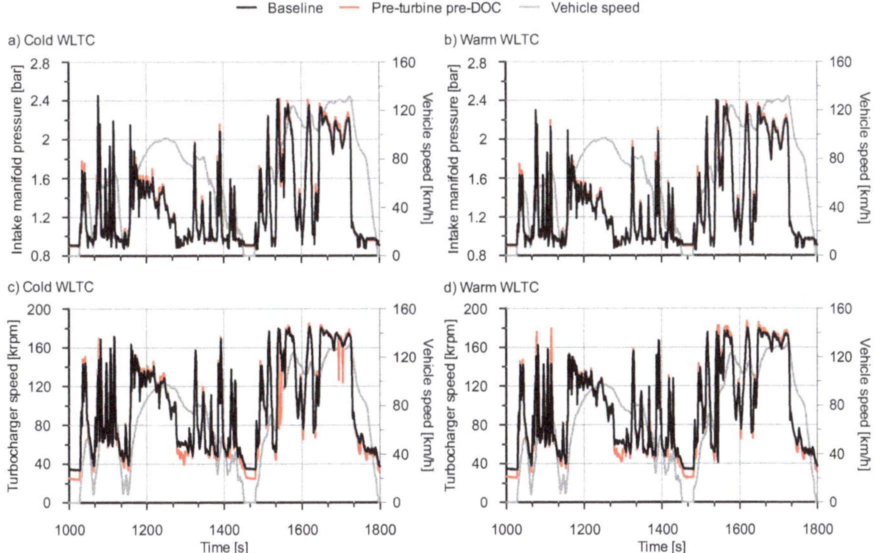

Figure 6. Intake manifold pressure and turbocharger speed in cold and warm WLTCs with baseline ATS and pre-turbine pre-DOC configurations during the high- and extra-high-speed phases. (**a**) Intake manifold pressure in cold WLTC, (**b**) intake manifold pressure in warm WLTC, (**c**) turbocharger speed in cold WLTC, (**d**) turbocharger speed in warm WLTC.

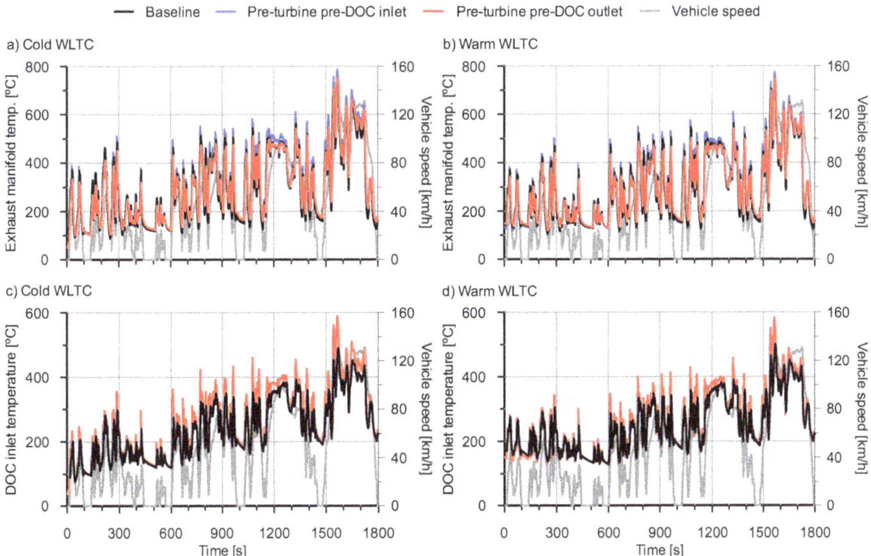

Figure 7. Pre-DOC, turbine and DOC inlet temperature in cold and warm WLTCs with baseline ATS and pre-turbine pre-DOC configurations. (**a**) Exhaust manifold temperature in cold WLTC, (**b**) exhaust manifold temperature in warm WLTC, (**c**) DOC inlet temperature in cold WLTC, (**d**) DOC inlet temperature in warm WLTC.

Table 4. Cumulative engine-out CO and HC emission in cold and warm WLTCs with baseline ATS and pre-turbine pre-DOC configurations.

	Cold WLTC		Warm WLTC	
	Baseline	**Pre-Turbine Pre-DOC**	**Baseline**	**Pre-Turbine Pre-DOC**
CO [g]	34.1	50.2 (+47.2%)	29.3	44.1 (+50.5%)
HC [g]	3.9	6.2 (+58.9%)	3.2	4.9 (+53.1%)

To understand better the new emission pattern, Figure 9 shows the cumulative CO emission in the relevant locations of the exhaust line normalized against the total engine-out emission in every test. Top charts are referred to the baseline ATS configuration, distinguishing between cold and warm WLTC in plots (a) and (b) respectively. The cumulative engine-out CO emission presents almost identical distribution along the speed phases between cold and warm tests. The main discrepancies appeared in the low-speed phase, especially in cold WLTC. Nevertheless, the medium- and high-speed phases involved almost 60% of the total engine-out CO emissions due to their higher duration and concurrence of sharp engine load demands, which avoid complete in-cylinder CO oxidation. Tailpipe CO emissions showed similar pattern than engine-out ones but with more marked differences between cold and warm tests during the low-speed phase because of the faster DOC warm-up in warm WTLC. Consequently, the tailpipe CO emission reached its lowest contribution during this first cycle phase in the warm test in contrast to the cold-start case. The remainder phases showed a progressive decrease of the cumulative tailpipe CO emission from medium- to extra-high-speed phase with increasing differences with respect to the engine-out cumulative emission.

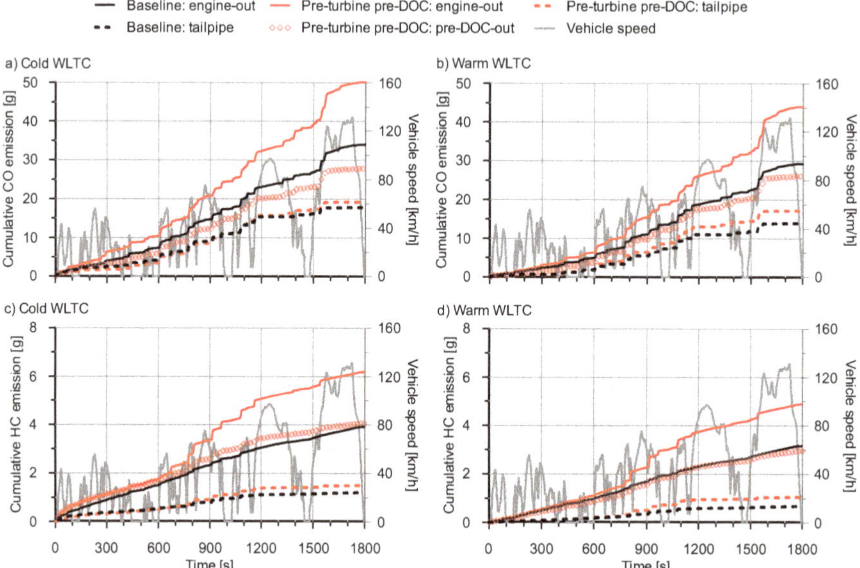

Figure 8. Cumulative CO and HC emission along the exhaust line in cold and warm WLTCs with baseline ATS and pre-turbine pre-DOC configurations. (**a**) Cumulative CO emission in cold WLTC, (**b**) cumulative CO emission in warm WLTC, (**c**) cumulative HC emission in cold WLTC, (**d**) cumulative HC emission in warm WLTC.

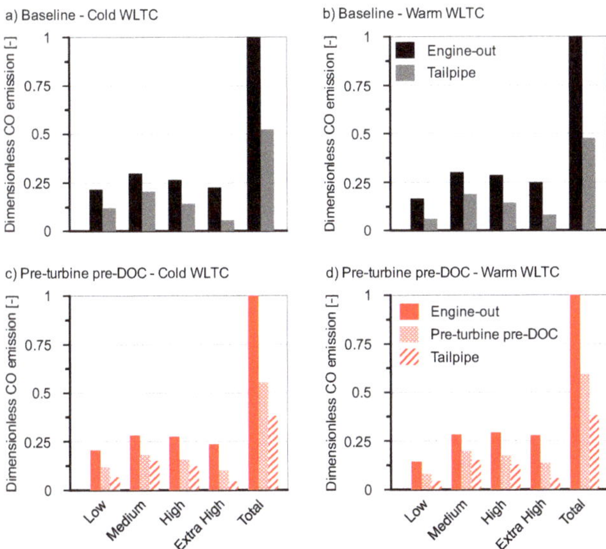

Figure 9. Normalized cumulative CO emission along the exhaust line in every WLTC phase in cold and warm tests with baseline ATS and pre-turbine pre-DOC configuration. (**a**) Baseline configuration in cold WLTC, (**b**) Baseline configuration in warm WLTC, (**c**) Pre-turbine pre-DOC configuration in cold WLTC, (**d**) Pre-turbine pre-DOC configuration in warm WLTC.

This last trend means a CO conversion efficiency increase, as depicted in Figure 10. The black series represents the CO conversion efficiency in baseline ATS configuration (DOC). It is interesting to note that the CO conversion efficiency in the low-speed phase was not the lowest one even in cold WLTC. Despite the warm-up, the DOC inlet temperature exceeded 150 °C most of time during this phase (Figure 7), which enabled the reaching of the CO light-off. Together with the smooth accelerations during this first cycle period, i.e., lack of engine-out CO emission peaks, the result was a CO conversion efficiency of 44% and 62% for cold and warm tests. These values were higher than those reached in the medium- and high-speed phases, despite the lower DOC temperature during the low-speed phase. The reason lies on the CO emission peaks along the medium- and high-speed phases during fast acceleration because of the mass flow increase along with high CO emission. As a result, mass transfer and inhibition condition the catalyst performance [30]. In fact, the gradual DOC temperature increase governs the recovery of the conversion efficiency till reaching its maximum in the extra-high-speed phase.

Complementary, Figure 9c,d represent the normalized cumulative CO emission distribution for the pre-turbine pre-DOC configuration. The main trends in CO engine-out emission and abatement for the baseline case were also found with the pre-turbine pre-DOC configuration. The cumulative CO emission in every exhaust region behaved as their counterpart in the baseline ATS configuration, both in cold and warm WLTCs. The most relevant result is that the percentage tailpipe CO emission was lower than the one obtained in baseline ATS configuration in every cycle phase. A detailed analysis of the CO conversion efficiency with pre-turbine pre-DOC shown in Figure 10 reveals that the presence of the pre-DOC enhanced the overall CO abatement throughout the cycle, especially during the low-speed phase in cold WLTC. The pre-DOC contributed with very high conversion efficiency, which was comparable to that of the DOC in baseline ATS configuration during the cold WLTC. The reason of the high pre-turbine pre-DOC reaction rate was two-fold. First, this catalyst operated at higher temperature than the DOC. In addition, the reactants partial pressure in pre-turbine location was also higher than in the DOC because of the higher gas pressure prior to the turbine. Higher CO and HC mole fraction was also found along with slightly lower O_2 amount due to the higher equivalence ratio.

Concerning the CO and HC mole fraction increase, this contribution is positive for base-level engine emissions (low mole fraction) but can turn negative during CO emission peaks related to accelerations because of the inhibition [31]. By contrast, the DOC performance was worsened with respect to the baseline ATS configuration, in which it was dealing with all the engine-out CO emission and operated at lower temperature (Figure 7). In fact, the DOC provided lower CO conversion efficiency than the pre-DOC when these two catalysts worked together.

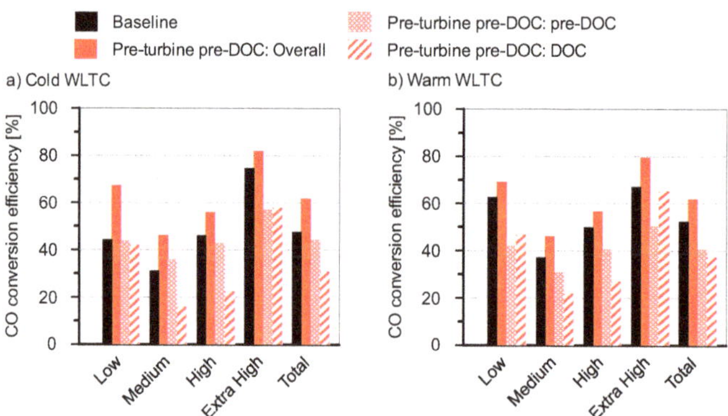

Figure 10. CO conversion efficiency in each WLTC phase in (**a**) cold and (**b**) warm tests: baseline (DOC) and pre-turbine pre-DOC (overall and contributions from pre-DOC and DOC) configurations.

The cause of this unexpected result was found in the different pattern of CO mass flow that the two catalysts must face. It is exemplified in Figure 11, where the CO mass flow and conversion efficiency in pre-DOC (plot (a)) and DOC (plot (b)) are shown during the medium speed phase of the cold WLTC. According to Figure 11a, the pre-DOC abated most of the base-level engine-out CO emissions, with instantaneous efficiencies ranging 60–80%. However, it was unable to deal with the CO emission peaks due to the previously mentioned mass transfer and inhibition limitations. According to this response, the CO emissions at the DOC inlet were mostly concentrated in the peaks related to the fast accelerations, as depicted in Figure 11b. Consequently, and despite the higher instantaneous conversion efficiency against base-level emissions (80–95%) than the pre-DOC, the CO conversion efficiency of the DOC decreased with respect to the baseline ATS configuration and was even lower than the one of the pre-DOC. The dependence on the flow pattern was also demonstrated by the fact that the CO conversion efficiency was better in the DOC than in the pre-DOC during the low- and extra-high-speed phases, where the emission peaks are less prominent and frequent. By contrast, the DOC performance was minimum during the medium- and high-speed phases penalized by the high dynamics of these periods.

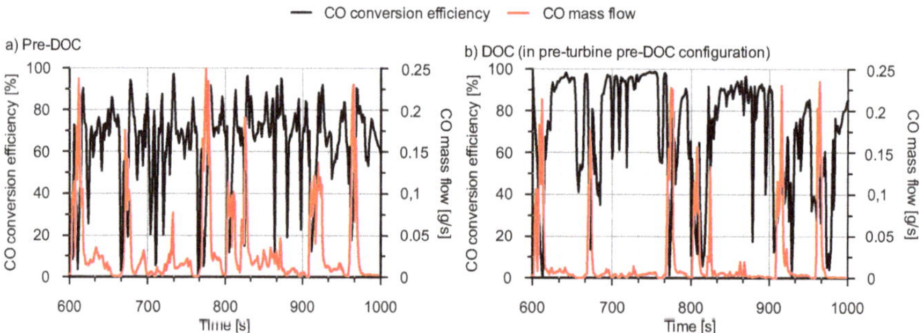

Figure 11. Instantaneous CO conversion efficiency and mass flow in (**a**) pre-DOC and (**b**) DOC (pre-turbine pre-DOC configuration) during the medium speed phase of the cold WLTC.

The results corresponding to HC conversion efficiency are shown in Figure 12. Unlike CO case, the pre-turbine pre-DOC configuration scarcely improved the overall HC conversion efficiency, especially under warm conditions. This was due to the pre-DOC characteristics and the HC oxidation behavior. On the one hand, the pre-DOC was not coated with zeolites, so that its HC conversion efficiency during the low-speed phase was very poor, especially in cold WLTC. By contrast, the DOC showed very high HC conversion efficiency during the cold start emphasizing the importance of the adsorption mechanism. On the other hand, the small size of the pre-DOC involved additional limitations to mass transfer, which are more relevant in HC than CO because of the higher diffusion volume of species of high molecular weight [32]. Despite the better fluid-dynamic conditions in pre-turbine location to increase the reaction rate and the advantage to face base-level HC emission first than the DOC, the pre-DOC HC conversion efficiency was almost constant along the WLTC, both cold and warm tests. In fact, a slight decrease in HC conversion efficiency was found from medium- to extra-high-speed phases due to the residence time decrease. By contrast, the higher size of the DOC allowed it to deal better with HC mass transfer limitations and took advantage of the temperature increase throughout the cycle to increase its HC conversion efficiency from medium- to extra-high-speed phases. Nevertheless, its HC conversion efficiency was damaged with respect to the baseline ATS configuration, as for CO abatement, and the benefits of the pre-DOC were partially degraded when evaluating the entire exhaust line layout.

Despite the increase in overall CO and HC conversion efficiency with pre-turbine pre-DOC configuration, the parallel increase of the engine-out CO and HC emissions canceled these improvements out. Figure 13 shows the variation of the cumulative engine-out and tailpipe CO and HC emissions brought by the pre-turbine pre-DOC configuration in each WLTC. Even though the increase of the engine-out emissions were completely countered in low-speed phases of cold WLTC, the total cumulative tailpipe emissions increased for CO and HC. Nevertheless, there are two relevant remarks. First, the percentage penalty was always lower in tailpipe than engine-out emissions. In addition, the penalties were lower in cold conditions (8.1% for CO and 24.4% for HC) than in warm WLTC (20.8% for CO and 37.6% for HC) despite the similar percentage increase of engine-out emissions. These two trends confirm the potential for cold-start conversion efficiency improvement of pre-turbine ATS provided that the effects on the engine operation were minimized.

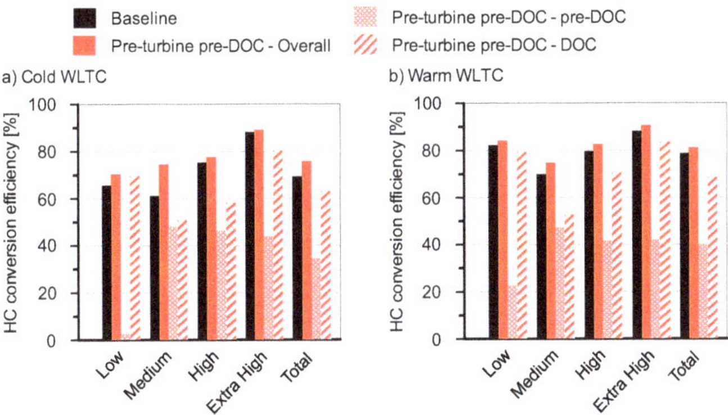

Figure 12. HC conversion efficiency in each WLTC phase in (**a**) cold and (**b**) warm tests: baseline (DOC) and pre-turbine pre-DOC (overall and contributions from pre-DOC and DOC) configurations.

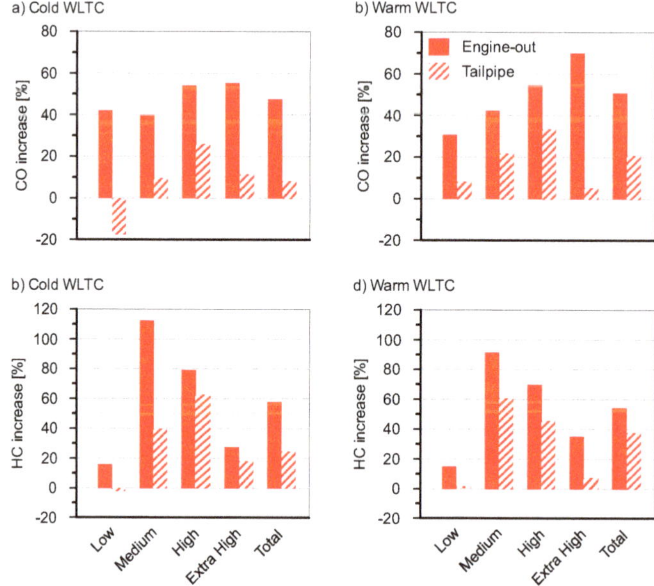

Figure 13. Increment of the CO and HC engine-out and tailpipe emission of the pre-turbine pre-DOC configuration with respect to the baseline ATS configuration in cold and warm WLTCs. (**a**) CO increment in cold WLTC, (**b**) CO increment in warm WLTC, (**c**) HC increment in cold WLTC, (**d**) HC increment in warm WLTC.

4. Conclusions

The influence of a pre-turbine pre-DOC on the engine performance and emissions has been analyzed experimentally from driving cycle tests in cold and warm conditions. The pre-DOC consisted of a small-sized metallic monolith placed at the exhaust manifold outlet and whose diameter coincided with that of the turbine inlet.

The results indicate that the addition of the pre-turbine pre-DOC had relevant effects on the engine performance and emission abatement. Although the small monolith volume avoided the turbocharger lag, the reduced cross-section generated a high pressure drop upstream of the turbine affecting the gas exchange process. The direct effect was the deterioration of the combustion leading the engine-out CO, HC and soot emissions to increase as well as a small NOx emission decrease because of the increase of internal EGR.

Concerning the engine performance, the increase in soot emission led the SCRF pressure drop to increase causing, in turn, an additional increase of the exhaust manifold pressure with respect to the baseline ATS configuration. As a result, the fuel consumption was penalized in 3.2% and 3.9% in cold and warm WLTCs, respectively.

The increase of the engine-out CO and HC emissions with respect to the baseline ATS were ranging 50% in cold and warm WTLCs. This huge increase was not able to be balance out by the presence of the pre-DOC upstream of the turbine. The CO tailpipe emissions increased by 8% and 21% in cold and warm WLTCs, respectively. In the same conditions, HC did by 24% and 37%. The tailpipe percentage increments were lower than engine-out ones because of the high pre-DOC conversion efficiency. This was promoted by the positive fluid-dynamic conditions upstream of the turbine concerning high temperature and pressure, the last increasing the partial pressure of the reactant species and, hence, the reaction rate. However, these advantages were only evident to abate base level emissions. The pre-DOC conversion efficiency fell against engine-out emission peaks generated during fast accelerations due to the high mass transfer limitations of small-sized monoliths. The main concern of this behavior is that the close-coupled DOC also reduced its conversion efficiency compared to the baseline ATS configuration. The higher emission peaks with respect to this configuration and the low base-level emissions because of their removal by the pre-DOC compromised the DOC performance. As a result, the overall pollutants conversion efficiency was improved when using the pre-DOC but was not high enough to offset the increase in engine-out emissions.

Considering these results as a part of the literature context, the potential of pre-turbine ATS configuration to increase the pollutants conversion efficiency under driving conditions is confirmed even for very small monolith sizes. This kind of geometry is also positive to reduce the damage on the turbocharger lag, the main concern of pre-turbine ATS layouts. However, the effects on the engine fluid-dynamics and, hence, on the fuel consumption and pollutants formation, penalizes this solution with respect to full-size pre-turbine ATS approaches. These alternatives can improve further the conversion efficiency because of the inherent lower mass transfer limitations at the same time the fuel consumption is highly reduced if the wall-flow filter is placed upstream of the turbine (high reduction in engine backpressure). The high thermal inertia of the monoliths and its management based on e-turbocharger technology would remain as the main challenge to combine high pollutants conversion with fuel savings.

Author Contributions: Conceptualization, J.R.S. and P.P.; Formal analysis, P.P. and M.J.R.; Funding acquisition, J.R.S.; Investigation, J.R.S., P.P., J.D.l.M. and M.J.R.; Methodology, P.P. and J.D.l.M.; Project administration, J.R.S. and P.P.; Resources, J.R.S.; Supervision, J.R.S. and P.P.; Visualization, P.P., J.D.l.M. and M.J.R.; Writing—original draft, P.P. and J.D.l.M.; Writing—review and editing, J.R.S., P.P., J.D.l.M. and M.J.R. All authors have read and agreed to the published version of the manuscript.

Funding: This research has been partially supported by FEDER and the Government of Spain through project TRA2016-79185-R and by Universitat Politècnica de València under a grant with reference number FPI-2018-S2-10 to the Ph.D. student María José Ruiz.

Conflicts of Interest: The authors declare no conflict of interest.

Abbreviations

The following abbreviations are used in this manuscript:

ATS	Aftertreatment system
BEV	Battery electric vehicle
CAC	Charge air cooling
CLS	Chemi-luminescence detector
DOC	Diesel oxidation catalyst
ECU	Electronic control unit
EGR	Exhaust gas recirculation
HFID	Heated flame ionization detector
HP-EGR	High-pressure exhaust gas recirculation
HSDI	High-speed direct injection
ICE	Internal combustion engine
LP-EGR	Low-pressure exhaust gas recirculation
NDIR	Nondispersive infrared detector
NEDC	New European driving cycle
OFA	Open frontal area
PHEV	Plug-in hybrid electric vehicle
SCR	Selective catalytic reduction
SCRF	Selective catalytic reduction on filter
VGT	Variable geometry turbocharger
WLTC	Worldwide harmonized light vehicles test cycle
ZLEV	Zero- and low-emission vehicle

References

1. Serrano, J.R.; Novella, R.; Piqueras, P. Why the development of internal combustion engines is still necessary to fight against global climate change from the perspective of transportation. *Appl. Sci.* **2019**, *9*, 4597. [CrossRef]
2. Road Transport: Reducing CO_2 Emissions from Vehicles. European Commission. 2019. Available online: https://ec.europa.eu/clima/policies/transport/vehicles/cars (accessed on 29 September 2020).
3. Mock, P. *CO_2 Emission Standards for Passenger Cars and Light-Commercial Vehicles in the European Union*; Policy Update; ICCT: Washington, WA, USA 2019.
4. Joshi, A. *Review of Vehicle Engine Efficiency and Emissions*; SAE Technical Paper 2020-01-0352; SAE International: Warrendale, PA, USA, 2020; doi:10.4271/2020-01-0352. [CrossRef]
5. Gohil, D.B.; Pesyridis, A.; Serrano, J.R. Overview of clean automotive thermal propulsion options for India to 2030. *Appl. Sci.* **2020**, *10*, 3604. [CrossRef]
6. Jain, A.; Krishnasamy, A.; Pradeep, V. Computational optimization of reactivity controlled compression ignition combustion to achieve high efficiency and clean combustion. *Int. J. Engine Res.* **2020**. [CrossRef]
7. Claßen, J.; Pischinger, S.; Krysmon, S.; Sterlepper, S.; Dorscheidt, F.; Doucet, M.; Reuber, C.; Görgen, M.; Scharf, J.; Nijs, M.; et al. Statistically supported real driving emission calibration: Using cycle generation to provide vehicle-specific and statistically representative test scenarios for Euro 7. *Int. J. Engine Res.* **2020**. [CrossRef]
8. Di Maio, D.; Beatrice, C.; Fraioli, V.; Napolitano, P.; Golini, S.; Rutigliano, F.G. Modeling of three-way catalyst dynamics for a compressed natural gas engine during lean–rich transitions. *Appl. Sci.* **2019**, *9*, 4610. [CrossRef]
9. Reitz, R.D.; Ogawa, H.; Payri, R.; Fansler, T.; Kokjohn, S.; Moriyoshi, Y.; Agarwal, A.K.; Arcoumanis, D.; Assanis, D.; Bae, C.; et al. IJER editorial: The future of the internal combustion engine. *Int. J. Engine Res.* **2019**, *21*, 3–10. [CrossRef]
10. Angerbauer, M.; Grill, M.; Bargende, M.; Inci, F. Fundamental research on pre-turbo exhaust gas aftertreatment systems. In Proceedings of the 20th Internationales Stuttgarter Symposium, Stellenbosch, South Africa, 17–20 March 2020; Springer Vieweg: Wiesbaden, Germany, 2020.
11. Choi, D.; Kang, Y.S.; Cheong, I.C.; Kang, H.K.; Chung, Y.R.; Lee, J.K. The new Hyundai in-line 4-cylinder 2.2 L diesel engine - Smartstream D2.2 FR. In Proceedings of the 40th International Vienna Motor Symposium, Vienna, Austria, 15–17 May 2019.
12. Schmitz, T. Light duty gasoline catalyst development for Euro 7 type legislation. In Proceedings of the 19th Hyundai-Kia International Powertrain Conference, Daejeon, Korea, 22–23 October 2019.
13. Kawaguchi, A.; Wakisaka, Y.; Nishikawa, N.; Kosaka, H.; Yamashita, H.; Yamashita, C.; Iguma, H.; Fukui, K.; Takada, N.; Tomoda, T. Thermo-swing insulation to reduce heat loss from the combustion chamber wall of a diesel engine. *Int. J. Engine Res.* **2019**, *20*, 805–816; doi:10.1177/1468087419852013. [CrossRef]
14. Luján, J.M.; Serrano, J.R.; Piqueras, P.; Diesel, B. Turbine and exhaust ports thermal insulation impact on the engine efficiency and aftertreatment inlet temperature. *Appl. Energy* **2019**, *240*, 409–423. [CrossRef]
15. Arnau, F.J.; Martín, J.; Pla, B.; Au nón, A. Diesel engine optimization and exhaust thermal management by means of variable valve train strategies. *Int. J. Engine Res.* **2020**. [CrossRef]
16. Maniatis, P.; Wagner, U.; Koch, T. A model-based and experimental approach for the determination of suitable variable valve timings for cold start in partial load operation of a passenger car single-cylinder diesel engine. *Int. J. Engine Res.* **2019**, *20*, 141–154. [CrossRef]
17. Lújan, J.M.; Bermúdez, V.; Piqueras, P.; García-Afonso, O. Experimental assessment of pre-turbo aftertreatment configurations in a single stage turbocharged Diesel engine. Part 1: Steady-state operation. *Energy* **2015**, *80*, 599–613. [CrossRef]
18. Lújan, J.M.; Serrano, J.R.; Piqueras, P.; García-Afonso, O. Experimental assessment of a pre-turbo aftertreatment configuration in a single stage turbocharged Diesel engine. Part 2: Transient operation. *Energy* **2015**, *80*, 614–627. [CrossRef]
19. Serrano, J.R.; Climent, H.; Piqueras, P.; Angiolini, E. Analysis of fluid-dynamic guidelines in diesel particulate filter sizing for fuel consumption reduction in post-turbo and pre-turbo placement. *Appl. Energy* **2014**, *132*, 507–523. [CrossRef]

20. Joergl, V.; Keller, P.; Weber, O.; Mueller-Haas, K.; Konieczny, R. Influence of pre turbo catalyst design on diesel engine performance, emissions and fuel economy. *SAE Int. J. Fuels Lubr.* **2009**, *1*, 82–95. [CrossRef]

21. Serrano, J.R.; Bermúdez, V.; Piqueras, P.; Angiolini, E. On the impact of DPF downsizing and cellular geometry on filtration efficiency in pre-and post-turbine placement. *J. Aerosol Sci.* **2017**, *113*, 20–35. [CrossRef]

22. Christmann, R. Advanced boosting technologies for future emission legislations. In Proceedings of the 19th Hyundai-Kia International Powertrain Conference, Seoul, Korea, 22–23 October 2019.

23. Lindemann, B.; Schönen, M.; Schaub, J.; Robb, L.; Fiebig, M.; Sankhla, H.; Klein, R. White eco diesel powertrain with pre-turbine exhaust aftertreatment and mild-hybrid concept for lowest NOx emission under urban driving conditions. In Proceedings of the 40th International Vienna Motor Symposium, Vienna, Austria, 15–17 May 2019.

24. Amar, P.; Li, J. Volvo SuperTruck 2, pathway to cost-effective commercialized freight efficiency. In Proceedings of the US Department of Energy Annual Merit Review, Washington, DC, USA, 8–10 June 2019.

25. Kröcher, O.; Elsener, M.; Bothien, M.R.; Dölling, W. Pre-turbo SCR - Influence of pressure on NOx reduction. *MTZ Worldw.* **2014**, *75*, 46–51. [CrossRef]

26. Bermúdez, V.; Serrano, J.R.; Piqueras, P.; García-Afonso, Ó. Assessment by means of gas dynamic modelling of a pre-turbo diesel particulate filter configuration in a turbocharged hsdi diesel engine under full-load transient operation. *Proc. Inst. Mech. Eng. Part D J. Automob. Eng.* **2011**, *225*, 1134–1155. [CrossRef]

27. Payri, F.; Serrano, J.R.; Piqueras, P.; García-Afonso, O. *Performance Analysis of a Turbocharged Heavy Duty Diesel Engine with a Pre-Turbo Diesel Particulate Filter Configuration*; SAE Technical Paper 2011-37-0004; SAE International: Warrendale, PA, USA, 2011; doi:10.4271/2011-37-0004. [CrossRef]

28. Serrano, J.R.; Guardiola, C.; Piqueras, P.; Angiolini, E. *Analysis of the Aftertreatment Sizing for Pre-Turbo DPF and DOC Exhaust Line Configurations*; SAE Technical Paper 2014-01-1498; SAE International: Warrendale, PA, USA, 2014; doi:10.4271/2014-01-1498. [CrossRef]

29. Schönen, M.; Schaub, J.; Robb, L.; Fiebig, M.; Reinwald, H.; Lindemann, B. Emissions and fuel consumption potential of a mild-hybrid-diesel-powertrain with a pre-turbine exhaust aftertreatment. In Proceedings of the 28th Aachen Colloquium Automobile and Engine Technology, Aachen, Germany, 7–9 October 2019.

30. Klaewkla, R.; Arend, M.; Hoelderich, W.F. A review of mass transfer controlling the reaction rate in heterogeneous catalytic systems. In *Mass Transfer—Advanced Aspects*; Chapter 29; Nakajima, H., Ed.; Intech Open: London, UK 2011. [CrossRef]

31. Piqueras, P.; García, A.; Monsalve-Serrano, J; Ruiz, M.J. Performance of a diesel oxidation catalyst under diesel-gasoline reactivity controlled compression ignition combustion conditions. *Energy Convers. Manag.* **2019**, *196*, 18–31. [CrossRef]

32. Sampara, C.S.; Bissett, E.J.; Chmielewski, M. Global kinetics for a commercial diesel oxidation catalyst with two exhaust hydrocarbons. *Ind. Eng. Chem. Res.* **2008**, *47*, 311–322. [CrossRef]

Publisher's Note: MDPI stays neutral with regard to jurisdictional claims in published maps and institutional affiliations.

Article

Development of a Computationally Efficient Tabulated Chemistry Solver for Internal Combustion Engine Optimization Using Stochastic Reactor Models

Andrea Matrisciano [1,2,*], Tim Franken [3], Laura Catalina Gonzales Mestre [3], Anders Borg [1] and Fabian Mauss [3]

[1] Lund Combustion Engineering LOGE AB, Scheelevägen 17, 22370 Lund, Sweden; anders.borg@logesoft.com
[2] Department of Mechanics and Maritime Sciences, Division of Combustion and Propulsion Systems, Chalmers University of Technology, Horsalvägen 7a, 41296 Gothenburg, Sweden
[3] Thermodynamics and Thermal Process Engineering, Brandenburg University of Technology Cottbus, Senftenberg, Siemens-Halske-Ring 8, 03046 Cottbus, Germany; Tim.Franken@b-tu.de (T.F.); LauraCatalina.GonzalezMestre@b-tu.de (L.C.G.M.); FMauss@b-tu.de (F.M.)
* Correspondence: andmatr@chalmers.se; Tel.: +46-7-0863-0560

Received: 28 October 2020; Accepted: 8 December 2020; Published: 16 December 2020

Abstract: The use of chemical kinetic mechanisms in computer aided engineering tools for internal combustion engine simulations is of high importance for studying and predicting pollutant formation of conventional and alternative fuels. However, usage of complex reaction schemes is accompanied by high computational cost in 0-D, 1-D and 3-D computational fluid dynamics frameworks. The present work aims to address this challenge and allow broader deployment of detailed chemistry-based simulations, such as in multi-objective engine optimization campaigns. A fast-running tabulated chemistry solver coupled to a 0-D probability density function-based approach for the modelling of compression and spark ignition engine combustion is proposed. A stochastic reactor engine model has been extended with a progress variable-based framework, allowing the use of pre-calculated auto-ignition tables instead of solving the chemical reactions on-the-fly. As a first validation step, the tabulated chemistry-based solver is assessed against the online chemistry solver under constant pressure reactor conditions. Secondly, performance and accuracy targets of the progress variable-based solver are verified using stochastic reactor models under compression and spark ignition engine conditions. Detailed multicomponent mechanisms comprising up to 475 species are employed in both the tabulated and online chemistry simulation campaigns. The proposed progress variable-based solver proved to be in good agreement with the detailed online chemistry one in terms of combustion performance as well as engine-out emission predictions (CO, CO_2, NO and unburned hydrocarbons). Concerning computational performances, the newly proposed solver delivers remarkable speed-ups (up to four orders of magnitude) when compared to the online chemistry simulations. In turn, the new solver allows the stochastic reactor model to be computationally competitive with much lower order modeling approaches (i.e., Vibe-based models). It also makes the stochastic reactor model a feasible computer aided engineering framework of choice for multi-objective engine optimization campaigns.

Keywords: tabulated chemistry; chemical kinetics; 0-D stochastic reactor models

---

1. Introduction

The ever-stringent regulations on carbon dioxide and criteria pollutants (i.e., nitrogen oxides or particulate matter) for internal combustion engine vehicles (ICEV), as well as the original equipment manufacturers (OEM) needing to reduce the technology development times, are among the key drivers of modern computer aided engineering (CAE) for engine development toolchains. To thoroughly study and optimize engine fuel efficiency and reduce pollutant formation, an experimentally driven campaign generally requires the deployment of expensive and highly complex techniques. Moreover, the ever-increasing hardware complexities being introduced in modern powertrains (i.e., pre-chamber or advanced multi-stage aftertreatment) make the experimental engine development process even more challenging. Each new technology introduces a new degree of freedom in the parameter range of a combustion engine development. In this scenario, numerical methods represent an attractive tool to aid the engine development process, more so if they are capable of accounting for both the chemical and physical phenomena occurring in an internal combustion engine-based powertrain. Provided that such numerical methods deliver an acceptable level of accuracy, engine development costs can be substantially reduced by running engine optimization campaigns within a virtual framework. However, modeling of the in-cylinder combustion process poses many challenges, such as (1) Turbulence-Chemistry Interaction (TCI), (2) fuel injection and mixture formation and (3) gaseous pollutants and particulates formation mechanisms. On top of the numerical complexity, computational cost is also among the major decisive factors of whether a certain modeling approach shall be deployed in the development stage or not. The methods need to be fast to deliver information in time during an engine or vehicle development process.

Among the different reactive flow simulation frameworks currently adopted for engine development, 3-D Computational Fluid Dynamics (CFD) allows to model flow, turbulence and combustion chemistry processes with high level of detail. However, depending on the chosen model parameters such as computational grid/time-step size, numerical differencing scheme and the number of species/reactions in the chosen chemical kinetic model, 3-D CFD may require an unfeasible computational cost. This is particularly true when engine and fuel chemistry effects are to be considered across a large set of operating points or during transient operations. In this respect, lower order tool chains (i.e., 0-D and 1-D frameworks) require a small fraction of the computational times compared to advanced 3-D CFD analyses. The price for this benefit is the limited numerical accuracy and frequently a loss of chemical and physical information. The treatment of the combustion chemistry and the turbulence-chemistry interaction effects are among the main aspects to be addressed in order to achieve high accuracy and feasible simulation times. Both these phenomena become particularly important in the development and optimization of novel internal combustion engine concepts which may include, for instance, dual-fuel, highly premixed fuel/oxidizer mixtures or complex exhaust gas recirculation (EGR) strategies.

Numerous 0-D methods have been proposed to describe the Rate of Heat Release (RoHR) and the turbulence for both Spark Ignition (SI) and Compression Ignition (CI) engines to with different level of complexity [1–6].

In addition to the turbulence/burn rate interaction, the computational treatment of in-homogeneities in the combustion chamber can strongly affect the predictive capability of a 0-D model especially under Diesel engine conditions. The most common approach is to discretize the trapped mass into several computational zones, which vary depending on the number of physical regions included in the model formulation (i.e., flame front, cylinder wall area, crevice [7–11]). While such models present a remarkable advantage against single or two zone models, a mean temperature and gas composition within each zone has still to be imposed by definition. This implies that the calculation of the chemical source terms is done assuming negligible variations in enthalpy and composition spaces within each zone and hence no TCI effects are considered. These simplifications, together with the lack of detailed chemistry sub-models, impact the quality of engine-out emission predictions.

An alternative method to consider turbulent mixing as well as detailed chemistry in 0-D is the Stochastic Reactor Model (SRM). The SRM discretizes the mixture in the combustion chamber in a given number of notional particles under the assumption of statistical homogeneity, as opposed to special homogeneity in multi-zone models. Each notional particle features a realization of possible temperature and mixture compositions. TCI is mimicked by stochastic mixing of particles, stochastic heat exchange with the walls and detailed chemistry evaluations. However, as it is the case for detailed chemistry-based 3-D CFD methods, depending on the size of the chemical mechanism considered, the chemistry step may take up to 99% of the simulation time. In addition to mechanism reduction techniques [12], chemistry storage and run-time retrieval methods are viable solutions to reduce computational costs.

Different tabulated chemistry-based methods have been proposed primarily for 3-D CFD applications. These methodologies are based on the decoupling of flow and chemistry. While the flow is computed during run-time, the chemistry solution, usually intended as auto-ignition and/or emission formation processes, is computed in a pre-processing step typically performed on a one-time basis for a given fuel. The two major combustion modeling concepts used to decouple flow and chemistry are various flamelet methods, including Conditional Moment Closure (CMC) and the Well-Stirred Reactor (WSR) methods. Remarkable efforts have been made towards formulation of predictive flamelet-based tabulated chemistry solvers for auto-ignition [13–17] as well as advanced soot and NO_x emission formation modeling [18–21]. The present article is based on the WSR approach. This is the best choice for Probability Density Function (PDF) methods, such as the herein employed SRM. These methods use operator splitting loops to separate vaporization, mixing, compression, heat transfer and chemistry. In these processes, each particle is considered as a well stirred reactor. The chemistry storage is constructed by means of 0-D adiabatic constant pressure/volume reactors across wide ranges of initial pressure, temperature and equivalence ratio. Pires da Cruz et al. [22] proposed a method where both the high and low temperature ignition delay times are stored in the look-up table. The validation was done under 0-D adiabatic reactor conditions (assuming constant pressure/volume) as well as under diesel engine conditions in 3-D CFD. An improved version of such model was later proposed by Colin et al. [23] where a progress-variable-based approach was used rather than a tabulation of the ignition delays. In this configuration, an additional tabulation dimension is introduced as the tabulation is done across a predefined set of grid points, varying between unburned and fully burned mixture. During run-time, the progress variable source term was retrieved for each cell, and it was used to reconstruct the chemical state. Later improvements of such method proposed by Knop and co-authors [24] also incorporated a turbulence-chemistry interaction term during the tabulation process. Thanks to an additional tabulation dimension, TCI effects could be accounted for in 3-D CFD reactive flow simulations within the Extended Coherent Flame Model (ECFM) framework. The model (referred as ECFM-TKI in [24]) has been applied to predict the ignition process of Diesel and homogeneous charge compression ignition (HCCI) engines and is currently implemented in various commercial engine CFD software.

With respect to modeling of spray flames and Diesel engines, numerous models, such as the Partially Stirred Reactor (PaSR) concept [25], have been formulated and implemented in OpenFOAM [26,27].

When it comes 0-D/1-D frameworks, the number of studies featuring tabulated chemistry-based methods is rather limited. Its implementation, however, is potentially very useful as it allows to include detailed chemistry effects, as opposed to the commonly used empirically derived correlations for emission predictions. Leicher et al. [28] proposed a table look-up approach based on mixture fraction and reaction entropy as progress variable. Their methodology was implemented in an SRM and tested under constant pressure reactor conditions. Bernard and co-authors [10] simulated heat release and pollutant formation by means of a Flame Prolongation of Intrinsic Low Dimensional Manifold (FPI/ILDM initially proposed in [29]) as well as a timescale-based sub-model for NO formation. The tabulation method used constant volume reactors and a CO-CO_2-based progress variable definition. Upon table generation, CO, CO_2, H, H_2 and the fuel molecule were stored as key species for the reconstruction of the thermodynamic state during engine simulation. The method was validated for a wide range of Diesel engine conditions. Within the SI engine simulation framework, Bougrine et al. [30] proposed a two-zone 0-D model (referenced as One-Dimensional Coherent Flame Model-Tabulated Chemistry (CFM1D-TC)) where the chemical part of the combustion process was tabulated using laminar 1-D premixed flame solutions. In addition, a time-scale model was formulated to better represent the relaxation towards equilibrium of CO and NO species with the help of homogeneous reactor calculations. Bozza et al. [31] implemented the previously mentioned ECFM-TKI [24] approach within a 0-D/1-D phenomenological combustion model for better knock prediction in spark ignited engines compared to the traditionally used Livengood-Wu [32] approach. Validation under both 0-D reactors and knocking SI engine simulations under stoichiometric conditions showed promising results when compared to the online chemistry solutions.

Motivation

The present work aims to (1) introduce and validate an improved version of the tabulated chemistry solver initially presented in [33]; (2) assess the solvers' combustion and emissions predictions under Diesel and gasoline engine conditions within the 0-D SRM framework; (3) demonstrate the computational efficiency of the method, which allows large design optimization studies while taking detailed chemistry effects into account. With respect to the analyses discussed in [33], the present work includes numerous performance and accuracy refinements on both the tabulation and the engine simulation codes, phenomenological formulations of the turbulence models for CI and SI engines, a novel approach for thermal NO source terms tabulation and a method validation under SI engine conditions in addition to a broader CI engine simulation campaign. To the authors' knowledge, a comprehensive SRM-based tool chain featuring tabulated chemistry, physical turbulence models and computational efficiency comparable to multi-zone models (i.e., [7]) has not been demonstrated before.

The paper is structured as follows: In Section 2, a concise description of the SRM framework is given together with a description of the turbulence models and progress variable-based solver. In Section 3, an accuracy assessment investigation of the refined storage/retrieval strategy under zero-dimensional constant pressure reactors in shown. The validation of the development is demonstrated in Sections 4–6. Online and tabulated chemistry-based solvers are compared using the SRM under Diesel (Section 4) and spark ignited engine (Section 5) conditions. Finally, the accuracy of the introduced solvers is discussed with focus on combustion and emissions predictions along with remarks on computational performances in Section 6. For validation, the following criteria were defined: (1) Is the solver accuracy consistent with the general model accuracy? (2) Is the CPU time of the developed software acceptable for engine development, engine optimization and driving cycle analysis?

2. Simulation Methods

The SRM has been coupled with two different chemistry solvers: (1) online, where chemical source terms are calculated during run-time; (2) Combustion Progress Variable (CPV), which uses a pre-calculated look-up table to retrieve the source terms for combustion as well as thermal NO formation. In the first sub-section, a brief description of the SRM engine modelling framework is presented for better readability. In the following sub-sections, aspects of the different phenomenological turbulence models adopted for the SI and CI simulation campaigns are briefly presented as well as a description of the tabulated chemistry solver.

2.1. The Stochastic Reactor Model

In the SRM computational domain, the in-cylinder trapped mass is discretized as an ensemble of notional particles that can mix with each other and exchange heat with the cylinder walls. A schematic visualization of the concept, along with an exemplary distribution of the particles' property (i.e., enthalpy or gas composition) is shown in Figure 1. Depending on the initial conditions assigned at Intake Valve Closure (IVC), a given chemical composition, temperature, and mass are assigned to each particle, while pressure is assumed to be the same across all particles.

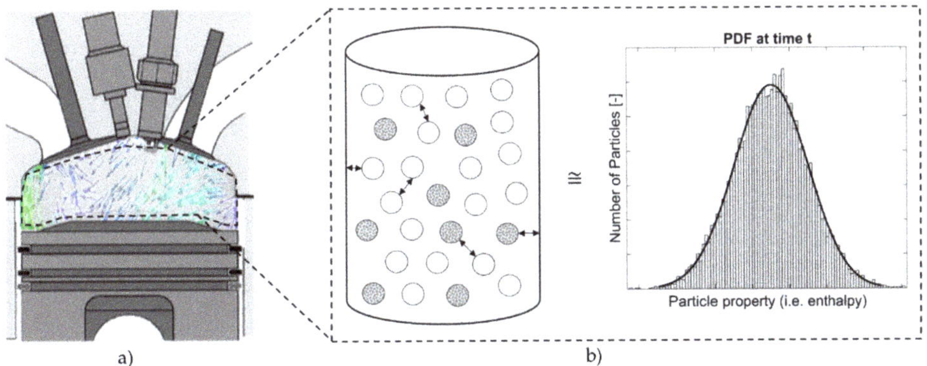

Figure 1. Schematic visualization of the Stochastic Reactor Model (SRM) concept; (**a**) engine physical space; (**b**) 0-D particles in the SRM computational space and an exemplary distribution of particle properties at a given time-step t.

A probability density function is used to describe the in-cylinder content and each particle contributes to the realization of the given PDF at each time-step. Since all stochastic particles in the SRM represent a portion of the in-cylinder mass, a Mass Density Function (MDF) is used to solve the main transport equation of the 0-D SRM. The joint scalar vector of the MDF can be expressed as reported in Equation (1).

$$F_{\Phi}(\psi, t) = F_{\Phi}\left(\psi_1, \ldots, \psi_{N_{Sc}}; t\right) \tag{1}$$

The vector ψ is the realization of the vector of local scalar variables noted herein as Φ, while N_{Sc} is the number of scalars transported in the computational domain. Depending on the chemistry solver used in the simulation framework, the number and type of transported scalars varies significantly.

$$\text{Online chemistry solver } \Phi(t) = \left(Y_1, Y_2, \ldots, Y_{N_{Sp}}, h; t\right) \tag{2}$$

$$\text{Tabulated chemistry solver (CPV) } \Phi(t) = \left(\phi, h, w_q, \mathbf{T_p}, yEGR, h_{298}; t\right) \tag{3}$$

In the online chemistry solver formulation (see Equation (2)), the specific enthalpy (h) as well as the full vector of species mass fractions $\mathbf{Y}$ must be transported in order to correctly compute the chemical reactions during run time. The size of the mass fraction vector is therefore dependent on the number of species (N_{Sp}) defined in the chemical mechanism adopted. With respect to the tabulated chemistry solver, only a reduced set of quantities need to be transported in the computational domain independently on the size of the mechanism used to generate the look-up table. These are the equivalence ratio (ϕ), the specific enthalpy (h), the mean molecular weight (w_q), the vector ($\mathbf{T_p}$) containing the thermodynamic polynomial coefficients, the EGR mass fraction (yEGR) and the latent enthalpy (h_{298}) of the in-cylinder mixture. Once initial conditions are assigned, the transport equation of the joint scalar vector of the MDF is numerically solved using a Monte Carlo particle method (e.g., Pope [34]) with the operator splitting technique as previously presented by Maigaard et al. [35]. A series of sub-models are employed to sequentially solve the different physical and chemical processes occurring in the combustion chamber, as summarized in Equation (4):

$$\frac{\partial F_{\Phi}}{\partial t} = \frac{\partial F_{\Phi}}{\partial t}\bigg|_{\Delta V} + \frac{\partial F_{\Phi}}{\partial t}\bigg|_{inj} + \frac{\partial F_{\Phi}}{\partial t}\bigg|_{FP} + \frac{\partial F_{\Phi}}{\partial t}\bigg|_{mix} \frac{\partial F_{\Phi}}{\partial t}\bigg|_{chem} + \frac{\partial F_{\Phi}}{\partial t}\bigg|_{ht} \tag{4}$$

In the equation above, the subscript ΔV represents the piston movement, inj fuel injection, FP flame propagation (considered only for SI engine simulations), mix the turbulence and particle interaction process, chem the chemical reactions and ht heat transfer to the walls. Detailed descriptions of the sub-models used for piston movement, pressure correction, online chemistry solver and fuel injection can be found in the work from Tuner [36]. The treatment of the flame propagation has been previously introduced by Bjerkborn et al. [37] and broadly validated by Pasternak et al. [38] and Netzer [39]. The Woschni model [40] is used to evaluate total wall heat transfer, while the distribution of the heat transfer over the SRM particles is calculated using a stochastic approach, explained by Tuner [36] and further validated by Franken et al. [41] using Direct Numerical Simulation (DNS) results from Schmitt et al. [42]. A short overview of the turbulence models adopted in the present work is given in the following sub-section.

2.2. Phenomenological Turbulence Models and Particle Interaction

The expression of the mixing term in Equation (4) is presented below:

$$\left.\frac{\partial F_{\Phi}}{\partial t}\right|_{mix} = \frac{C_{\Phi}\beta}{\tau}\left[\int_{\Delta\psi} F_{\Phi}(\psi - \Delta\psi, t)F_{\Phi}(\psi + \Delta\psi, t)d(\Delta\psi) - F_{\Phi}(\psi, t)\right] \tag{5}$$

C_{Φ} and β are two model parameters that in the present study have been set to 2.0 and 1.0, respectively. The mixing time history, τ in Equation (5), is the main input parameter of the SRM. The mixing time is needed to model the turbulent mixing occurring in the combustion chamber due to various phenomena such as: fuel injection, swirl motion, tumble motion, etc. The shape and magnitude of the mixing time profile controls how intense the SRM particle mixing process is. Since τ influences mixture inhomogeneities in both composition and enthalpy spaces, a strong influence on the auto-ignition process, the local rates of heat release and pollutant formation are seen when the mixing time is changed.

A simple approach to construct the mixing time history is by using a set of empirical constants as done by Pasternak et al. [43,44] for Diesel combustion studies. A more comprehensive approach is to calculate τ during run-time by employing a k-ε or K-k based turbulence model. Depending on whether Diesel or gasoline combustion are considered, different approaches have been implemented in the present work. For Diesel combustion, the approach initially proposed by Kožuch [45] was adopted and validated by Franken et al. [46] across a large set of operating points. For gasoline engine conditions, the K-k model proposed by Dulbecco et al. [47] was implemented and validated in the SRM framework for different gasoline fuel surrogate mixtures in [48,49]. Examples of the mixing time histories that are computed with the mentioned turbulence models are shown in Figure 2 as function of crank angle degrees (CAD) assuming 0 as firing top dead center (TDCf).

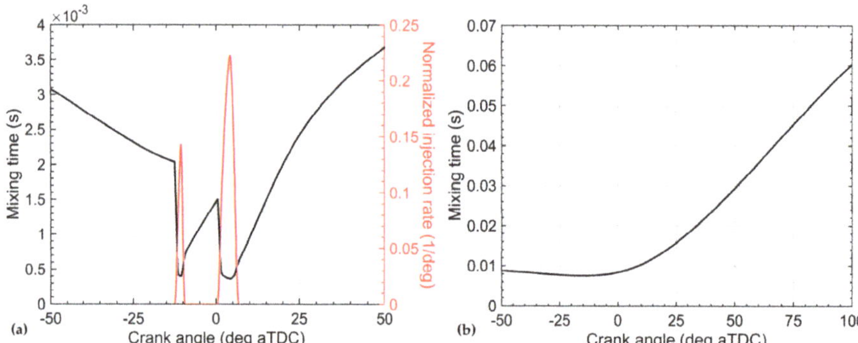

Figure 2. Exemplary mixing time histories computed using turbulence models for (**a**) Diesel engine conditions [47] and (**b**) gasoline engine conditions [49].

Once the mixing time history has been computed, an additional sub-model is employed to decide which and how many particles are selected in each mixing step. In this work the Curl model [50] has been used to describe the particle interaction for both the Diesel and the gasoline engine simulations.

2.3. Combustion Progress Variable Model

The CPV model relies on the pre-tabulation of the auto-ignition and emission formation processes for a broad range of initial conditions in terms of exhaust gas recirculation rates (yEGR), pressure (p), unburned temperature (T_u) and equivalence ratio (ϕ). The reconstruction of the thermo-chemical state during run-time is then performed by means of an appropriately chosen progress variable (herein noted as C). As discussed in detail by Niu et al. [51] and Ihme et al. [52], any progress variable-based method needs to fulfill the following conditions to be mathematically sound: (1) strict monotonicity with time; (2) non-negativity and normalization (C = 0 for unreacted mixture and C = 1 for fully burned mixture); (3) if reactive scalars (i.e., chemical species) are used in the definition of the progress variable, said scalars should evolve on comparable time scales [52].

Several progress variable definitions have been proposed by different groups for both homogeneous and flamelet-based tabulation frameworks [10,13,21]. The most common approach consists in formulating the progress variable via a combination of chemical species (typically including the surrogate fuel molecule, O_2, CO, CO_2, H_2O, OH, CH_2O and others). The choice of the species to use and their weighting factors within the progress variable definition is usually optimized so that both low and high temperature combustion regimes are captured [51,52]. In the present work, latent enthalpy (enthalpy of formation at standard state, h_{298}) is used to define the reaction progress variable C as reported below in a normalized fashion.

$$C(t) = \frac{h_{298}(t) - h_{298,u}}{h_{298,max} - h_{298,u}} \tag{6}$$

In Equation (6), h_{298} is the latent enthalpy calculated at 298 K, and summed over all species, subscript u denotes unburned state, and max denotes the most reacted state, which is assumed to be where the maximum of the accumulated chemical heat is released. As discussed in [15,33], this progress variable can be used to track both low and high temperature reactions, which is important when tabulation methods are developed for fuels exhibiting cool flames. Figure 3 shows an exemplary h_{298} profile together with temperature for a constant pressure reactor calculation.

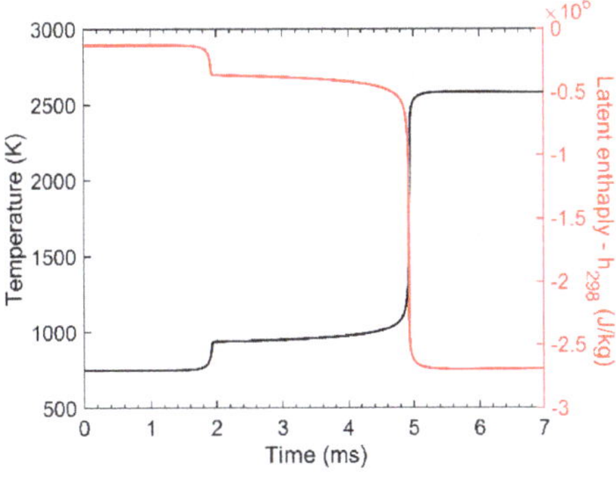

Figure 3. Temperature (black) and latent enthalpy at 298 K (red) as function of time for a constant pressure reactor calculation at 10 bar and 750 K for stochiometric *n*-heptane/air mixture.

It can be discussed that in a constant pressure reactor, progress variables based on temperature (i.e., as proposed by Knop et al. [24]) or on latent enthalpy will be equal. However, unlike temperature the latent enthalpy will not be influenced by pressure changes, fuel vaporization or mixing. As h_{298}

is a conserved scalar, a transport equation, as well as a spray source term, can be easily formulated. For the majority of the tabulated conditions, the maximum of the accumulated chemical heat release coincides with the chemical equilibrium state. However, for fuel-rich conditions as well as for states that lead to pronounced endothermic reactions (i.e., high pressure or high EGR conditions), the mixture relaxation towards equilibrium is not taken into account by Equation (6). Nevertheless, this definition was considered acceptable for engine applications in the SRM since it affects a small fraction of the particles for a limited portion of the cycle in direct injected-engine simulations. Over the whole range of computed conditions, this definition was found to be convenient to correctly capture both low and high temperature combustion regimes when generating the table through adiabatic homogeneous constant pressure reactor calculations. To keep the table size within a feasible range for engine applications, storage of the progress variable source terms was done over 25 points for each reactor simulation.

As introduced in [33], the coupling of the CPV solver with the SRM required substantial changes to the pre-existing online chemistry-based code. As discussed in the previous section, the tabulated chemistry solver requires only a reduced set of scalars as opposed to the full vector of species mass fraction needed for the online solver. It is important to notice that the normalized reaction progress, as reported in (6), is not transport directly. The latent enthalpy (h_{298}) is transported instead and the progress variable is calculated, whenever a table look-up is needed. This practice facilitates the formulation of the fuel injection model and the resulting evaporation source term and the particle–particle interaction sub-model. Beside the treatment of the chemistry step, the same set of sub-models noted in Equation (4) and their relative constants were applied for both the tabulated chemistry runs the online chemistry simulations. Hence potential differences between the two solver solutions are to be interpreted as due to the chemistry step only.

3. Look-Up Table Generation and Testing

Prior to the engine simulation campaigns, a grid density investigation together with an assessment of the interpolation error was performed so that the CPV model could be verified. In the following sub-sections, the details of the table construction process as well as the outcome of the interpolation error analysis are discussed.

3.1. Look-Up Table Generation

Detailed chemistry simulations under adiabatic constant pressure reactor conditions were performed using the software LOGEtable [53]. Initial mixture unburned temperature (T_u), pressure (p), equivalence ratio (ϕ) and mass fraction EGR (yEGR) are the input variables of the look-up table generator. Table grid points were chosen so that the typical range of thermodynamic conditions encountered in conventional direct injected engines that can be expected to be found in conventional direct injected engines. The table grid adopted in this work relies consists of 205,200 points which are distributed across tabulation parameters as outlined in Table 1.

Table 1. Tabulation grid adopted for all tabulated chemistry runs.

Parameter	Lower Bound	Upper Bound	Number of Points
Unburned Temperature (K)	300	1400	76
Pressure (bar)	1.0	200	18
Equivalence Ratio (-)	0.05	6.0	30
EGR mass fraction (-)	0.0	0.4	5

Except for the EGR mass fraction space, the points are distributed in a non-equidistant manner so that typical engine relevant conditions (i.e., around stoichiometry in equivalence ratio space) are better resolved. The EGR stream is assumed to have a fixed composition comprising combustion products (CO_2, H_2O and N_2) of the given fuel mixture at stoichiometry.

For each of the table grid points, the progress variable C introduced in Equation (6) is used to trace the reaction trajectory, and its source terms (dC/dt) are stored in the table at 25 different points between unburned C = 0 and fully burned C = 1.0 state. In addition, mean molecular weight, the thermodynamic polynomial coefficients and any chemical species that the user decides to monitor are also stored across the mentioned progress variable grid. In the present work, the following species have been included in the tabulation process so that major engine-out emissions and standard engine performance analyses could be done: fuel molecule, i.e., n-C_7H_{16}, O_2, N_2, CO_2, H_2O, CO, H_2, C_2H_2, C_2H_4 and unburned hydrocarbons (uHC) (defined as the sum of all species containing an H and C or H, C and O atoms). The resulting size on disk was approximately 1 GB.

3.2. Table Density/Accuracy Investigation

Before applying the CPV solver to engine simulations in the SRM, the interpolation strategy for the progress variable source term retrieval was verified. For each of the data entry points included in the tabulation grid (see Table 1), the auto-ignition solution from the constant pressure reactor calculation using the online chemistry solver has been compared to the respective solution retrieved from the table using linear interpolation. The discrepancy between the two solution has been quantified based on the mean relative difference of the vector composed by the errors at 5, 10, 50 and 90 percent of energy released location (noted as $E_{5\%}$, $E_{10\%}$, $E_{50\%}$ and $E_{90\%}$ in Figure 4) at all given reactor points.

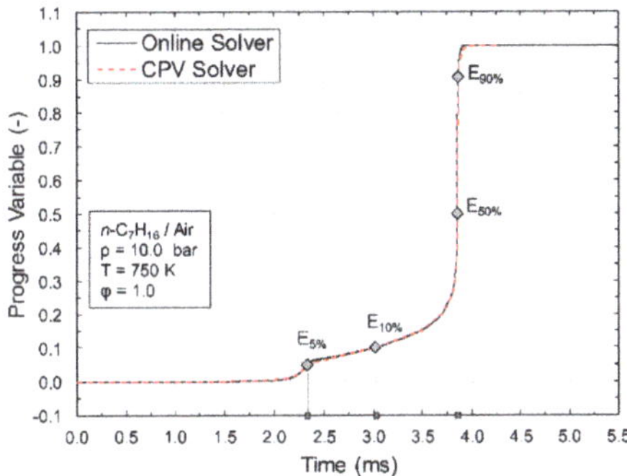

Figure 4. Online and tabulated chemistry solutions for the progress variable (based on Equation (6)) evolution of a stoichiometric *n*-heptane/air mixture at 10 bar and 750 K.

The global interpolation error has been then defined based on the mean square difference (MSD) of the location of $E_{5\%}$, $E_{10\%}$, $E_{50\%}$ and $E_{90\%}$ points. In Equations (7), where i = 5, 10, 50, 90, the exact definition of the interpolation error is reported.

$$\varepsilon_i = \frac{E_{i,\text{Tabulated}} - E_{i,\text{Online}}}{E_{i,\text{Online}}} \times 100 \quad \bar{\varepsilon} = \sum_{i=1}^{N_{\text{errors}}} |\varepsilon_i| / N_{\text{errors}} \quad \varepsilon_{\text{MSD}} = \sqrt{\sum_i (\bar{\varepsilon} - |\varepsilon_i|)^2} \tag{7}$$

The interpolation errors have been assessed for a large set of operating conditions covering lean to rich (φ = 0.6–3.0) conditions as well as from low to high unburned temperatures (T = 600–1400 K) under constant pressure reactor conditions. *n*-Heptane was used as fuel molecule, and the mechanism developed by Zeuch et al. [54] was employed. The main goal of this campaign was to quantify the accuracy of the interpolation routine under the simplest reactor conditions, so to have a higher confidence in results assessment during the subsequent engine simulation campaign. In the SRM simulation, numerous sub-models are coming into play, and hence, error compensation effects may arise. In Figure 5, exemplary results of the error quantification campaign are displayed in pixel plot format for conditions at 1 and 35 bar pressure.

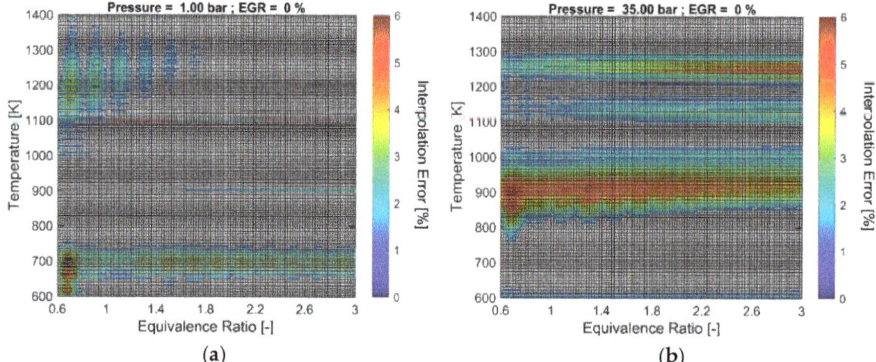

Figure 5. Error maps for the interpolation error (ε_{MSD} as defined in Equation (7)) between online and tabulated chemistry solver under constant pressure reactor conditions (**a**) at 1 bar and (**b**) 35 bar, both at 0% exhaust gas recirculation using *n*-heptane as fuel.

For a more effective error visualization as well as to better highlight the exact areas where interpolation errors may arise, pixel contours have been conditionally formatted so that every case having an error below 0.5% is displayed in white. All cases presenting an error above 0.5% will have contours consistent with the RGB color bar instead. For the whole range of computed conditions, the solution retrieved from the table did not exceed a 6% discrepancy on a single point basis, while the overall average error is approximately 1.5%. As expected, interpolation errors are more visible in the Negative Temperature Coefficient (NTC) regimes, which is the most challenging area to parametrize when progress variable methods are concerned [24,55]. While a tighter tabulation grid in temperature space could have delivered even lower errors, the authors considered this level of accuracy to be acceptable. This decision was based on the globally low errors but also on method usability considerations.

A tighter table grid results in a larger file size on disk, which in turn affects the random-access memory (RAM) requirement for the engine simulations during run-time. Hence, the present configuration was considered to be a good trade-off.

4. Compression Ignition Engine Simulations

In this section, results of a heavy-duty Diesel engine simulation campaign are presented and discussed. In the first sub-section, experimental data and simulation setups are presented. Secondly, a result comparison between experimental data and the two chemistry solvers is shown and discussed with respect to engine performance parameters and engine-out emissions. All the simulation results presented in this section were obtained using the commercial software LOGEengine version 3.2.1 [53].

4.1. Engine Data and Simulation Setup

The simulation setups were constructed based on experimental data from a heavy-duty (HD) 13.0 L Diesel engine that features a direct injection system capable of injection pressures up to 2000bar. Although the engine has an in-built external EGR system, the analyzed engine conditions did not include external EGR. A total of 10 operating points from 1000 rpm to 1700 rpm and 6 bar to 22 bar indicated mean effective pressure (IMEP) are used for the present simulation campaign. Details of the operating conditions are outlined in Table 2.

Table 2. Heavy-duty engine operating conditions.

Case Name	Speed (rpm)	IMEP (bar)	EGR (mass%)	Injection Pulses (#)
HD01	1700	19.0	4.0	1
HD02	1300	22.0	4.0	1
HD03	1300	14.5	4.0	1
HD04	1300	6.0	4.0	1
HD05	1200	6.0	4.0	1
HD06	1200	14.5	4.0	1
HD07	1200	22.0	4.0	1
HD08	1000	6.0	4.0	1
HD09	1000	11.0	4.0	1
HD10	1000	22.0	4.0	2

As presented in Figure 6, operating point 10 features a double injection rate profile (pilot + main) while the rest have a single injection event. The crank-angle resolved pressure profiles were measured for one cylinder and used to calibrate the SRM engine model.

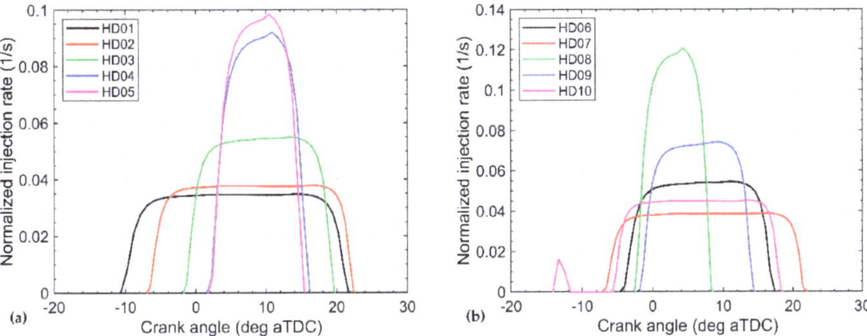

Figure 6. Fuel injection rates as a function of crank angle degree (assuming 0 as firing TDC) for operating conditions (**a**) from 1–5 and (**b**) 6–10, as listed in Table 2.

Commercial Diesel fuel was used during experiments, whereas in the simulations a blend of *n*-decane and α-methylnaphthalene (75–25% mass fractions respectively, also noted as IDEA surrogate fuel) was employed as surrogate fuel model. A liquid fuel properties comparison is summarized in Table 3.

Table 3. Liquid properties of the experimental and surrogate fuel mixture used in the heavy-duty Diesel engine simulation campaign.

Fuel Name	Lower Heating Value (MJ/kg)	Density at 15 °C (kg/m^3)	Cetane Number (-)	C (w%)	H (w%)	O (w%)
EU Diesel	41.6	820.0	49	86.0	13.4	0.6
IDEA surrogate fuel	42.94	783.0	52.3	86.62	13.38	0.0

The chemical kinetic scheme employed in this study has been taken from the LOGEfuel database [53]. The mechanism is an improved version of the detailed model from Wang [56]. It features oxidation models for *n*-decane, α-methyl-naphthalene and methyl decanoate as main fuel species as well as a detailed PAH growth mechanism [57] and thermal NO_x model. The detailed reaction scheme was reduced to a size of 189 species using the method described in [12].

The measured pressure history was analyzed using the thermodynamic analysis of LOGEengine [53]. This procedure provided chemical kinetics-based estimations of wall temperature, in-cylinder temperature at IVC, internal EGR fraction and the apparent rate of heat release (RoHR). In Tables 4 and 5, the main SRM model parameters and the calibrated k-ε model constants are presented, respectively [45].

Table 4. SRM main model settings for the heavy-duty Diesel engine simulation campaign.

Parameter	Value
Number of particles (-)	500
Simulation time-step (CAD)	0.5
Number of consecutive cycles (-)	30
Woschni constant C1	2.28
Woschni constant C2	0.0035
Stochastic heat transfer constant (-)	15

Table 5. Calibrated constants for the k-ε turbulence model.

C_{squish}	$C_{injection}$	C_{swirl}	$C_{dissipation}$	C_{tau}
1.0	11.0	136.0	9.1	0.35

The SRM model calibration for the presented operating conditions was carried out using the procedure described in [58], and the Curl [50] particle interaction sub-model was used. To ensure consistency during chemistry solver comparisons, the same set of model parameters and constants were applied to both the online and tabulated chemistry solver runs without any re-calibration.

4.2. Simulation Results

In Figure 7, comparisons of experimental and simulated in-cylinder pressure histories, rate of heat release, combustion phasing parameters and normalized emissions are presented for a low-load (HD04), a mid-load (HD06) and a high-load (HD07) operating points. The quantities noted in the following figures as CA05, CA10, CA50 and CA90 represent the crank angle location at which 5, 10, 50 and 90 per cent of the total heat has been released, respectively.

Detailed comparisons by means of pressure, RoHR and crank angle resolved emissions for all the operating conditions listed in Table 2 can be found in Appendix A. In Figure 8, comparisons of the CA50, peak cylinder pressure location in CAD (PCP_{CAD}), as well as CO, CO_2, unburned hydrocarbons (uHC) and NO at EVO are shown for all operating points. To comply with data confidentiality restrictions, all the results shown in this simulation campaign are presented in a normalized fashion. With respect to engine out emissions, different normalization strategies have been applied to ensure meaningfulness of the shown comparisons. More in detail, for CO_2 and NO, the simulated ppm values have been normalized with respect to the experimental measurements. For CO and uHC instead, the normalization has been computed based on the difference in ppm between simulated and experimental data with a threshold value set to approximately 100 ppm. In other words, if simulated uHC or CO presents a normalized factor of 2.0, it means that the absolute difference between experimental and simulated values is approximately 200 ppm.

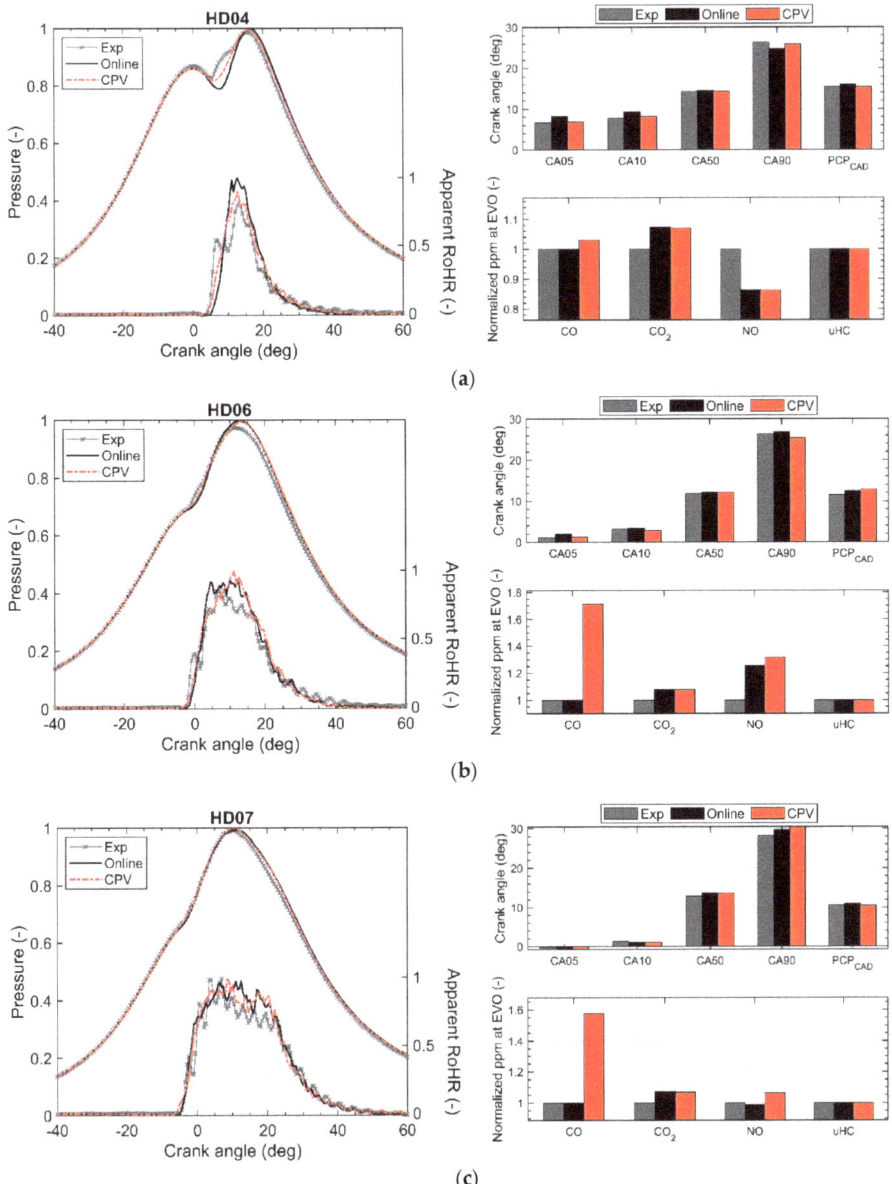

Figure 7. In-cylinder pressure history and apparent rate of heat release (left), combustion phasing parameters (top right) and normalized engine-out emissions (bottom right) comparisons between experimental data and SRM simulations for operating points (**a**) HD04, (**b**) HD06 and (**c**) HD07.

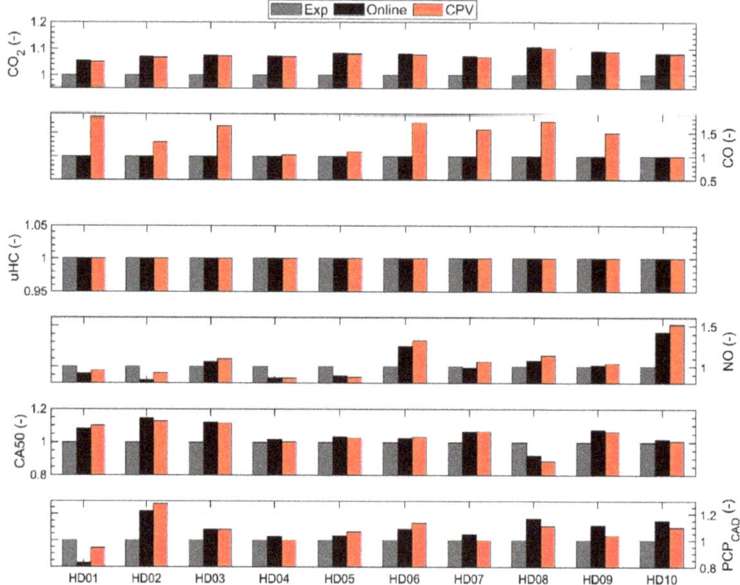

Figure 8. Experimental and simulated engine-out emissions as well as performance parameters ($CA50$ and PCP_{CAD} location) for all operating conditions of the heavy-duty engine simulation campaign. Data have been normalized with respect to experimental values.

On the other hand, if simulated and experimental HC or CO differ by less than 100 ppm, then the factor is set to 1.0, so to underline an acceptable agreement. Such formulation was considered necessary to cope with the fact that the measured CO and uHC are in absolute terms very low. Hence a standard normalization would have resulted in a set of misleadingly high factors for CO and uHC from the engineering standpoint. For uHC, in particular, the difference between experimental and simulated engine-out ppm values never exceeded 30 ppm across all the operating points, and therefore, the set of comparison factors for uHC in Figure 8 is homogeneously set to 1.0.

By means of combustion phasing, the SRM results are in good agreement with experimental data for the majority of the analyzed operating conditions. Visible discrepancies can be observed by means of peak cylinder pressure predictions in HD08 and HD09. However, such result was considered acceptable, considering that on the experimental side, these operating points showed a strong cycle-to-cycle variability, as it can be noticed via the large fluctuations of the heat release rate between 0 and 40 crank angle degrees after TDC (see Figures A8 and A9 in Appendix A).

As for the online versus CPV simulation results, both simulations resulted to be in close agreement with each other across the whole range of simulated data. The tabulated chemistry solver predicted a combustion phasing within less than 0.5 CAD difference with the detailed online chemistry solver at the mid and high loads. For the low load points (HD04 and HD05), a slightly more noticeable difference (≈ 2.0 CAD) between online and tabulated chemistry solutions can be seen when comparing the predicted start of combustion (See Figures A4 and A5 in Appendix A). At low loads, combustion initiates while the mixture is the NTC region, which, as discussed in Section 3.2, is the most challenging regime for progress-variable-based models. It is therefore likely to happen that under these conditions the interpolation error starts to play a visible role. Nevertheless, a 2 CAD discrepancy in start of combustion is well within a range typically considered acceptable for engine performance studies and considering accuracy of the sensors used during the experimental campaigns.

With respect to engine-out emissions, both solvers showed good agreement with experimental data for CO_2, uHC and NO. Differences in CO_2 are explained by differences in the C/H ratio of the real fuel blend and the surrogated fuel blend. These differences also influence the CO emission calculation. With respect to NO, the differences between tabulated and detailed chemistry are lower than differences between the SRM predictions and the engine measurements. For these species, it can be stated that the accuracy of the tabulated chemistry solver is not influencing the accuracy of the tool chain. This statement is also true for the prediction of CA50 and PCP. For carbon monoxide emissions, the online chemistry solver showed a noticeably closer match (less than 100 ppm difference) with experimental data. While the CO predictions from the CPV solver lie within a more than acceptable range from the engineering point of view, it is important to note that correct tabulation of CO during the expansion phase is another challenging area when progress variable models are concerned. Unlike methods proposed in [23,24], the present method does not account for a time-scale dependent retrieval of the CO emissions from the table. This means that the accuracy of the final CO yield depends on how the close to (or far from) equilibrium the value stored in the table at progress 1 is. In the present study, the presented level of accuracy between online and tabulated chemistry solver-based CO values ($\pm$ 200 ppm) was considered to be acceptable. In future studies, however, a time-scale dependent CO retrieval strategy will be considered.

5. Gasoline Engine Simulations

In this section, results of a spark-ignited engine simulation campaign are shown and discussed. In the first sub-section, a brief description of the experimental data and computational setups are presented. Secondly, engine performance parameters and engine-out emissions are compared between experimental data, online and tabulated chemistry solver. All the simulation results presented in this section were obtained using the commercial software LOGEengine version 3.2.1 [53].

5.1. Engine Data and Simulation Setups

The experimental data were measured on a single cylinder research engine at the TU Berlin [48]. Cylinder bore and stroke are 82.0 and 71.9 mm, respectively, while the compression ratio is 10.75:1. The single cylinder engine is specifically designed for combustion investigations and features both port and direct fuel injection systems. The present engine experiments were conducted using the centrally mounted direct fuel injector. The start of fuel injection is at −270 CAD aTDC. Eight fired operating points were selected and are summarized in Table 6. For each operating condition, in-cylinder as well as manifold pressure were recorded for 250 consecutive cycles by means of a low- and high-pressure sensors. More details on the experimental setup and measuring equipment used can be found in Kauf et al. [59].

Table 6. Spark-ignition engine operating conditions.

Case Name	Speed (rpm)	IMEP (bar)	EGR (mass%)	Spark timing (CAD aTDC)
SI01	1500	15.0	1.7	−1.5
SI02	2000	5.0	9.1	−4.0
SI03	2000	10.0	4.9	−6.0
SI04	2000	15.0	2.0	−3.0
SI05	2000	20.0	1.1	2.0
SI06	2500	5.0	9.4	−9.0
SI07	2500	10.0	4.9	−5.0
SI08	2500	15.0	2.0	−5.0

The fuel used during the experimental campaign was a RON95 E10 commercial gasoline, at 150 bar injection pressure. A four-component mixture comprising mole percentages of 31.9% *iso*-octane, 11.4% *n*-heptane 35.6% toluene and 20.8% Ethanol was used in the simulation campaign instead. Comparisons of the major fuel properties are listed in Table 7. The adopted reaction mechanism is based on the detailed ETRF scheme developed by Seidel consisting of 475 species and 5160 reactions. The detailed reaction scheme was validated for different experiments and engine relevant conditions for both auto-ignition and laminar flame speed in several previous works [39,60].

Table 7. Liquid properties of the experimental and surrogate fuel mixture used in the SI engine simulation campaign.

Fuel Name	Lower Heating Value (MJ/kg)	Density at 15 °C (kg/m^3)	RON/MON (-)	C (n)	H (n)	O (n)
E10 Gasoline	41.78	748.7	96.7/85.8	6.6	12.8	0.21
ETRF mixture	41.14	756.4.0	96.7/87.4	6.3	11.8	0.21

The SRM model calibration for the presented operating conditions was carried out using the procedure described in [61]. In Tables 8 and 9, the main SRM model parameters and the calibrated K-k model constants [47] are presented, respectively. As done for the compression ignition engine campaign, the same set of model parameters and constants were applied to both the online and tabulated chemistry solver runs without any re-calibration.

Table 8. SRM main model settings for the spark-ignition engine simulation campaign.

Parameter	Value
Number of particles (-)	500
Simulation time-step (CAD)	0.5
Number of consecutive cycles (-)	30
Woschni constant C1	2.28
Woschni constant C2	0.0035
Stochastic heat transfer constant (-)	15

Table 9. Calibrated constants for the K-k turbulence model.

$C_{injection}$	$C_{compression}$	$C_{dissipation}$	C_{TKE}	C_{length}	C_{β}	$C_{\Delta,1}$	$C_{\Delta,2}$	C_{tau}
0.005	0.67	1.0	0.85	0.30	0.25	0.073	0.1313	4.3

5.2. Simulation Results

In Figure 9, comparisons of experimental and simulated in-cylinder pressure histories, rate of heat release, combustion phasing parameters and normalized emissions are presented for a low-speed mid-load (SI01), a mid-speed high-load (SI05) and a high-speed low-load (SI06) operating points. Extended comparisons by means of pressure, RoHR and crank angle resolved emissions for all the operating conditions listed in Table 6 can be found in Appendix B. Overall, the SRM simulation results show a close match with experiment by means of in-cylinder pressure for different operating conditions. Slight deviations can be seen for the operating points SI05 (2000 rpm and 20 bar IMEP) and SI06 (2500 rpm and 5 bar IMEP) in Figure 9b,c by means of start of combustion and peak cylinder pressure. However, the overall agreement is considered to be acceptable, especially considering the typical cycle-to-cycle variability. Compared to the Diesel engine campaign, a much closer match between online and CPV solver can be seen in the SI cases.

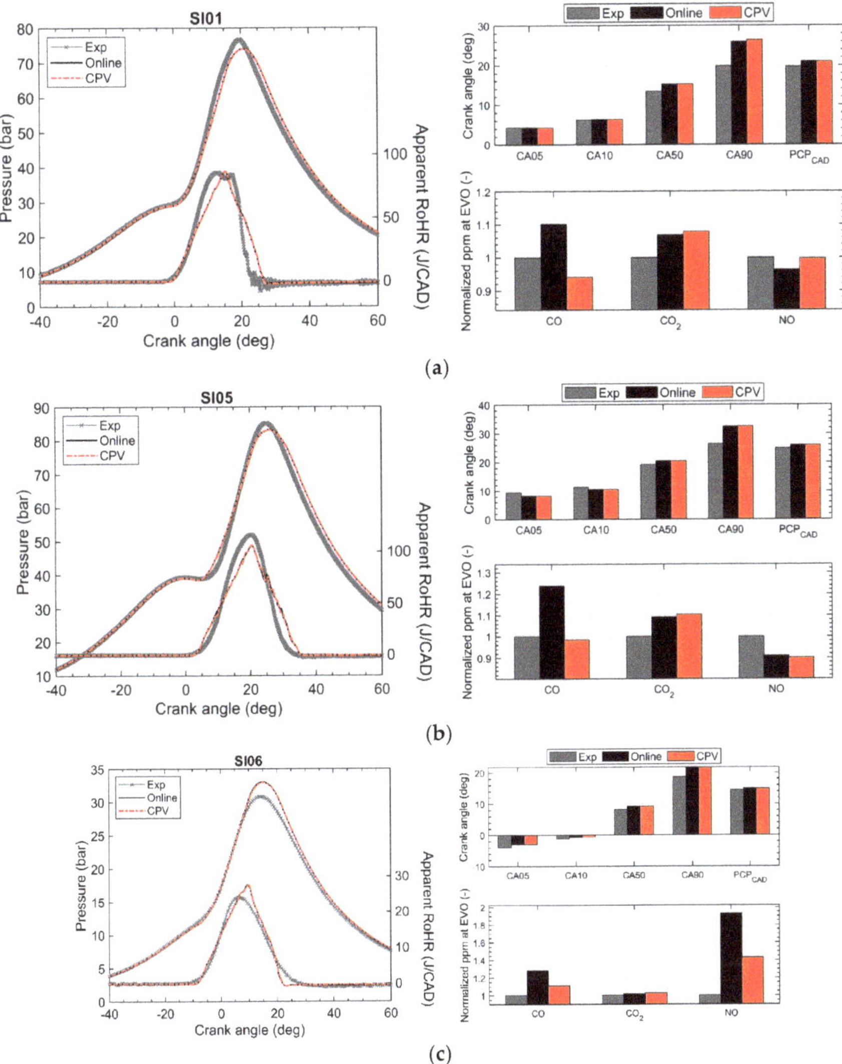

Figure 9. In-cylinder pressure history and apparent rate of heat release (left), combustion phasing parameters (top right) and normalized engine-out emissions (bottom right) comparisons between experimental data and SRM simulations for operating points SI01 (**a**), SI05 (**b**) and SI06 (**c**).

This is explained by the fact that in SI mode, the dominant phenomenon is the flame propagation rather than mixing controlled combustion, where particles reach fully burned state (C = 1) much faster and are moved to the burned zone. In addition, given the early start of injection, the mixture is assumed to be homogeneous and close to stoichiometry. The homogeneity in lambda, together with the particles quickly reaching C = 1, makes the interpolation particularly accurate. Stochastic effects are still present due to the SRM treatment of the heat transfer; however, hardly any difference can be seen in terms of pressure and rate of heat release histories as well as in terms of combustion phasing parameters and peak cylinder pressure location.

In Figure 10, engine-out emissions and major combustion phasing parameters are summarized for all the investigated operating conditions. All values are normalized to experimental data. By means of engine-out emissions, while simulations and experiments agree acceptably well, for most of the cases, noticeable differences can be seen for CO and NO between online and tabulated chemistry solver. Regarding NO, the differences are related to the fact that in the detailed scheme a more advanced prompt and thermal formation mechanism for NOx is accounted for, while in the tabulated chemistry solver only thermal NO source terms are considered. As for CO, a similar discrepancy as in the Diesel simulation campaign can be seen. The operating point SI06 (at 2500 rpm and 2 bar IMEP) shows the largest difference (approximately 17%) for NO emissions against experimental data. This may be explained by the noticeable in predicted peak cylinder pressure which results in a different in-cylinder temperature. The comparison of the tool chain accuracy and the tabulated chemistry solver accuracy is leading to similar conclusions as for the Diesel engine test case in Section 4.

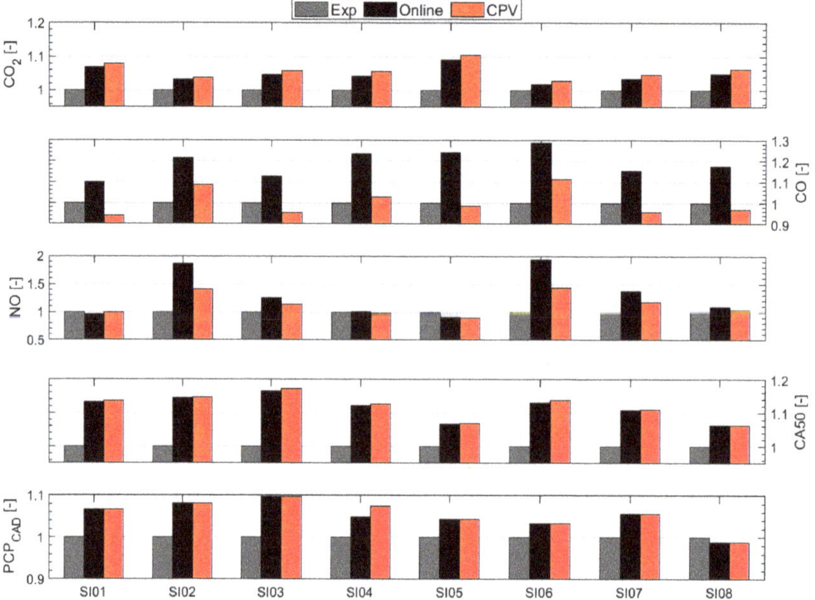

Figure 10. Experimental and simulated engine-out emissions as well as performance parameters (CA50 and PCP$_{CAD}$ location) for all operating conditions of the SI engine simulation campaign. Data have been normalized with respect to experimental values.

6. Computational Aspects

With respect to the SI engine campaign (see model settings in Table 8), a single SRM cycle simulation employing the 475 species chemical mechanism using the online chemistry solver takes approximately 19 min to complete on 24 parallel cores (Intel Xeon E5-2687W v4 @ 3.00GHz processors from the year 2016). The simulation of thirty consecutive cycles results in a total CPU time per operating point of 9.4 h. Although these figures are a small fraction of the CPU cost of a RANS 3-D CFD multi-cycle simulation, typical 0-D/1-D simulation frameworks (i.e., based on a multi-zone Vibe combustion model) usually require a few tenths of a second to run and, in some cases (i.e., Mean Value Engine Models MVEMs), real-time simulation capability is easily reached. In addition, considering that the turbulence model calibration procedure [49,58] relies on a few thousands of Genetic Algorithm (GA) driven SRM simulations to find the optimal constants (see Tables 5 and 9), the application of the online chemistry solver with large mechanisms (i.e., more than 150 species) becomes unfeasible for engine development studies such as driving cycle simulations or engine performance mapping. With respect to the present simulation campaigns, a summary of the CPU times obtained with both solvers and the reported model settings (see Tables 4 and 8) are reported in Table 10.

Table 10. Computational performance summary assuming SRM model settings listed in Tables 4 and 8, on an Intel Xeon E5-2687W v4 @ 3.00GHz CPU from the year 2016.

Simulation Setup	Online Chemistry on 24 Parallel Cores (s/cycle)	Tabulated Chemistry on 1 Core (s/cycle)
SI engine simulation	1136.5	4.1
CI engine simulation	631.7	1.6

Considering that the SRM with CPV tabulated chemistry solver can be easily run on a single core, as opposed to the online chemistry solver that requires multiple cores per run, one can conclude that the present solver delivers a speed-up of at least three orders of magnitude. The size of the auto-ignition table valid for a wide range of typical engine relevant conditions including EGR variations (between 0 and 40%) requires about 1.0 GB of RAM memory. These level resource requirements allow usage of the SRM with CPV not only on dedicated high-performance computing (HPC) systems but also on modern industry grade laptops. Moreover, given the high degree of physical and chemistry models included in its formulation, engine parameter optimization campaigns can be performed within feasible engineering times. To put the computational results shown in Table 10 in a broader prospective, in Figure 11 are shown the extrapolated computational costs of two relevant engine development simulation campaigns: an engine performance mapping and a WLTP cycle.

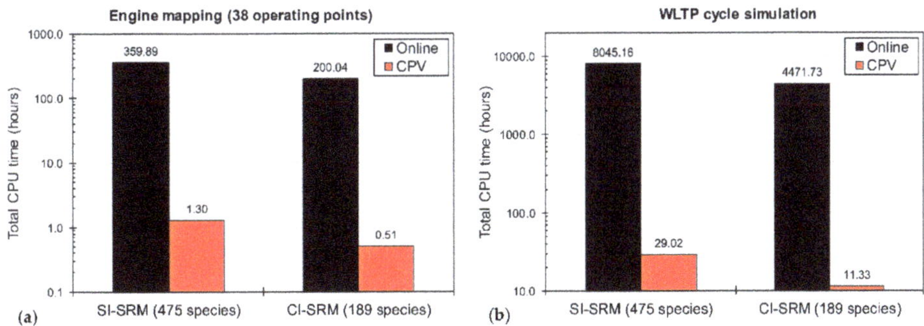

Figure 11. Comparison between online and tabulated chemistry solver computational performances for (**a**) a full engine performance mapping simulation campaign comprising a total of 38 operating conditions and (**b**) the simulation of the full WLTP cycle (30 min) in terms of combustion and emission simulations only.

Both results have been extrapolated considering only the CPU time needed by in-cylinder combustion model. Additional system components (i.e., intake and exhaust air paths, aftertreatment systems) and their contribution to the total simulation time are not considered. Nevertheless, it can be stated that the tabulated chemistry allows to include detailed chemistry effects in a number of applications that are, in most cases, unfeasible for the online chemistry solver.

The developed CPV tabulated chemistry solver was recently applied in two publications for the engine and fuel co-optimization of Diesel and gasoline engines. In the work of Franken et al. [58] a heavy-duty Diesel engine was optimized to find the best set of engine parameters to reduce fuel consumption and NO_x emissions at different speeds and loads. The authors reported optimization times of 20 h to 40 h for one operating point. For a single-cylinder, research on gasoline engine optimization campaign was published by Franken and co-authors [48,61]. A dual fuel tabulated chemistry approach, based on CPV, was used to find the best set of engine parameters in terms of water/fuel-ratios to reduce the knock tendency at a high load operating point and improve the engine efficiency. The optimization times reported in [48,61] are within 10h for one operating point using 4 cores (Intel i7-7820HQ @ 2.90 GHz, from the year 2017) while an equivalent run with the online chemistry solver would have taken several days.

7. Conclusions

This article reported on the application and comparisons of two different chemistry solvers for in-cylinder combustion simulations of compression and spark-ignited engines using a zero-dimensional PDF-based framework. A well-stirred reactor-based online chemistry solver was compared to the developed progress-variable-based solver noted as CPV. Latent enthalpy is chosen as the only parameter for the formulation of the reaction progress variable while a dedicated source term-based method is applied to thermal NO formation. Verification of the newly introduced CPV solver was first assessed under homogeneous constant pressure reactor conditions in order to optimize the table density/interpolation accuracy trade-off. Subsequently, a stochastic reactor model was used to validate the interaction of chemistry and flow during engine combustion processes. Model performance with respect to experimental data and the two solvers was assessed under heavy-duty Diesel engine as well as passenger-car SI engine conditions.

The SRM was shown to be capable of predicting the mixing controlled as well as flame propagation driven combustion processes independently of the chemistry solver. Main engine-out emissions (CO, CO_2, NO and uHC) as well as combustion phasing parameters (CA50, PCP location) are in good agreement with experimental data. With respect to the results obtained with the online and tabulated chemistry solvers, minor differences have been noticed for the start of combustion location and CO emissions. Although still limited to a magnitude of 2.0 CAD, more noticeable discrepancies between the two solvers, in terms of combustion onset, were seen at low-load low-speed in both the Diesel and the gasoline engine simulation campaigns. Under these conditions, the fuel undergoes the NTC behavior where the interpolation error becomes more evident due to highly non-linear reactivity trajectories of the reacting mixture. Despite the thorough table grid optimization study performed to minimize interpolation error, when low temperature combustion is the dominant phenomenon, a tighter, if not adaptive, tabulation grid point distribution may be needed to match even better the start of combustion location predicted by the online chemistry solver. Nevertheless, given the accuracy level shown in the present work, it was concluded that the predictive capabilities of the 0-D SRM is well within commonly noted uncertainty ranges caused by, for instance, sensors inaccuracy or cycle-to-cycle variability.

From the computational cost standpoint, the CPV solver was found to be at least three orders of magnitude faster than the online chemistry solver while keeping the same order of chemical and physical models. The proposed approach is therefore a competitive tool, in terms of CPU time, to lower order methods (i.e., multizone Vibe models) widely used in 0-D/1-D engine performance studies. Generally, CPU cost is one of the main burdens when deployment of detailed chemical mechanisms in 0-D and 3-D CFD simulations is concerned. In particular, if simulations aim to an accurate prediction of exhaust emissions, it often comes a point where a trade-off has to be made between computational performances and size of the chemical mechanisms. Employing a tabulated chemistry solver has the potential to break this tread-off, by using the large mechanism only during table generation (a one-time process) while keeping the high-fidelity combustion and emission predictive capability. In conclusion, it can be stated that the present validation of CPV tabulated chemistry solver allows the SRM to be a useful CAE tool, which holds the accuracy of the overall model tool chain.

Author Contributions: Conceptualization, A.M. and T.F.; methodology, A.M., T.F. and L.C.G.M. Gonzales Mestre.; software, A.M., A.B. and T.F.; validation, A.M., T.F. and L.C.G.M. Gonzales Mestre.; formal analysis, A.M., T.F. and F.M.; investigation, A.M. and T.F.; resources, A.M.; data curation, A.M. and T.F.; writing—original draft preparation, A.M.; writing—review and editing, A.M. and T.F.; visualization, A.M.; supervision, A.B. and F.M.; project administration, A.M.; funding acquisition, A.M. All authors have read and agreed to the published version of the manuscript.

Funding: The research leading to these results has received funding from the People Programme (Marie Curie Actions) of the European Union's Seventh Framework Programme FP7/2007-2013/under REA grant agreement n° 607214. Funding from the Swedish Energy Agency (P39368-2-F-Flex2) is gratefully acknowledged.

Conflicts of Interest: The authors declare no conflict of interest.

Nomenclature

0-D	Zero-Dimensional
3-D	Three-Dimensional
AMN/1-MN	Alpha Methylnaphthalene, $C_{11}H_{10}$
aTDC	After Top Dead Centre
BEV	Battery Electric Vehicle
bTDC	Before Top Dead Centre
CAE	Computer Aided Engineering
CAD	Crank Angle Degree
CFD	Computational Fluid Dynamics
CFM1D-TC	One-Dimensional Coherent Flame Model-Tabulated Chemistry
CI	Compression Ignition
CMC	Conditional Moment Closure
CO	Carbon Monoxide
CO_2	Carbon Dioxide
CPV	Combustion Progress Variable
DI	Direct Injection
ECFM-3Z	Extended Coherent Flame Model 3 Zones
EGR	Exhaust Gas Recirculation
EOI	End of Injection
FPI	Flame Prolongation of Intrinsic Low Dimensional Manifold
ETRF	Ethanol Toluene Reference Fuel
GHG	Green House Gases
HCCI	Homogeneous Charge Compression Ignition
ICE	Internal Combustion Engine
ICEV	Internal Combustion Engine Vehicles
IFP-EN	French Institute of Petroleum
ILDM	Intrinsic Low Dimensional Manifold
KLSA	Knock Limit Spark Advance
LES	Large Eddy Simulation
LHV	Lower Heating Value
MD	Methyl Decanoate, $C_{11}H_{22}O_2$
MDF	Mass Density Function
NOx	Nitrogen Oxides
NTC	Negative Temperature Coefficient
OEM	Original Equipment Manufacturer
PAH	Polycyclic Aromatic Hydrocarbons
PaSR	Partially Stirred Reactor
RDE	Real Driving Emissions
RoHR	Rate of Heat Release
RANS	Reynolds Averaged Navier Stokes
SI	Spark Ignition
SOI	Start of Injection
SRM	Stochastic Reactor Model

TCI	Turbulence Chemistry Interaction
TKI	Tabulated Kinetics of Ignition
TDC	Top Dead Centre
TRF	Toluene Reference Fuel
uHC	Unburned Hydrocarbons
WLTP	Worldwide harmonized Light vehicle Test Procedure
WSR	Well Stirred Reactor

Appendix A Heavy-Duty Diesel Engine Simulation Campaign

In this section, in-cylinder pressure, apparent rate of heat release as well as CO, CO_2, uHC and NO profiles comparisons between experiments and simulations for both solvers are presented for all heavy-duty Diesel engine conditions outlined in Table 2.

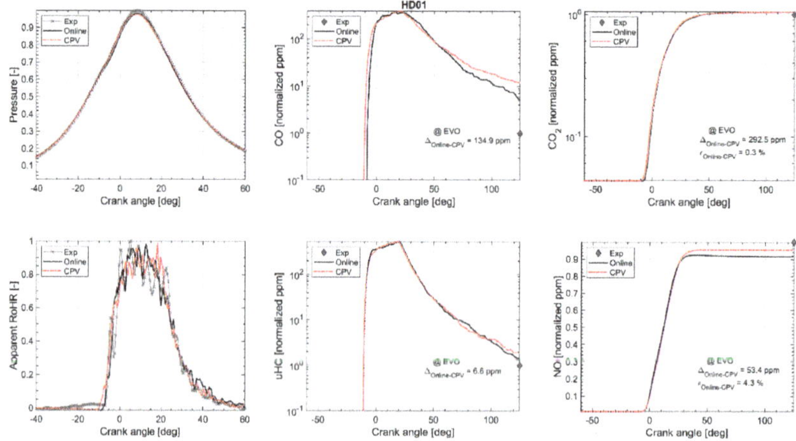

Figure A1. Engine performance and emissions for operating point HD01 (1700 rpm, 19.0 bar IMEP).

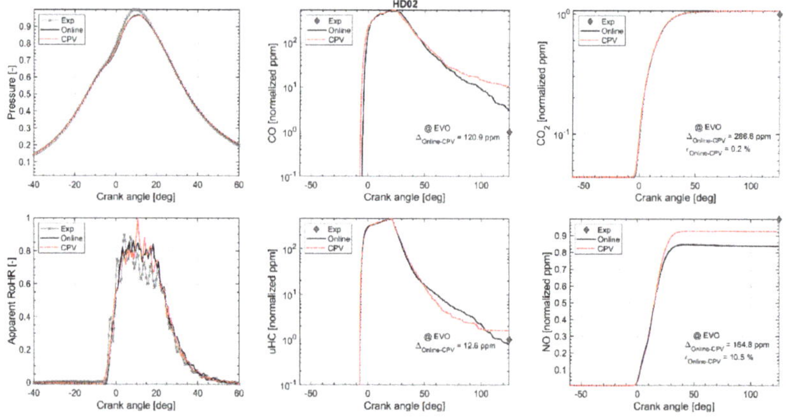

Figure A2. Engine performance and emissions for operating point HD02 (1300 rpm, 22.0 bar IMEP).

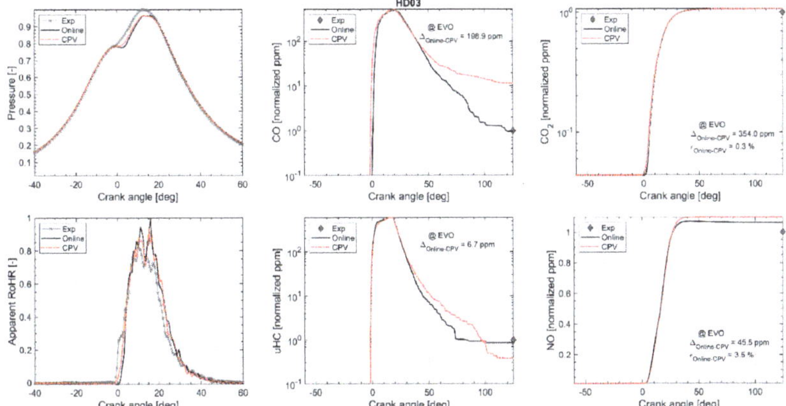

Figure A3. Engine performance and emissions for operating point HD03 (1300 rpm, 14.5 bar IMEP).

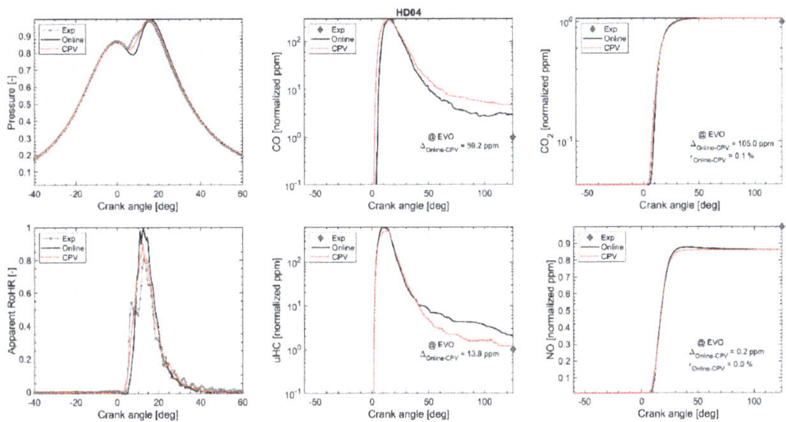

Figure A4. Engine performance and emissions for operating point HD04 (1300 rpm, 6.0 bar IMEP).

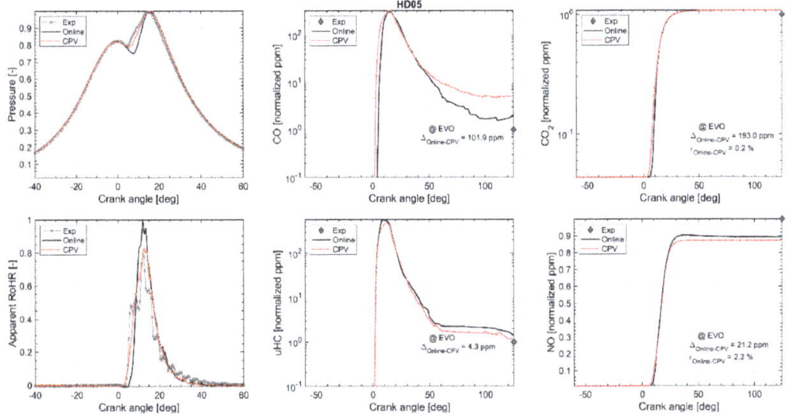

Figure A5. Engine performance and emissions for operating point HD05 (1200 rpm, 6.0 bar IMEP).

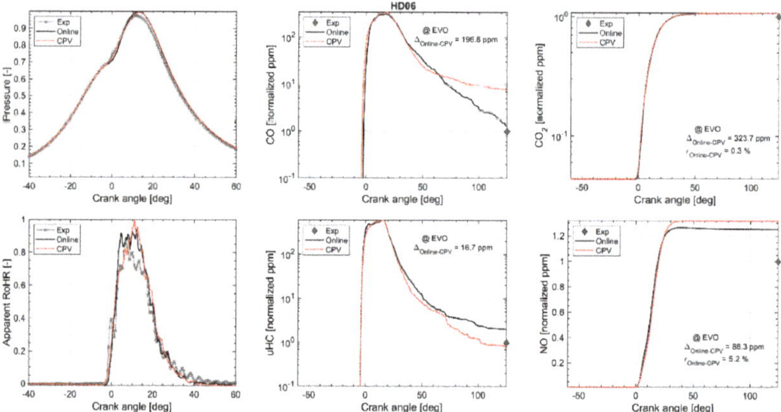

Figure A6. Engine performance and emissions for operating point HD06 (1200 rpm, 14.5 bar IMEP).

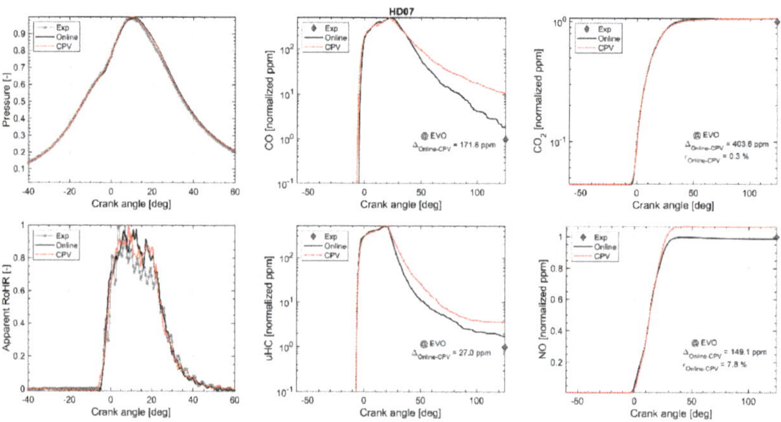

Figure A7. Engine performance and emissions for operating point HD07 (1200 rpm, 22.0 bar IMEP).

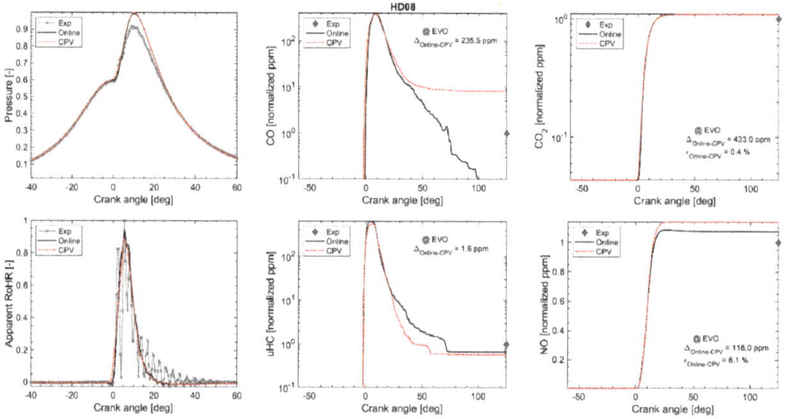

Figure A8. Engine performance and emissions for operating point HD08 (1000 rpm, 6.0 bar IMEP).

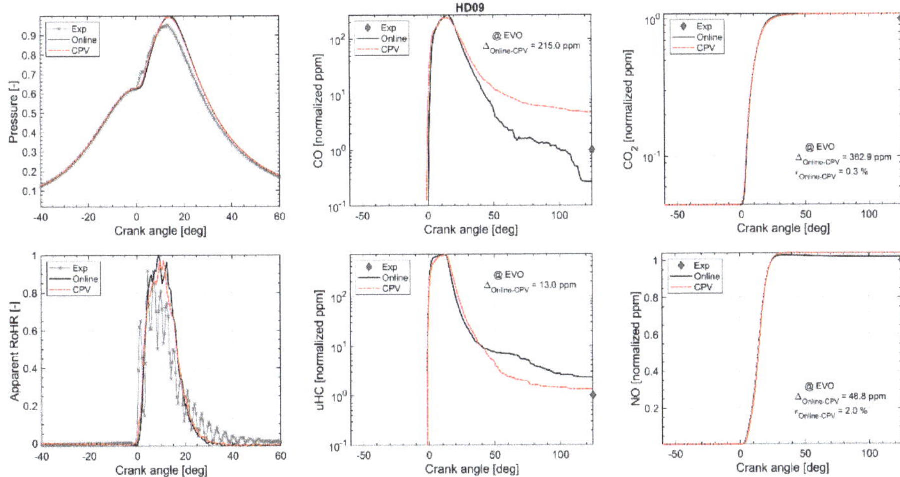

Figure A9. Engine performance and emissions for operating point HD09 (1000 rpm, 11.0 bar IMEP).

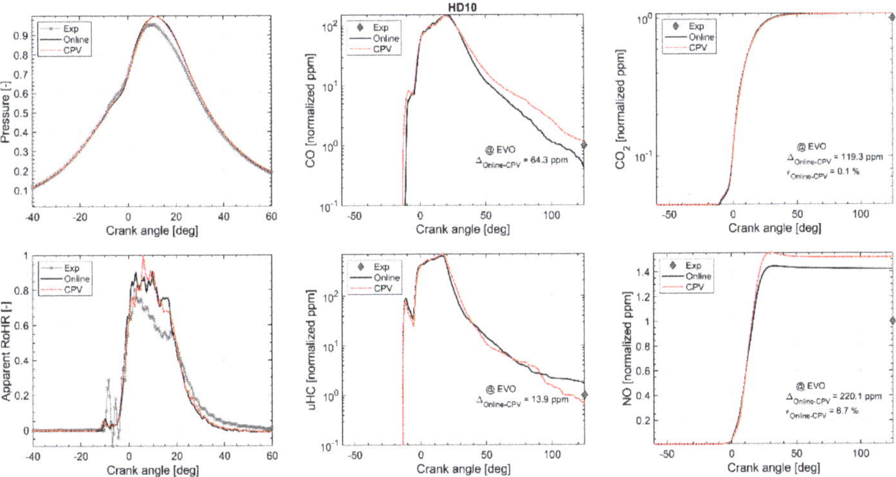

Figure A10. Engine performance and emissions for operating point HD10 (1000 rpm, 22.0 bar IMEP).

Appendix B Spark Ignition Engine Simulation Campaign

In this section, in-cylinder pressure, apparent rate of heat release as well as CO, CO_2 and NO profiles comparisons between experiments and simulations for both solvers are presented for all the spark ignition engine conditions outlined in Error! Reference source not found.

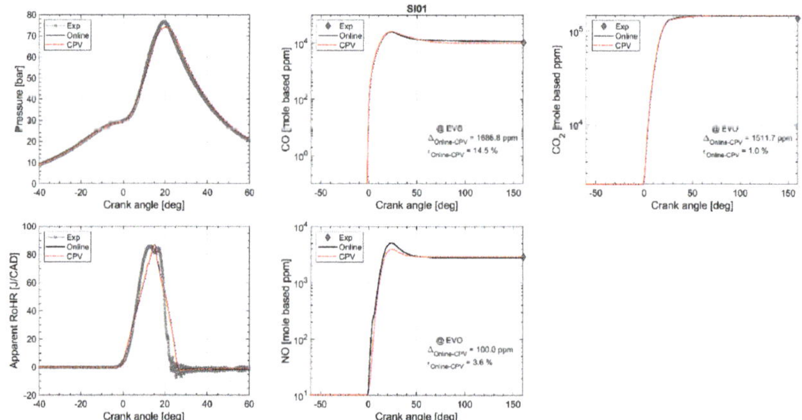

Figure A11. Engine performance and emissions for operating point SI01 (1500 rpm, 15.0 bar IMEP).

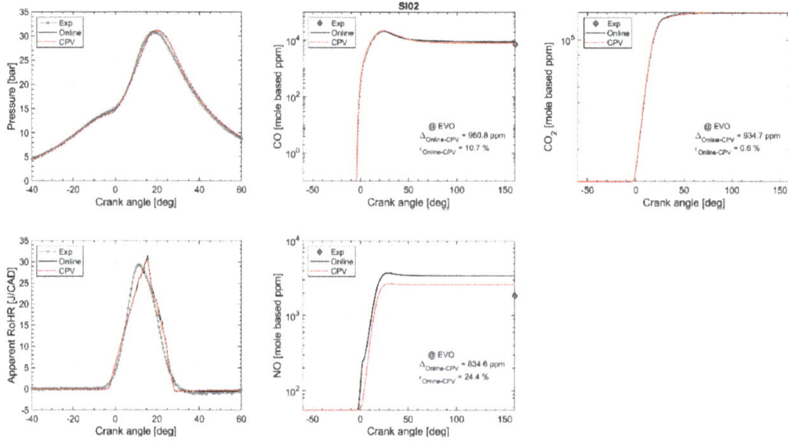

Figure A12. Engine performance and emissions for operating point SI02 (2000 rpm, 5.0 bar IMEP).

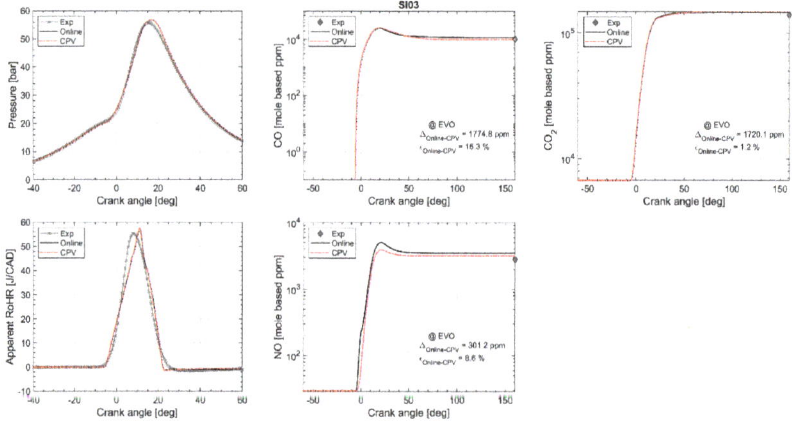

Figure A13. Engine performance and emissions for operating point SI03 (2000 rpm, 10.0 bar IMEP).

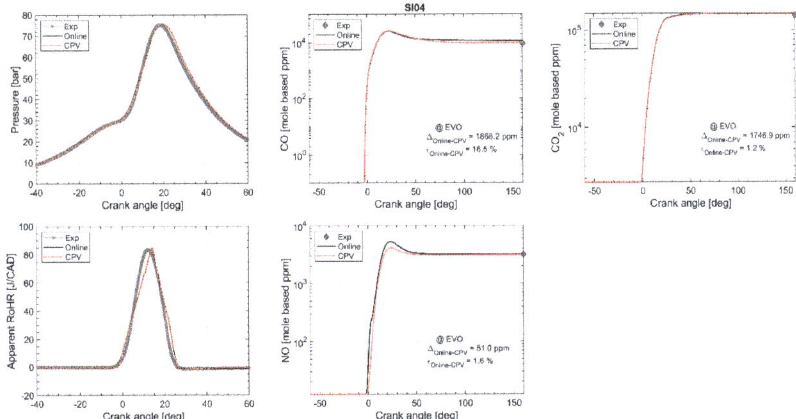

Figure A14. Engine performance and emissions for operating point SI04 (2000 rpm, 15.0 bar IMEP).

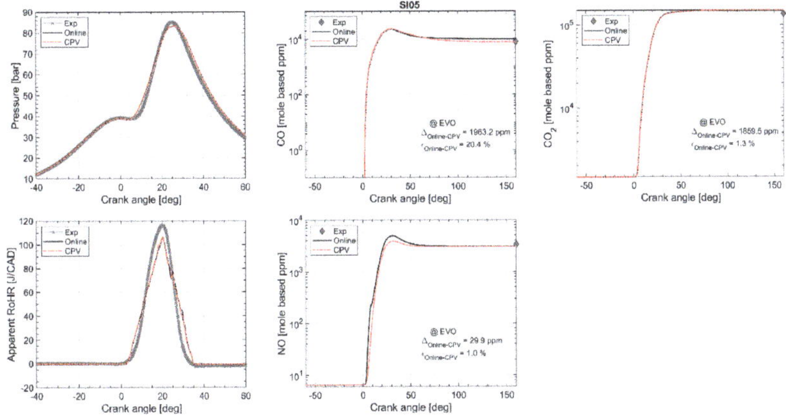

Figure A15. Engine performance and emissions for operating point SI05 (2000 rpm, 20.0 bar IMEP).

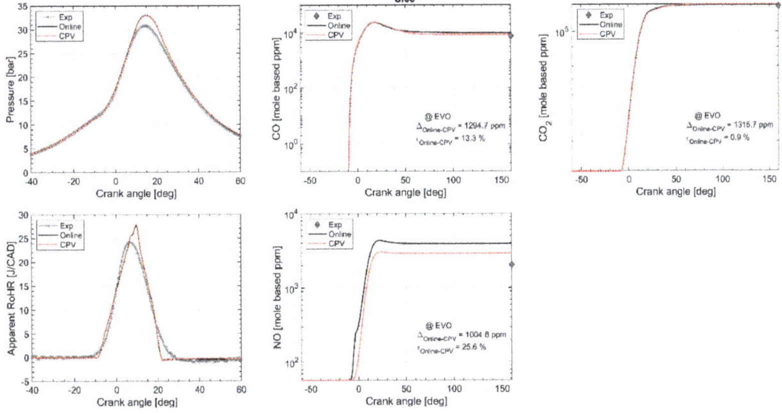

Figure A16. Engine performance and emissions for operating point SI06 (2500 rpm, 5.0 bar IMEP).

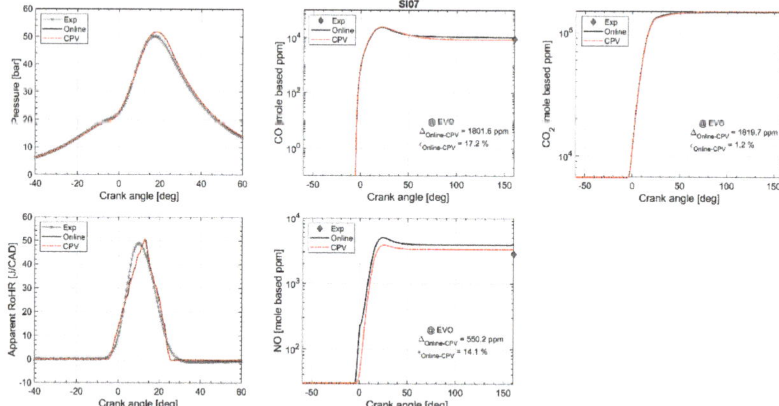

Figure A17. Engine performance and emissions for operating point SI07 (2500 rpm, 10.0 bar IMEP).

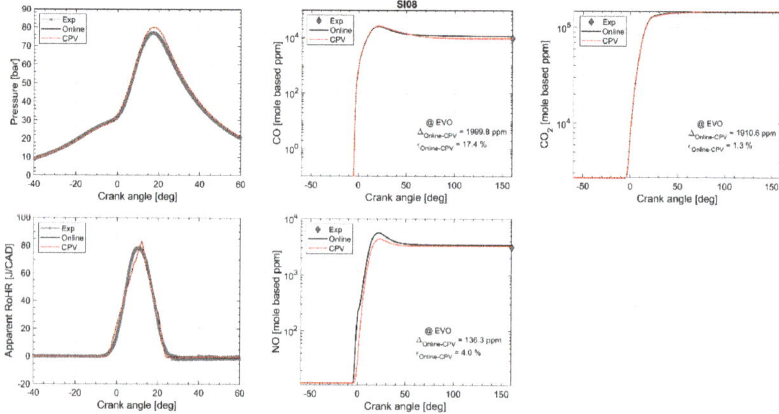

Figure A18. Engine performance and emissions for operating point SI08 (2500 rpm, 15.0 bar IMEP).

References

1. Poulos, S.G.; Heywood, J. The Effect of Chamber Geometry on Spark-Ignition Engine Combustion. In *SAE Technical Paper 830334*; SAE International: Detroit, MI, USA, 1983.
2. Vibe, I. *Halbempirische Formel fur die Verbrennungs-Geschwindigkeit*; Verl. Akad. Wiss. Vd SSR: Moscow, Russia, 1956.
3. Hiroyasu, H.; Kadota, T.; Arai, M. Development and Use of a Spray Combustion Modeling to Predict Diesel Engine Efficiency and Pollutant Emissions: Part 1 Combustion Modeling. *Bull. JSME* **1983**, *26*, 569–575. [CrossRef]
4. Chmela, F.G.; Orthaber, G.C. Rate of Heat Release Prediction for Direct Injection Diesel Engines Based on Purely Mixing Controlled Combustion. In *SAE Technical Paper 1999-01-0186*; SAE International: Detroit, MI, USA, 1999.
5. Nishiwaki, K. A hybrid fractal flame model for si engine combustion comprising turbulent dissipation and laminar flamelets. *COMODIA JSME* **2008**, *8*, 202.
6. Bozza, F.; Gimelli, A.; Merola, S.; Vaglieco, B. Validation of a Fractal Combustion Model through Flame Imaging. In *SAE Technical Paper 2005-01-1120*; SAE International: Detroit, MI, USA, 2005.
7. Jia, M.; Xie, M.; Peng, Z. A Comparative Study of Multi-zone Combustion Models for HCCI Engines. In *SAE Technical Paper 2008-01-0064*; SAE International: Detroit, MI, USA, 2008.

8. Ogink, R.; Golovitchev, V. Gasoline HCCI Modeling: An Engine Cycle Simulation Code with a Multi-Zone Combustion Model. In *SAE Technical Paper 2002-01-1745*; SAE International: Detroit, MI, USA, 2002.

9. Ofner, H.; Schutting, E. Select Assessment of a Multi Zone Combustion Model for Analysis and Prediction of CI Engine Combustion and Emissions. In *SAE Technical Paper 2011-01-1439*; SAE International: Detroit, MI, USA, 2011.

10. Demoulin, F.-X.; Bernard, B.; Lebas, R. A 0D Phenomenological Model Using Detailed Tabulated Chemistry Methods to Predict Diesel Combustion Heat Release and Pollutant Emissions. In *SAE Technical Paper 2011-01-0847*; SAE International: Detroit, MI, USA, 2011.

11. Aceves, S.M.; Flowers, D.L.; Espinosa-Loza, F.; Babajimopoulos, A.; Assanis, D.N. Analysis of Premixed Charge Compression Ignition Combustion with a Sequential Fluid Mechanics-Multizone Chemical Kinetics Model. In *SAE Technical Paper 2005-01-0115*; SAE International: Detroit, MI, USA, 2004.

12. Seidel, L.; Netzer, C.; Hilbig, M.; Mauss, F.; Klauer, C.; Pasternak, M.; Matrisciano, A. Systematic Reduction of Detailed Chemical Reaction Mechanisms for Engine Applications. *J. Eng. Gas Turbines Power* **2017**, *139*, 091701. [CrossRef]

13. Bekdemir, C.C.; Somers, L.B.; De Goey, L.P.; Tillou, J.; Angelberger, C. Predicting diesel combustion characteristics with Large-Eddy Simulations including tabulated chemical kinetics. *Proc. Combust. Inst.* **2013**, *34*, 3067–3074. [CrossRef]

14. Lehtiniemi, H.; Zhang, Y.; Rawat, R.; Mauss, F. Efficient 3-D CFD Combustion Modeling with Transient Flamelet Models. In *SAE Technical Paper 2008-01-0957*; SAE International: Detroit, MI, USA, 2008.

15. Lehtiniemi, H.; Mauss, F.; Balthasar, M.; Magnusson, I. Modeling diesel spray ignition using detailed chemistry with a progress variable approach. *Combust. Sci. Technol.* **2006**, *178*, 1977–1997. [CrossRef]

16. Bo, T.; Mauss, F.; Beck, L.M. Detailed Chemistry CFD Engine Combustion Solution with Ignition Progress Variable Library Approach. In *SAE Technical Paper 2009-01-1898*; SAE International: Detroit, MI, USA, 2009.

17. Fiorina, B.; Gicquel, O.; Vervisch, L.; Carpentier, S.; Darabiha, N. Premixed turbulent combustion modeling using tabulated detailed chemistry and PDF. *Proc. Combust. Inst.* **2005**, *30*, 867–874. [CrossRef]

18. Nakov, G.; Mauss, F.; Wenzel, P.; Steiner, R.; Krüger, C.; Zhang, Y.; Rawat, R.; Borg, A.; Perlman, C.; Fröjd, K.; et al. Soot Simulation under Diesel Engine Conditions Using a Flamelet Approach. *SAE Int. J. Engines* **2009**, *2*, 89–104. [CrossRef]

19. Nakov, G.; Mauss, F.; Wenzel, P.; Krüger, C. Application of a stationary flamelet library based CFD soot model for low-NOx diesel combustion. In Proceedings of the THIESEL Conference on Thermo- and Fluid Dynamic Processes in Diesel Engines, Valencia, Spain, 14–17 September 2010.

20. Karlsson, A.; Magnusson, I.; Balthasar, M.; Mauss, F. Simulation of Soot Formation Under Diesel Engine Conditions Using a Detailed Kinetic Soot Model. In *SAE Technical Paper 981022*; SAE International: Detroit, MI, USA, 1998.

21. Wenzel, P.; Steiner, R.; Krüger, C.; Schießl, R.; Hofrath, C.; Maas, U. 3D-CFD Simulation of DI-Diesel Combustion Applying a Progress Variable Approach Accounting for Detailed Chemistry. In *SAE Technical Paper 2007-01-4137*; SAE International: Detroit, MI, USA, 2007.

22. Da Cruz, A.P. Three-dimensional modeling of self-ignition in hcci and conventional diesel engines. *Combust. Sci. Technol.* **2004**, *176*, 867–887. [CrossRef]

23. Colin, O.; Da Cruz, A.P.; Jay, S. Detailed chemistry-based auto-ignition model including low temperature phenomena applied to 3-D engine calculations. *Proc. Combust. Inst.* **2005**, *30*, 2649–2656. [CrossRef]

24. Knop, V.; Michel, J.-B.; Colin, O. On the use of a tabulation approach to model auto-ignition during flame propagation in SI engines. *Appl. Energy* **2011**, *88*, 4968–4979. [CrossRef]

25. Kosters, A.; Golovitchev, V.; Karlsson, A. A Numerical Study of the Effect of EGR on Flame Lift-off in n-Heptane Sprays Using a Novel PaSR Model Implemented in OpenFOAM. *SAE Int. J. Fuels Lubr.* **2012**, *5*, 604–610. [CrossRef]

26. Lucchini, T.; Della Torre, A.; D'Errico, G.; Onorati, A. Modeling advanced combustion modes in compression ignition engines with tabulated kinetics. *Appl. Energy* **2019**, *247*, 537–548. [CrossRef]

27. Zhou, Q.; Lucchini, T.; D'Errico, G.; Hardy, G.; Lu, X. Modeling heavy-duty diesel engines using tabulated kinetics in a wide range of operating conditions. *Int. J. Engine Res.* **2020**. [CrossRef]

28. Leicher, J.; Wirtz, S.; Scherer, V. Evaluation of an Entropy-Based Combustion Model using Stochastic Reactors. *Chem. Eng. Technol.* **2008**, *31*, 964–970. [CrossRef]

29. Gicquel, O.; Darabiha, N.; Thévenin, D. Laminar premixed hydrogen/air counterflow flame simulations using flame prolongation of ILDM with differential diffusion. *Proc. Combust. Inst.* **2000**, *28*, 1901–1908. [CrossRef]

30. Bougrine, S.; Richard, S.; Michel, J.-B.; Veynante, D. Simulation of CO and NO emissions in a SI engine using a 0D coherentflame model coupled with a tabulated chemistry approach. *Appl. Energy* **2014**, *113*, 1199–1215. [CrossRef]

31. Bozza, F.; De Bellis, V.; Teodosio, L. A Tabulated-Chemistry Approach Applied to a Quasi-Dimensional Combustion Model for a Fast and Accurate Knock Prediction in Spark-Ignition Engines. In *SAE Technical Paper 2019-01-0471*; SAE International: Detroit, MI, USA, 2019.

32. Livengood, J.C.; Wu, P.C. Correlation of Autoignition Phenomena in Internal Comustion Engines and Rapid Compression Machines. *Symp. (Int.) Combust.* **1955**, *5*, 347–356. [CrossRef]

33. Matrisciano, A.; Franken, T.; Perlman, C.; Borg, A.; Lehtiniemi, H.; Mauss, F. Development of a Computationally Efficient Progress Variable Approach for a Direct Injection Stochastic Reactor Model. In *SAE Technical Paper 2017-01-0512*; SAE International: Detroit, MI, USA, 2017.

34. Pope, S. PDF methods for turbulent reactive flows. *Prog. Energy Combust. Sci.* **1985**, *11*, 119–192. [CrossRef]

35. Maigaard, P.; Mauss, F.; Kraft, M. Homogenous charge compression ignition engine: A simulation study on the effect of in-homogeneities. *ASME J. Eng. Gas Turb. Power* **2003**, *125*, 466–471. [CrossRef]

36. Tuner, M. *Stochastic Reactor Models for Engine Simulations*; Lund University: Lund, Sweden, 2008.

37. Bjerkborn, S.; Fröjd, K.; Perlman, C.; Mauss, F. A Monte Carlo Based Turbulent Flame Propagation Model for Predictive SI In-Cylinder Engine Simulations Employing Detailed Chemistry for Accurate Knock Prediction. *SAE Int. J. Engines* **2012**, *5*, 1637–1647. [CrossRef]

38. Pasternak, M.; Mauss, F.; Sens, M.; Riess, M.; Benz, A.; Stapf, K.G. Gasoline engine simulations using zero-dimensional spark ignition stochastic reactor model and three-dimensional computational fluid dynamics engine model. *Int. J. Engine Res.* **2015**, *17*, 76–85. [CrossRef]

39. Netzer, C.; Franken, T.; Seidel, L.; Lehtiniemi, H.; Mauss, F. Numerical Analysis of the Impact of Water Injection on Combustion and Thermodynamics in a Gasoline Engine Using Detailed Chemistry. *SAE Int. J. Engines* **2018**, *11*, 1151–1166. [CrossRef]

40. Heywood, J. *Internal Combustion Engine Fundamentals*; McGraw-Hill: New York, NY, USA, 1988.

41. Franken, T.; Klauer, C.; Kienberg, M.; Matrisciano, A.; Mauss, F. Prediction of thermal stratification in an engine-like geometry using a zero-dimensional stochastic reactor model. *Int. J. Engine Res.* **2019**, *21*, 1750–1763. [CrossRef]

42. Schmitt, M.; Frouzakis, C.E.; Wright, Y.M.; Tomboulides, A.G.; Boulouchos, K. Investigation of wall heat transfer and thermal stratification under engine-relevant conditions using DNS. *Int. J. Engine Res.* **2015**, *17*, 63–75. [CrossRef]

43. Pasternak, M. Simulation of the Diesel Engine Combustion Process Using the Stochastic Reactor Model. Ph.D. Thesis, Brandenburg University of Technology, Cottbus, Germany, 2015.

44. Pasternak, M.; Mauss, F.; Klauer, C.; Matrisciano, A. Diesel engine performance mapping using a parametrized mixing time model. *Int. J. Engine Res.* **2017**, *19*, 202–213. [CrossRef]

45. Kožuch, P. Phenomenological Model for a Combined Nitric Oxide and Soot Emission Calculation in di Diesel Engines. Ph.D. Thesis, University of Stuttgart, Stuttgart, Germany, 2004.

46. Franken, T.; Sommerhoff, A.; Willems, W.; Matrisciano, A.; Lehtiniemi, H.; Borg, A.; Netzer, C.; Mauss, F. Advanced Predictive Diesel Combustion Simulation Using Turbulence Model and Stochastic Reactor Model. In *SAE Technical Paper*; SAE International: Detroit, MI, USA, 2017.

47. Dulbecco, A.; Richard, S.; Laget, O.; Aubret, P. Development of a Quasi-Dimensional K-k Turbulence Model for Direct Injection Spark Ignition (DISI) Engines Based on the Formal Reduction of a 3D CFD Approach. In *SAE Technical Paper 2016-01-2229*; SAE International: Detroit, MI, USA, 2016.

48. Franken, T.; Mauss, F.; Seidel, L.; Gern, M.S.; Kauf, M.; Matrisciano, A.; Kulzer, A.C. Gasoline engine performance simulation of water injection and low-pressure exhaust gas recirculation using tabulated chemistry. *Int. J. Engine Res.* **2020**, *21*, 1857–1877. [CrossRef]

49. Franken, T.; Seidel, L.; Matrisciano, A.; Mauss, F.; Kulzer, A.C.; Schuerg, F. Analysis of the Water Addition Efficiency on Knock Suppression for Different Octane Ratings. In *SAE Technical Paper 2020-01-0551*; SAE International: Detroit, MI, USA, 2020.

50. Breda, S.; D'Adamo, A.; Fontanesi, S.; Giovannoni, N.; Testa, F.; Irimescu, A.; Merola, S.; Tornatore, C.; Valentino, G. CFD Analysis of Combustion and Knock in an Optically Accessible GDI Engine. *SAE Int. J. Engines* **2016**, *9*, 641–656. [CrossRef]

51. Lafossas, F.-A.; Castagne, M.; Dumas, J.P.; Henriot, S. Development and Validation of a Knock Model in Spark Ignition Engines Using a CFD code. In *SAE Technical Paper 2002-01-2701*; SAE International: Detroit, MI, USA, 2002.

52. Ihme, M.; Shunn, L.; Zhang, J. Regularization of reaction progress variable for application. *J. Comput. Phys.* **2012**, *231*, 7715–7721. [CrossRef]

53. Lund Combustion Engineering—LOGE AB. LOGEsoft Products. 2020. Available online: www.logesoft.com (accessed on 14 December 2020).

54. Zeuch, T.; Moréac, G.; Ahmed, S.S.; Mauss, F. A comprehensive skeletal mechanism for the oxidation of n-heptane generated by chemistry-guided reduction. *Combust. Flame* **2008**, *155*, 651–674. [CrossRef]

55. Bekdemir, C.C.; Somers, L.B.; De Goey, L.P. Modeling diesel engine combustion using pressure dependent Flamelet Generated Manifolds. *Proc. Combust. Inst.* **2011**, *33*, 2887–2894. [CrossRef]

56. Wang, X. *Kinetic Mechanism of Surrogates for Biodiesel*; BTU Cottbus-Senftenberg: Cottbus, Germany, 2017.

57. Mauss, F. Entwicklung Eines Kinetischen Modells der Russbildung mit Schneller Polymerisation. Ph.D. Thesis, Rheinisch-Westfälische Technische Hochschule Aachen, Aachen, Germany, 1997.

58. Franken, T.; Duggan, A.; Matrisciano, A.; Lehtiniemi, H.; Borg, A.; Mauss, F. Multi-Objective Optimization of Fuel Consumption and NOx Emissions with Reliability Analysis Using a Stochastic Reactor Model. In *SAE Technical Paper 2019-01-1173*; SAE International: Detroit, MI, USA, 2019.

59. Kauf, M.; Gern, M.; Seefeldt, S. Evaluation of Water Injection Strategies for NOx Reduction and Charge Cooling in SI Engines. In *SAE Technical Paper 2019-01-2164*; SAE International: Detroit, MI, USA, 2019.

60. Netzer, C.; Seidel, L.; Ravet, F.; Mauss, F. Impact of the Surrogate Fromulation on 3D CFD Egnien Knock Prediction Using Detailed Chemistry. *Fuel* **2019**, *254*, 115678. [CrossRef]

61. Franken, T.; Netzer, C.; Mauss, F.; Pasternak, M.; Seidel, L.; Borg, A.; Lehtiniemi, H.; Matrisciano, A.; Kulzer, A.C. Multi-objective optimization of water injection in spark-ignition engines using the stochastic reactor model with tabulated chemistry. *Int. J. Engine Res.* **2019**, *20*, 1089–1100. [CrossRef]

Publisher's Note: MDPI stays neutral with regard to jurisdictional claims in published maps and institutional affiliations.

Article

0D/1D Thermo-Fluid Dynamic Modeling Tools for the Simulation of Driving Cycles and the Optimization of IC Engine Performances and Emissions

Andrea Marinoni *, Matteo Tamborski, Tarcisio Cerri, Gianluca Montenegro, Gianluca D'Errico, Angelo Onorati, Emanuele Piatti and Enrico Ernesto Pisoni

Department of Energy, Politecnico di Milano, Via Lambruschini 4A, 20156 Milano, Italy; matteo.tamborski@polimi.it (M.T.); tarcisio.cerri@polimi.it (T.C.); gianluca.montenegro@polimi.it (G.M.); gianluca.derrico@polimi.it (G.D.); angelo.onorati@polimi.it (A.O.); emanuele.piatti@mail.polimi.it (E.P.); enricoernesto.pisoni@mail.polimi.it (E.E.P.)
* Correspondence: andreamassimo.marinoni@polimi.it

Abstract: The prediction of internal combustion engine performance and emissions in real driving conditions is getting more and more important due to the upcoming stricter regulations. This work aims at introducing and validating a new transient simulation methodology of an ICE coupled to a hybrid architecture vehicle, getting closer to real-time calculations. A one-dimensional computational fluid dynamic software has been used and suitably coupled to a vehicle dynamics model in a user function framework integrated within a Simulink® environment. A six-cylinder diesel engine has been modeled by means of the 1D tool and cylinder-out emissions have been compared to experimental data. The measurements available have been used also to calibrate the combustion model. The developed 1D engine model has been then used to perform driving cycle simulations considering the vehicle dynamics and the coupling with the energy storage unit in the hybrid mode. The map-based approach along with the vehicle simulation tool has also been used to perform the same simulation and the two results are compared to evaluate the accuracy of each approach. In this framework, to achieve the best simulation performance in terms of computational time over simulated time ratio, the 1D engine model has been used in a configuration with a very coarse mesh. Results have shown that despite the high mesh spacing used the accuracy of the wave dynamics prediction was not affected in a significant way, whereas a remarkable speed-up factor was achieved. This means that a crank angle resolution approach to the vehicle simulation is a viable and accurate strategy to predict the engine emission during any driving cycle with a computation effort compatible with the tight schedule of a design process.

Keywords: diesel heavy-duty engine; engine transient simulation; longitudinal dynamics; fuel economy; hybrid; complex powertrains; real driving cycles

Citation: Marinoni, A.; Tamborski, M.; Cerri, T.; Montenegro, G.; D'Errico, G.; Onorati, A.; Piatti, E.; Pisoni, E.E. 0D/1D Thermo-Fluid Dynamic Modeling Tools for the Simulation of Driving Cycles and the Optimization of IC Engine Performances and Emissions. *Appl. Sci.* **2021**, *11*, 8125. https://doi.org/10.3390/app11178125

Academic Editors: Georgios Karavalakis and Ricardo Novella Rosa

Received: 3 May 2021
Accepted: 28 August 2021
Published: 1 September 2021

1. Introduction

Nowadays internal combustion engines play a crucial role in worldwide mobility. A general concern about climate change imposes the adoption of an environmentally sustainable policy, as well relevant efforts towards a reduction of fuel consumption and compliance with strict emission regulations [1,2]. In the past decades, analytical and numerical models provided fundamental support to the design of IC engines, in addition to the necessary experimental campaign on real engine prototypes. Today reliable 1D CFD tools are commonly used in the community of engineers and researchers to investigate the behavior and performance of new engines [3,4]. In general, state of the art 1D simulation models can calculate the steady-state engine maps (in terms of air mass flow rate, torque and power, max. cylinder pressure, fuel consumption and emissions), being able to represent every operating condition just starting from a few operating points analyzed in the experimental laboratory (map-based approach) [5,6]. This is a very powerful and fast simulation approach, but the hypothesis of steady state working condition represents a major limitation. In fact, the IC engine is an unsteady volumetric machine, and its behavior is deeply influenced by transient conditions due to mechanical, thermal and fluid-dynamics inertias, not considered in a map-based approach. A robust simulation method is required to account for the transient conditions of the IC engine coupled to the vehicle and its components, to obtain more realistic results in terms of fuel consumption and emissions [7].

The present work wants to highlight the application of the new simulation platform based on Gasdyn, for the 1D thermo-fluid dynamic engine simulation, and Velodyn, for the simulation of the vehicle longitudinal dynamics and powertrain composition, to cope with a typical transient condition simulation. The two simulation tools are integrated in a "co-simulation" framework, which allows the two software to interact in real-time within the MATLAB® Simulink® environment. The vehicle simulation tool interacts with the virtual engine to proceed forward in time and follow the prescribed test cycle.

In particular, the adoption of a specific numerical scheme, based on a staggered leapfrog, finite volume method applied to coarse meshes, allows to preserve the mass conservation and accuracy of the 1D fluid dynamic modeling of the IC engine, resulting in an optimal solution for the simulation of driving cycles in time frames comparable to real-time. In this way, the 1D simulation tool allows the investigation and comparison of different solutions of hybrid architectures in terms of performance, fuel consumption and emissions.

A six-cylinder, 210 kW diesel engine has been modeled in detail and then validated in steady-state conditions, on the basis of experimental engine map data. Then the 1D engine model was made compatible for the two-way coupling with the vehicle simulation tool, via an S-function block in the Simulink® framework.

Finally, the Co-Simulation environment has been applied and validated against the traditional map-based approach, which is based on the available experimental engine maps of the IC engine. The test case considered is a hybrid powertrain of a commercial bus. An analysis of fuel consumption in real driving emission (RDE) cycles is discussed, considering both the conventional and hybrid architectures of the bus, mainly focusing on the hybrid components design and hybrid control unit (HCU) logic settings.

2. 0D/1D Thermo-Fluid Dynamic Modeling of IC Engine

The 1D thermo-fluid dynamic tool used to carry out the simulation of the Diesel engine is based on the solution of the conservation equations of mass, momentum and energy along the IC engine domain, modelled by means of one-dimensional pipes, to represent the complete intake and exhaust duct systems, and zero-dimensional elements, to represent junctions of pipes, abrupt area changes, valves, turbocharger and cylinders. For what concerns the fluid dynamics, the gas flow is assumed to be one-dimensional, unsteady, compressible with friction and heat transfer at the pipe walls. Moreover, the transport equations for the tracking of chemical species along the pipes could have been considered as well, but in this study, this aspect was not considered. This comprehensive approach is extensively described in literature, see for example [5,6,8–10]. In this work, an innovative contribution is represented by the development and application of a fast simulation method (FSM) to reduce the computational time and reduce the CPU/real-time ratio. The FSM numerical solver developed can achieve satisfactory conservation of mass with coarse meshes adopted to discretize the ducts, up to 10 cm, to significantly reduce the computational effort, because of the decreased number of computational nodes and increased time step. Traditional numerical methods, based on finite difference techniques with 2nd order accuracy [11,12], behave as non-conservative when the calculation domain is so coarsely discretized [13]. Hence, an alternative numerical solver which ensures a good conservation with such large meshes has been applied, as briefly described below. The numerical method is a one-dimensional finite volume technique based on cells and connectors, named 1Dcell, explicit in time and 2nd order accurately derived from a quasi-

3D formulation [14]. The transported quantities are defined in the cell centers, while their fluxes are established on the connectors. The space-time grid is staggered, as shown in Figure 1. The centers of the computational control volumes for the solution of the momentum equation are represented as small squares, while the cell control volume centers for the solution of mass and energy equations are represented by circles. They are located at different positions in both space and time.

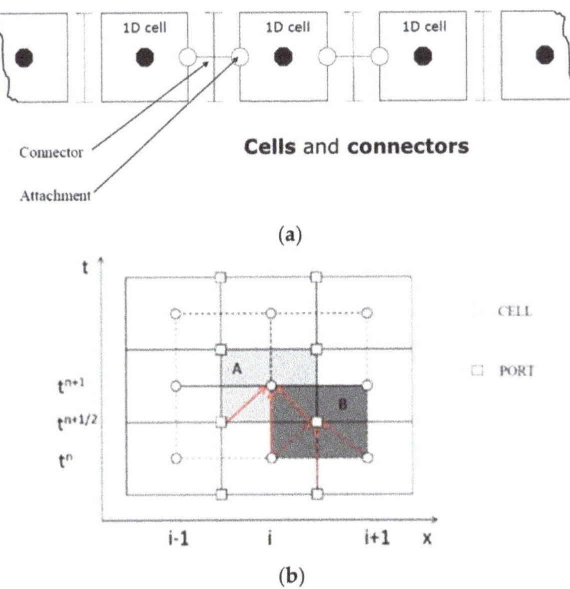

Figure 1. (**a**) One-dimensional 1Dcell approach, cells and connectors; (**b**) time marching procedure adopted for the solution of the governing equation in the 1Dcell method: (A) represents the cell control volume and (B) represents the port control volume.

With the above-mentioned reference frame, the continuity and energy conservation equations are solved with respect to the cell element, whereas the momentum equation is solved with respect to the connector elements between cells. The grid is also staggered in time as depicted by Figure 1. The closing equation is the perfect gas equation of state. The conservation equations of mass, energy and momentum are reported below:

$$\rho_{cell}^{t+1} = \rho_{cell}^{t} + \frac{\Delta t}{V} \sum_{P}^{Nports} (\rho v A)_P^{t+\frac{1}{2}} \tag{1}$$

$$\left(\rho e^0\right)_{cell}^{t+1} = \left(\rho e^0\right)_{cell}^{t} + \frac{\Delta t}{V} \sum_{P}^{Nports} \left(\rho v h^0 A\right)_P^{t+\frac{1}{2}} + S_e \tag{2}$$

$$(\rho v A)_{cell}^{t+\frac{1}{2}} = \left(\rho \vec{v} A\right)_{cell}^{t-\frac{1}{2}} + \frac{\Delta t}{\Delta L} \sum_{P}^{Nports} \left[\left(p + \rho v^2\right)_P A_P\right] + S_{M}, \tag{3}$$

where ρ is the gas density, p is the gas pressure, A is the pipe section and V is the cell volume; Δt and Δx the time step size and the distance between cell centers, respectively; h^0 is the total enthalpy of the cell; e^0 is the total internal energy of the cell; S_e and S_m terms are the sources of the corresponding equations which consider heat transfer and friction with the duct wall.

This 1Dcell method has been applied for the solution with coarse meshes, to achieve a robust fast simulation method (FSM). Regarding the accurate solution in the case of ordinary mesh sizes (1 cm), the Corberán–Gascon TVD, finite difference, symmetric, explicit method was applied [15]. The TVD feature of this explicit method allows the reduction of the intrinsic instabilities or spurious oscillations which may appear using the explicit formulation. For both numerical methods, the time step is computed according to the Courant–Friedrichs–Lewy condition described by [16].

Intra-pipe boundary conditions are treated resorting to a characteristic-based approach [11,16] to handle junctions, valves, sudden area changes, intercoolers and turbochargers. In particular, for pressure loss junctions experimentally derived loss coefficients are used. These data are available from the literature and are obtained mainly in steady-state flow conditions. Similarly, for poppet valves, measured flow coefficients have been used. Their applicability can be argued but, in any case, they allow a good representation of the physical phenomena with a reasonable approximation. The coupling with 3D models can be realized, improving the accuracy of the simulation, but this brings, as expected, an increase of computational burden that cannot be accepted when the calculation of a driving cycle is addressed [9]. A major concern in this type of simulation is also the approximation introduced by the modeling of the turbocharger. Usually, the compressor and the turbine are treated relying on performance maps, obtained experimentally in state-state flow conditions. The maps are provided by the supplier of the turbocharger or, alternatively, they can be obtained by measuring the turbocharger performances on the test bench [17]. In both cases, the result is a discrete map where information is available for specific revolution speeds. To extend the validity of the maps in regions outside of the experimental ones and between the single iso-velocity lines, interpolation techniques are used. The reliability of the procedure depends mainly on the interpolation technique adopted, but also on the pre-processing of the raw data of the map, which needs to be properly smoothed [18]. Alternatively, a few 1D models can solve the 1D conservation equations for the stator and the rotor, considering the effect of the rotation of the turbine and compressor wheel [19,20]. These are known as map-less approaches, which have a reduced level of approximation, but they need in any case a bit of calibration and detailed data about the turbo geometry. This approach, going in the direction of a better approximation, brings the increase of the computational burden, which, once again, cannot be accepted when the goal is to reach a real-time simulation of a driving cycle.

3. Engine Description

The engine considered is a 6.7 L, turbocharged Diesel engine for heavy-duty applications such as buses and trucks, developed by FPT Industrial. This engine belongs to a commercial family, with a maximum power ranging from 184 kW to 235 kW. In this work, the medium-size 210 kW engine has been investigated.

Specific information about the engines studied is summarized in Tables 1 and 2.

Table 1. Engine geometrical data.

Quantity	Value	Unit of Measure
Bore × Stroke	104 × 132	mm × mm
Rod Length	195	mm
Number of Cylinders	6 (In-Line)	-
Total Displacement	6728	cm3
Compression Ratio	17	-
Valves per Cylinder	4	-

Table 2. Engine performance data.

Quantity	Value	Unit of Measure
Rotational Speed Range	600 ÷ 2600	rpm
Maximum Brake Torque	1010	Nm
Maximum Brake Power	210	kW

The engine is equipped with a twin-entry turbocharger characterized by a fixed geometry and a wastegate valve.

A wide set of experimental data was available to validate the engine thermo-fluid dynamic model. All these data refer to different steady-state operating points, defined by engine speed and brake mean effective pressure (BMEP). The operating map of the 210 kW engine, investigated in this study, is represented by the red points in Figure 2.

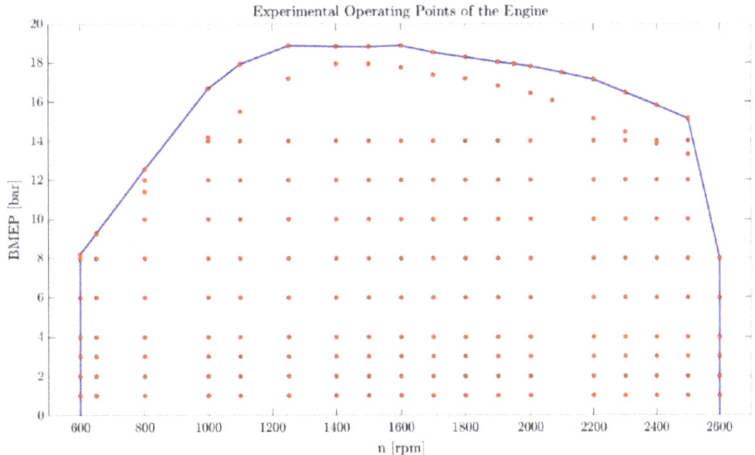

Figure 2. 210 kW (red points) and 235 kW (blue circles) engine operating maps (BMEP as function of engine speed).

4. Calculation Methodology

As anticipated, the final aim of this work is to carry out the transient simulation of the IC engine integrated with a complete vehicle model, considering the real driving conditions. The analysis conducted is summarized by the following steps:

1. development of the engine simulation model.
2. validation of the model at fixed point operating conditions.
3. vehicle and engine model integration.
4. vehicle dynamics simulation in real driving conditions.

During the first two steps of the activity, the 1D engine model is refined and then validated under steady-state conditions throughout the whole engine map. The simulation code iterates the calculation by studying all the operating points in sequence. Once the results obtained match the experimental data with minor discrepancy, the simulation of transient operation of the engine can be investigated through step 3 and step 4. The 1D model is reported in Figure 3.

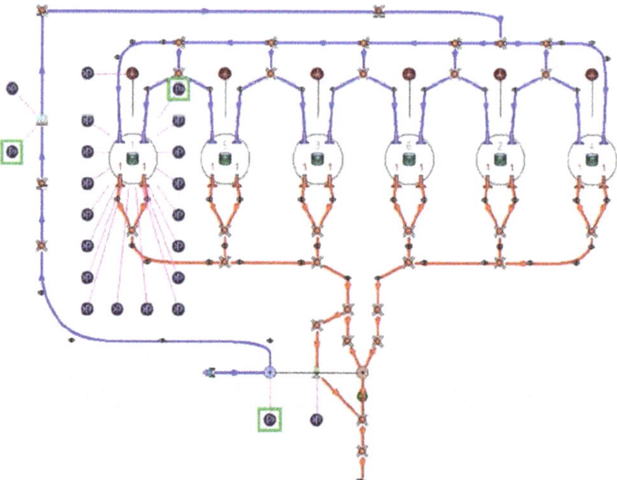

Figure 3. Gasdyn 1D schematic of the 6-cylinder, turbocharged Diesel engine investigated in this study. The "CoSim" input/output objects are clearly visible; "CoSim" input ports are represented in blue circular elements, whereas the "CoSim" output ports are highlighted in green.

Relying on the coupling functionality of the simulation tool, the 1D engine model can be exported to the Simulink® environment by means of an S-Function block. This possibility is exploited for the vehicle simulation, which is performed by means of the vehicle tool: this software is a Simulink® add-on that allows the simulation of the longitudinal dynamics of vehicles with a large variety of architectures [21]. The vehicle model consists of different sub-system blocks, each one in charge of a specific task, to represent the specific devices along the powertrain such as gearbox, electric motor, and battery as described with more details in [4].

The IC engine model operating in transient conditions is represented by the simplified scheme reported in Figure 4.

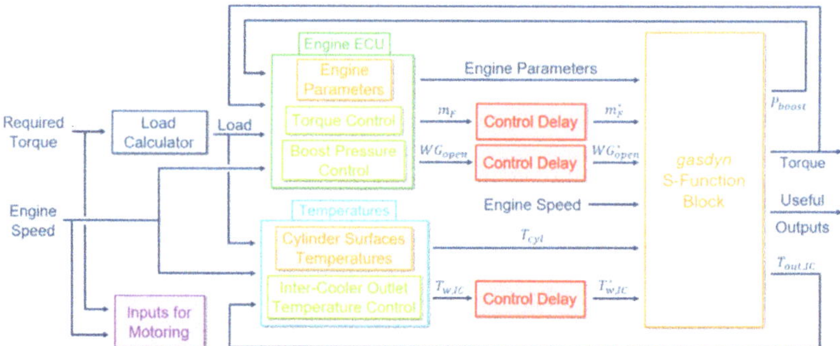

Figure 4. Schematic representation of the IC engine transient model.

In Figure 4 the S-Function block in light yellow contains the engine 0D/1D thermo-fluid dynamic model, which acts as virtual engine solver. All the parameters to be set or monitored and controlled while the engine is running are highlighted in the green and blue box.

The complete transient model receives as input the target torque and the engine speed as function of time: these are used to define the required engine parameters for the current engine speed and load by means of interpolation tables. The controllers (green blocks) determine the actuated quantities such as injected fuel mass, wastegate valve opening and inter-cooler wall temperature, which enter the virtual engine S-Function. Finally, the S-Function block provides the desired outputs.

This complete engine model, operating in transient conditions, can be coupled to the conventional vehicle model, improving the basic representation of the IC engine by means of measured steady-state operating maps. The standard vehicle simulation tool reproduces the behavior of the different vehicle components, including the engine, by pre-determined tables and via interpolation, every output quantity requested can be instantly calculated without considering the dynamics of the device.

To introduce the engine model by an S-Function block, the pre-existent vehicle velocity control is modified as reported in Figure 5, which shows the simulation architecture.

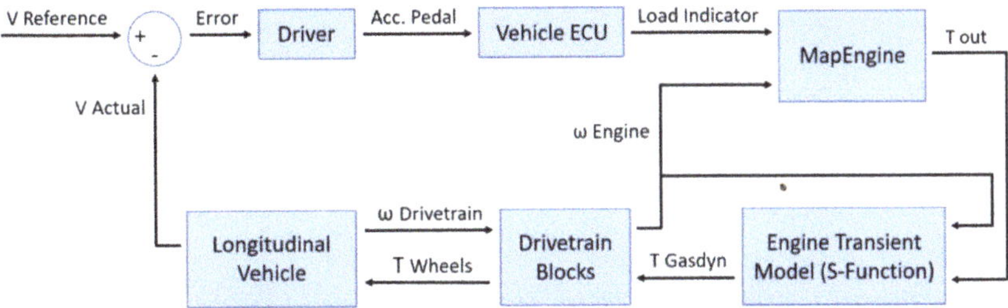

Figure 5. Updated vehicle velocity control; ω: angular speed, V: vehicle speed, T: torque.

While in the conventional approach, the output torque of the engine is simply determined via interpolation of the engine maps. The proposed methodology instead asks the running engine model to produce the required torque. This implies that the output torque of the engine might not be equal to the target but reflects the fluid dynamic solution obtained, hence it is affected by the intrinsic inertia and unsteadiness of the engine. It is the actual torque produced by the engine that is fed to the vehicle dynamics.

5. Calibration of Engine Model

An important step is related to the calibration of the 1D thermo-fluid dynamic model of the engine; in particular, the set up and consequent tuning of the combustion model is fundamental to achieve reliable results.

5.1. Description of Complete Engine Model

The 6-cylinder turbocharged Diesel engine schematic is reported in Figure 6, highlighting the typical components introduced for the simulation.

Figure 6 shows the air and exhaust gas paths, including the compressor and the fixed geometry twin-entry turbine with a waste-gate valve. Six fuel injectors are present, together with dedicated controllers for intercooler outlet temperature and engine BMEP, to make the engine work in the same operating points of the experimental campaign. In the case of steady-state operating points, the engine cycles calculations are performed until convergence to the targets is reached.

It is important to underline that the model does not feature any after-treatment system (ATS) and that the transport of chemical species along the ducts was not activated: this means that the emissions could be evaluated only as cylinder-out.

The calibration procedure of the model was carried out with a detailed 1D schematic, with a very refined mesh. After that, the subsequent simulations with the complete engine-vehicle system were based on the FSM approach with a coarse mesh, to achieve a significant reduction of the computational burden preserving the accuracy.

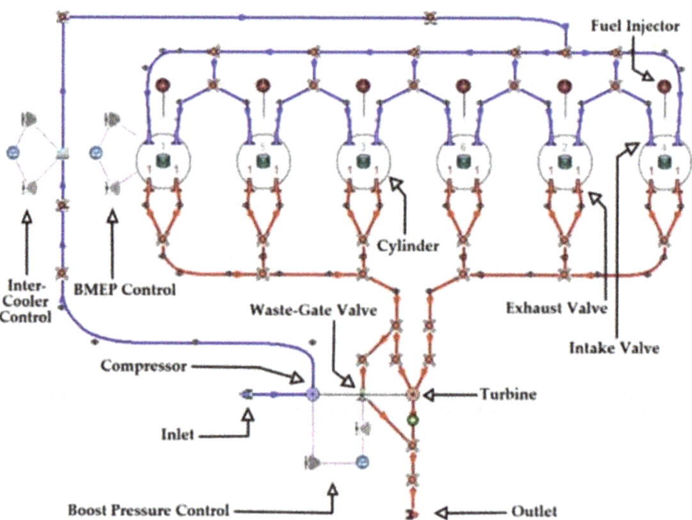

Figure 6. Six-cylinder turbocharged Diesel engine, 0D/1D schematic.

5.2. Engine Model Calibration

The CI engine combustion process is modeled with this approach: the cylinder volume is considered as a 0D system, in which the different chemical species are accounted for by a mixture of ideal gases. The process takes place with closed valves and is governed by the following energy equation (1st law of thermodynamics) [22]:

$$\frac{V}{\overline{\gamma}-1}\cdot\frac{dP}{d\theta} + \frac{\overline{\gamma}}{\overline{\gamma}-1}\cdot P\cdot\frac{dV}{d\theta} = m_F\cdot LHV\cdot\frac{dx_b}{d\theta}\cdot\eta_c - \frac{\dot{Q}_w}{\omega} \tag{4}$$

where V is the instantaneous in-cylinder volume, P the in-cylinder pressure (known from the experimental pressure traces), γ the average adiabatic constant of the equivalent gas mixture, θ the crank angle, m_F the injected fuel mass (per cylinder per cycle), LHV the lower heating value of the fuel, x_b the "fuel burnt mass fraction" (i.e., the mass fraction of fuel that has already taken part to the combustion, at a specific crank angle position), η_c the efficiency of the combustion process, $\dot{Q}_w$ the heat loss through the cylinder walls and ω the crankshaft angular speed. In the energy balance, the unknown to be computed is x_b.

In Equation (4), the left side member is the so-called "actual heat release rate" (AHRR), an energy contribution due to pressure and volume variation representing the energy effectively used to produce work. This term equals the difference between the "heat release rate" (HRR) by fuel combustion and the heat losses through the cylinder walls.

While the quantities AHRR and HRR have a known expression, the heat losses typically do not. The simple application of predictive formulas such as the Woschni model [23,24] ([6,7]) does not guarantee a reliable calculation of this term, since those models need to be recalibrated for each test case (as documented in [25,26]). Hence, by integrating Equation (4) during the combustion phase, it is possible to write that:

$$Q_{w,tot} = \int_{\theta_i}^{\theta_f} \frac{\dot{Q}_w}{\omega}d\theta = m_F\cdot LHV - \int_{\theta_i}^{\theta_f}\left(\frac{V}{\overline{\gamma}-1}\cdot\frac{dP}{d\theta} + \frac{\overline{\gamma}}{\overline{\gamma}-1}\cdot P\cdot\frac{dV}{d\theta}\right)d\theta \tag{5}$$

where $Q_{w,tot}$ is the total heat exchanged during combustion, by assuming the combustion efficiency η_c equal to 1. This quantity can be computed for all the engine map operating points, once the in-cylinder pressure is known by measurements and used to calibrate the

Woschni model point by point. In this work the tuning has been carried out by means of the two coefficients and (for more details please refer to [23,24]) and by minimizing the following error function for each operating point:

$$z(C_1, C_2) = \left| \frac{Q_{w,tot} - Q_w(C_1, C_2)}{\text{CHR}} \right| \tag{6}$$

where CHR is the cumulative heat release, and Q_w (C_1, C_2) is the computed wall heat exchanged for a given set of values (C_1, C_2). The distribution of C_1 and C_2 values on the engine map is shown in Figure 7. To retain a simpler set of C_1 and C_2 coefficients to be inserted in the 1D simulation model, their distribution on the engine map has been averaged along the BMEP axis, achieving a dependence only on engine speed.

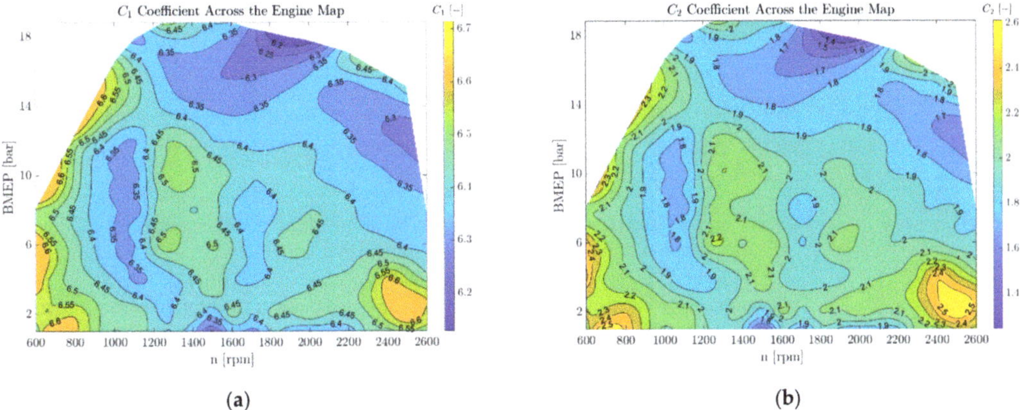

(a) (b)

Figure 7. (**a**) C_1 distribution across the engine map; (**b**) C_2 distribution across the engine map. C_1 and C_2 are the coefficients of the classical Woschni heat transfer model.

A little distortion in the distribution can be seen in some full load points: hence, these values have been reconstructed based on the adjacent points.

The total heat losses during the combustion process can be computed by applying the Woschni model, as reported in Figure 8, where the heat loss is expressed as a percentage of the total heat released by fuel combustion (i.e., simply, assuming η_c = 1).

As shown by the contour plots of Figure 8, the recalibration of Woschni's formula is fundamental to correctly evaluate the heat loss term, especially for medium-large displacements such as the one considered: without this tuning process, the heat losses would have been underestimated almost by a factor of two.

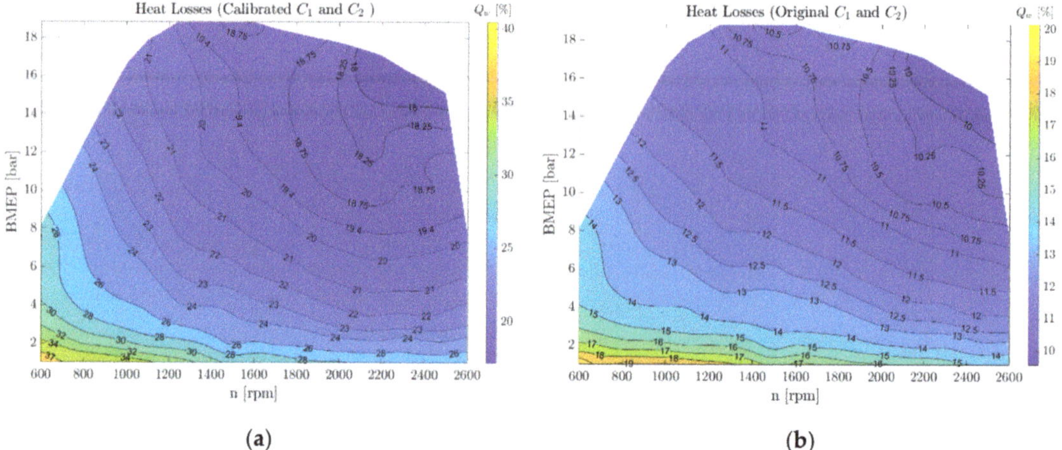

(a)   **(b)**

Figure 8. (**a**) Percentage heat losses with recalibrated Woschni model; (**b**) Percentage heat losses with original Woschni model (i.e., using the default coefficients of the model).

The fuel burnt mass fraction x_b can be computed once the energy balance has been verified. The calculation is performed by integrating Equation (5), which is implicit and non-linear, since γ depends on the chemical composition, which varies during the combustion. Once the x_b profiles are computed for each sampled operating point, they must be extended across the entire engine map, since the engine could run in whichever working condition. Instead of interpolating among the calculated profiles, which are functions defined over different domains, it is advisable to approximate them by an analytical function (in this case the "double Wiebe's" one) and then interpolate the fitting coefficients of these functions. The double Wiebe's function can be written as follows (Equation (7)) [27]:

$$x_b(\theta) = \beta \cdot \left(1 - e^{-a_1 \cdot \left(\frac{\theta - \theta_{i,1}}{\theta_{f,1} - \theta_{i,1}} \right)^{m_1+1}} \right) + (1-\beta) \cdot \left(1 - e^{-a_2 \cdot \left(\frac{\theta - \theta_{i,2}}{\theta_{f,2} - \theta_{i,2}} \right)^{m_2+1}} \right) \tag{7}$$

where β is the "weight factor", a_1 and a_2 are the "efficiency parameters", m_1 and m_2 the "shape factors", $\theta_{i,1}$ and $\theta_{i,2}$ the "starts of combustion", $\theta_{f,1}$ and $\theta_{f,2}$ the "ends of combustion". The fitting is performed by means of MATLAB® built-in algorithms and the only parameters used are β, $\theta_{f,1}$, m_1 and m_2; moreover, the following hypotheses are done:

- $a_1 = a_2 = 6.9078$ (complete combustion);
- $\theta_{i,1} = \theta_{i,2} = SOC_{main}$ (the start of combustion is set with respect to the ignition of the main injection, since very little fuel is injected in the operating points with a pilot injection);
- $\theta_{i,1} = \theta_{i,2} \leq \theta_{f,1} \leq \theta_{f,2} = EVO$ (exhaust valve opening);
- $0 \leq \beta \leq 1$;
- $0 \leq m_1, m_2 \leq 1$.

Once the fitting coefficients are defined for all the sampled points, their distribution across the engine map is further approximated with polynomial surfaces as function of engine speed and BMEP, to remove local distortions and have an easy computation of the coefficients for any combination of engine speeds and loads.

A comparison for the computed trends of the fuel burnt mass fraction x_b is represented in Figure 9, for two operating points as examples.

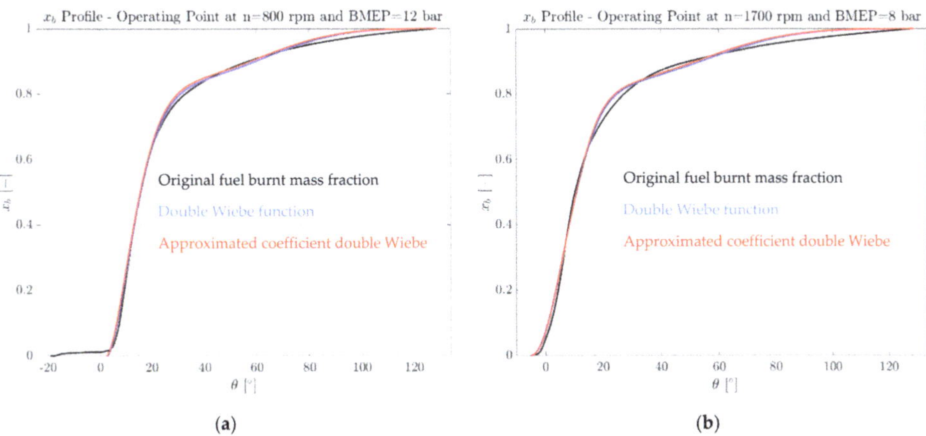

(**a**) (**b**)

Figure 9. (**a**) Fuel burnt mass fraction profile for the operating point at n = 800 rpm, BMEP = 12 bar. This is characterized by a double injection (a pilot and a main one), however the assumption of neglecting the pilot injection (for the fitting of the double-Wiebe function) is accurate; (**b**) Fuel burnt mass fraction profile for the operating point at n = 1700 rpm, BMEP = 8 bar (with single injection, in this case).

The profiles calculated by polynomial approximation are then considered for the remaining part of this study, given the satisfactory approximation they can provide.

6. Steady-State Map Results

After simulating all the map operating points, the predicted results can be collected and represented by means of contour plots, indexed by engine speed and load, which are used to identify the engine working conditions. These contour plots can be addressed as the steady-state maps of the IC engine, each one detailing a particular property. This kind of representation offers a quick, graphical way to verify the satisfactory correspondence between simulation results and experimental data.

As an example, Figure 10 shows a group of steady-state maps referring to the most important engine performance and emission characteristics: in particular, Figure 10a–f highlight the discrepancy between measured and calculated results in terms of percentage error, whereas Figure 10g,h directly show the measured and simulated (non-dimensional) cylinder-out NOx emissions.

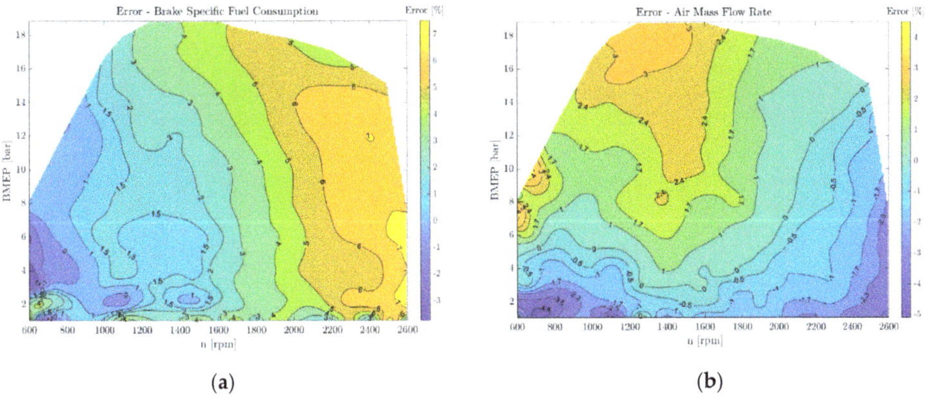

(**a**) (**b**)

Figure 10. *Cont.*

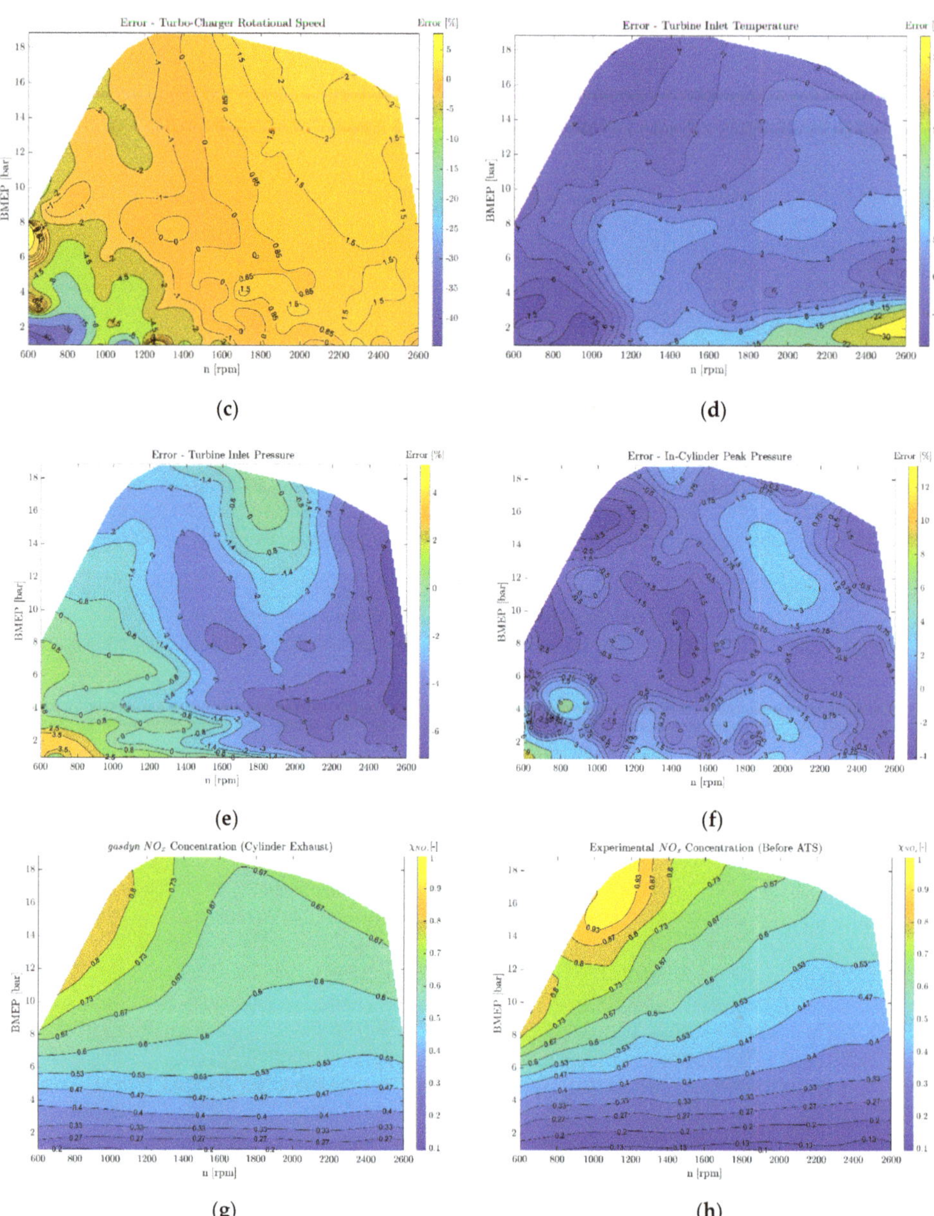

Figure 10. Steady-state maps of the six-cylinder Diesel engine predicted by *Gasdyn*: (**a**) % error—BSFC; (**b**) % error—air mass flow rate; (**c**) % error—turbocharger Speed; (**d**) % error—turbine inlet temperature; (**e**) % error—turbine inlet pressure; (**f**) % error—cylinder peak pressure; (**g**) cylinder-out NO_x concentration (non-dimensional); (**h**) cylinder-out experimental NO_x concentration (non-dimensional).

The prediction of specific fuel consumption by means of the 0D/1D code is very accurate at low and medium engine speeds, as reported in Figure 10a, whereas a larger discrepancy can be found at high regimes. However, considering the typical applications of the engine investigated, that part of the engine map is not frequently exploited, so that a reliable evaluation of fuel consumption is expected during an RDE test cycle.

Considering further important engine parameters (Figure 10b–f), the percentage errors are very small and local distortions can be acceptable at the edge of the maps (typically due to interpolation issues). Finally, the calculated NOx emission map (Figure 10g) agrees with the corresponding measured trends and values (Figure 10h), represented in normalized form with respect to the maximum experimental value. The satisfactory comparison confirms the good set-up of the combustion model, being NOx formation dependent on the correct prediction of in-cylinder temperature and local air-to-fuel ratio. The most important result of this activity is that the engine model can be run across non-mapped conditions reflecting the behavior of the real engine. When the model runs in transient, interpolations are performed to set up the engine model exploiting the know values of the nearest mapped operating conditions parameters.

7. Description of Vehicle and Test Cycles

The vehicle used for the simulations is an interurban hybrid bus, which is coupled to the six-cylinder Diesel engine analyzed for this investigation. The main characteristics of this vehicle (mass, size, gearbox ratios, target mission) with a classical architecture were taken from the literature. Moreover, it is assumed that the vehicle runs on medium payload, with an overall vehicle mass equal to 12,600 kg. In the simulations, the vehicle is equipped with a six-speed manual gearbox and, according to the software gear shifting strategy, the gear shift lines are placed to allow the engine to work in its high efficiency region and to prevent it from running at excessive speeds. These lines are represented by engine torque and speed curves in Figure 11.

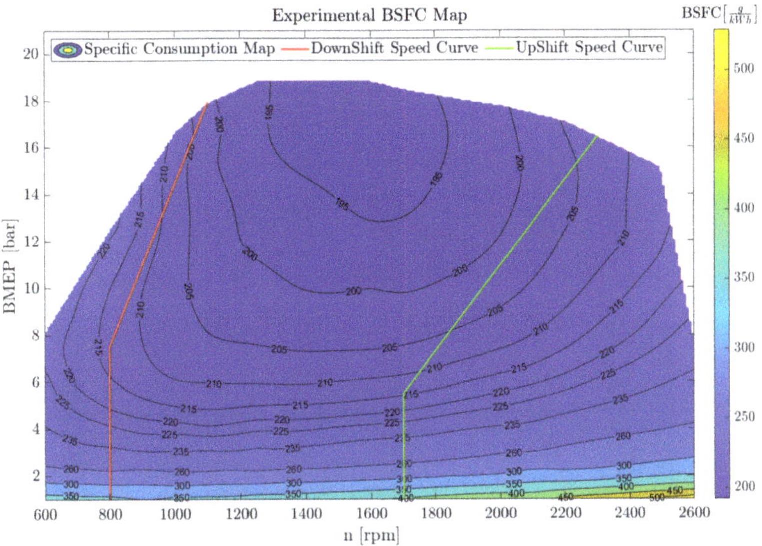

Figure 11. Gear shift lines corresponding to a six-speed manual gearbox mounted on the studied inter-urban bus, reported on the BSFC map of the six-cylinder Diesel engine.

To calculate the vehicle longitudinal dynamics, all the vehicle model blocks are provided with the requested architecture-layout data. Firstly, the simulations are carried out on the Diesel bus version with manual transmission. Then, a full hybrid parallel P2 configuration of the bus has been set up and investigated. The electric motor, placed between the engine flywheel and the gear box is a permanent magnet synchronous one. A "torque split" algorithm, together with a first-order thermal model, is used to test and validate the motor, which is simulated through steady-state maps. To match the different engine and motor speed ranges, a reduction gear connects the latter to the drive shaft between the clutches. The system is powered by a lithium-ion battery pack, characterized by a 1p110s (1 parallel, 110 series) arrangement working at 400 V, which has been designed taking into account the peak-power request, calculated by means of the "torque split" algorithm adopted, and of the overall energy storage requirements.

Once the electric components are chosen, a proper tuning of the hybrid control unit (HCU) is performed in the vehicle simulation tool, acting on the specific torque-speed curves that define the operating mode of the hybrid system, and the battery SoC (state of charge) threshold levels [28]. As mentioned, a first set of parameters to be defined is represented by the torque curves, which are indexed by the revolution speed, generalized to the drive shaft between the clutches. These curves characterize different regions on a torque-speed plane: depending on the total torque required by the vehicle and its actual speed, the position of such working condition on the torque-speed plane determines the hybrid system operating mode. For the hybrid architecture considered, the torque-speed curves are referred to the engine BSFC map: this allows their correct positioning, aimed at reaching the best overall efficiency.

In Figure 12, it is possible to distinguish four different curves, cited with their actual name in the vehicle simulation tool. To correctly place these curves, the underlying rationale of the hybrid powertrain must be pursued: letting the engine work in its high-efficiency regions.

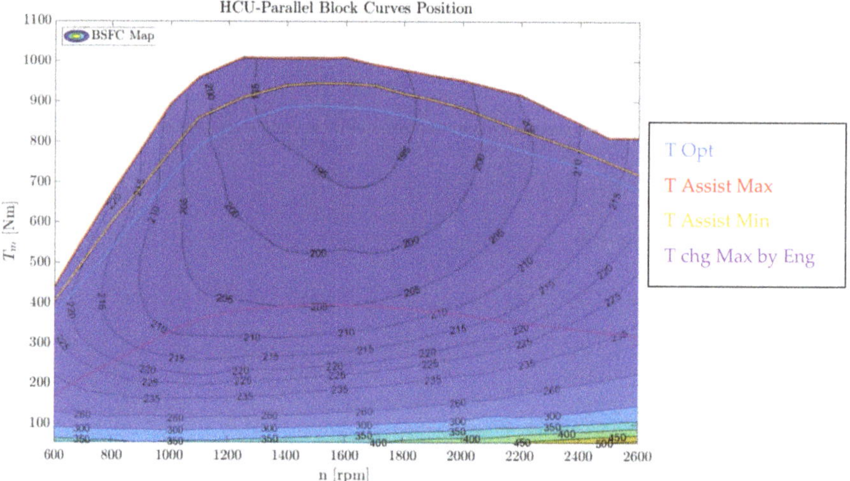

Figure 12. Position of the HCU-Parallel block curves on the engine BSFC map.

The lowest purple line defines a map region below which the engine is generally prevented from working: when the requested engine torque lies below this curve, the HCU determines an increase of torque request, so as to bring the operating point above the purple line. Then, the torque surplus is used to charge the battery through the motor generator mode. This curve has been placed to achieve a compromise: it prevents the engine from working in low-efficiency points, such as the bottom part of the map, so that above this curve an optimal operation of the engine is achieved.

The region between the purple line and the yellow one defines the area in which the engine can work. If the required torque is located between the yellow line and the red one the engine is supported by the electric motor to provide the requested torque. The blue curve represents the maximum efficiency curve for the engine, to achieve the minimum fuel consumption and CO_2 production. It is important to remark that, due to the characteristics of this engine, the maximum efficiency is often located around the full load in the engine map. If the overall requested torque exceeds the full load torque of the engine, the additional torque requested is provided by the electric motor. The vehicle control unit also decides when to turn off and on the engine. It is important to highlight that the engine model has the capability to be turned off and on, with the corresponding effects on the fluid dynamic solution.

To evaluate the bus performance under various scenarios, two different missions were chosen. The first is the orange county (OC) cycle, a chassis test for heavy-duty vehicles such as buses. Its speed profile is depicted in Figure 13a: with relatively low speeds, but frequent and steep accelerations and decelerations. This cycle is suitable to test the bus in urban driving conditions. The second one is the world harmonized vehicle cycle (WHVC), a generic test for heavy-duty vehicles, usually used for research purposes to compare the vehicle and engine emissions, derived from its corresponding engine dynamometer cycle: the WHTC. By looking at Figure 13b, this driving cycle consists of three different phases: an urban one, from the start of the cycle until the red dotted line; a rural phase, limited by the red and green dashed lines; a motorway phase, in the last part of the cycle. The composition of the cycle makes it suitable to test the bus in various conditions, including travelling on extra-urban roads, during the rural and motorway phases. Further information about these cycles can be found in [29,30].

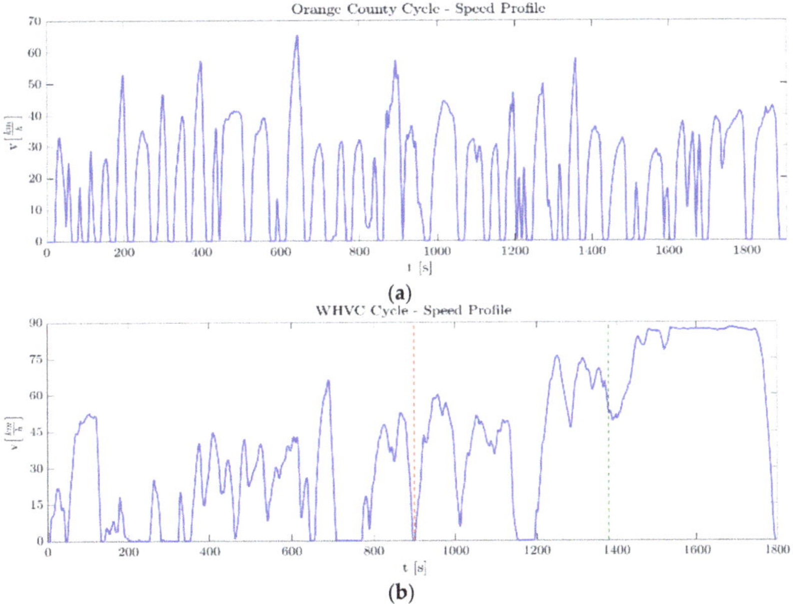

Figure 13. (a) Orange County cycle—speed profile; (b) WHVC—speed profile.

8. Vehicle Dynamics Simulations

The application and validation of this new co-simulation procedure are based on a direct comparison between two different approaches. In this section, for both bus versions (conventional Diesel and hybrid), the main outcomes of the map-based approach and the transient coupled model are reported and discussed. Only the results related to the WHVC cycle are shown since it represents a mixed mission able to test the new methodology into a wide range of accelerations and speeds, even if the full exploitation of a hybrid system can be reached in urban driving.

The fuel economy series for the different Diesel simulation models are collected in Table 3. The increase of fuel consumption for the transient model is probably due to the dynamics of the engine (inertia is considered with respect to the map-based steady-state model) as well as to a little overestimation of fuel consumption, as shown by Figure 10a and by the midway model (based on the map of fuel consumption, obtained by the analysis of steady-state operating conditions).

Table 3. WHVC: Diesel bus fuel economy comparison.

Diesel Bus—Fuel Economy	Value $\left[\frac{1}{100\ \text{km}}\right]$	Error [%]
Map-Based, Experimental [1]	24.88	Reference
Map-Based, calculated points [2]	25.57	2.77
Transient [3]	25.8	3.7

[1] Map-Based approach with fuel consumption map from experimental data. [2] Map-Based approach with fuel consumption map from engine model steady-state analysis. [3] Vehicle-engine transient model.

To carry out a fair comparison between the fuel consumption of the Diesel and hybrid versions, a normalization technique must be applied. The procedure devised by the "low carbon vehicle partnership" (LowCVP) and adopted in the United Kingdom is described in detail in [29]: intended just for buses, it perfectly fits the case considered in this work. According to this procedure, it is necessary to calculate a quantity called "net energy change" (NEC), expressed by the following formula:

$$NEC = (SoC_f - SoC_i) \cdot V_{sys} \cdot K_1 \tag{8}$$

where SoC_f and SoC_i are respectively the final and initial battery state of charge expressed in [Ah] (calculated multiplying the battery capacity by the percentage state of charge), V_{sys} is the nominal DC operating voltage of the battery, and K_1 is a constant equal to 3600 s/h, necessary to pass from hours to seconds. Besides the NEC, two further quantities must be calculated. These are the "total fuel energy" (TFE) and the "total cycle energy" (TCE):

$$TFE = m_{F,tot} \cdot LHV \tag{9}$$

$$TCE = TFE - NEC \tag{10}$$

where $m_{F,tot}$ is the total fuel mass consumed. The last formula (10) computes the total energy used to complete the driving cycle, provided by the fuel combustion and the battery. The final step is to calculate the "NEC variance" according to Equation (11) and check which range of Table 4 it falls into.

$$|NEC\ Variance| = \left|\frac{NEC}{TCE} \cdot 100\right| \tag{11}$$

Table 4. Possible |NEC Variance| ranges.

|NEC Variance|	Description
$\leq 1\%$	The fuel consumption measured in the cycle can be considered correct.
$>1\% \wedge \leq 5\%$	The fuel consumption measured in the cycle needs to be corrected.
$>5\%$	The vehicle run under analysis is not considered valid.

The correction requires to simulate the driving cycle multiple times, until getting three valid runs at least (i.e., each one with a |NEC Variance| $\leq 5\%$). Once this is achieved, each test run must be identified by an NEC Variance value and a fuel consumption; these points are then represented in a plane and processed with a linear regression (see Figure 14). If the "coefficient of linear correlation" of the regression is $R^2 \geq 0.8$, then the set of points constitutes a family of valid test runs and the corrected fuel consumption is given by the linear regression straight line, evaluated at NEC Variance = 0%. Hence, as indicated by the normalization procedure, three different SoC_i are chosen: 35%, 50% and 65% for both the map-based experimental procedure and the transient simulation. Note that all six runs are valid (i.e., |NEC Variance| $\leq$ 5%).

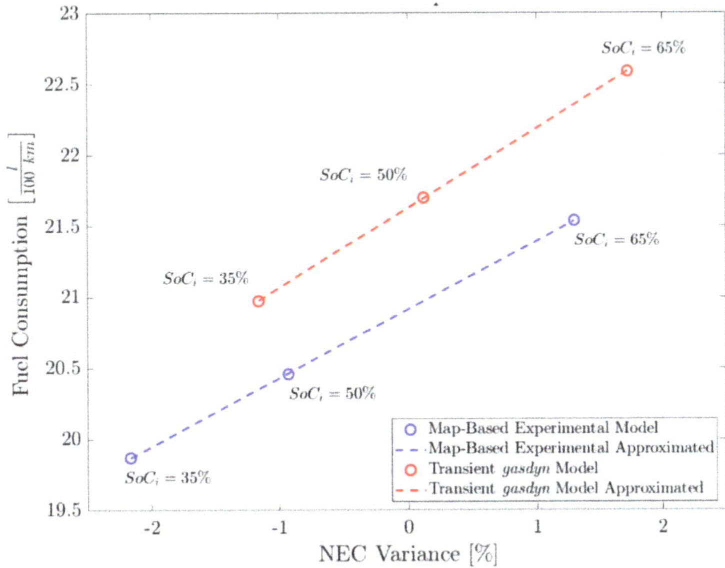

Figure 14. Calculation of WHVC fuel consumption by different models.

The normalized fuel consumption, calculated for the hybrid vehicle by both simulation approaches, is collected in Table 5 and directly compared to that for the conventional Diesel vehicle. As already discussed, the transient simulation model reveals slightly higher fuel consumption.

Table 5. WHVC: Fuel economy for Diesel and hybrid bus architectures.

Bus Version and Simulation Approach	Fuel Economy $\left[\frac{1}{100\,\text{km}}\right]$ (Normalized One in Case of the Hybrid Bus)
Map-Based Experimental [1] Diesel	24.88
Map-Based Experimental [1] hybrid	20.91
Transient [2] Diesel	25.8
Transient [2] hybrid	21.63

[1] Map-Based approach with fuel consumption from experimental data. [2] Vehicle with engine transient model.

In Figure 15a, the resulting cloud of engine operating points during the WHVC test cycle with a SoC_i of 50%, simulated by the transient approach, are shown together with the HCU curves of Figure 12. The engine operation follows the rules of the logic previously described: it works very rarely where the engine is not efficient, except when the engine is off or it is turned on trying to the target operating condition. Figure 15b shows the SoC of the battery during the same run, calculated by both simulation methods. The two curves are similar, but not identical: this depends on how the HCU manages and uses the two sources of torque (i.e., the e-motor and the IC engine). It bases the splitting on the required torque at that time instant: since the requested torque can be a little bit different between the two modeling approaches (due to the influence of engine dynamics), a particular engine condition can fall into different HCU regions on the map. Therefore, the managing of the required torque can be different, as well as the SoC profile.

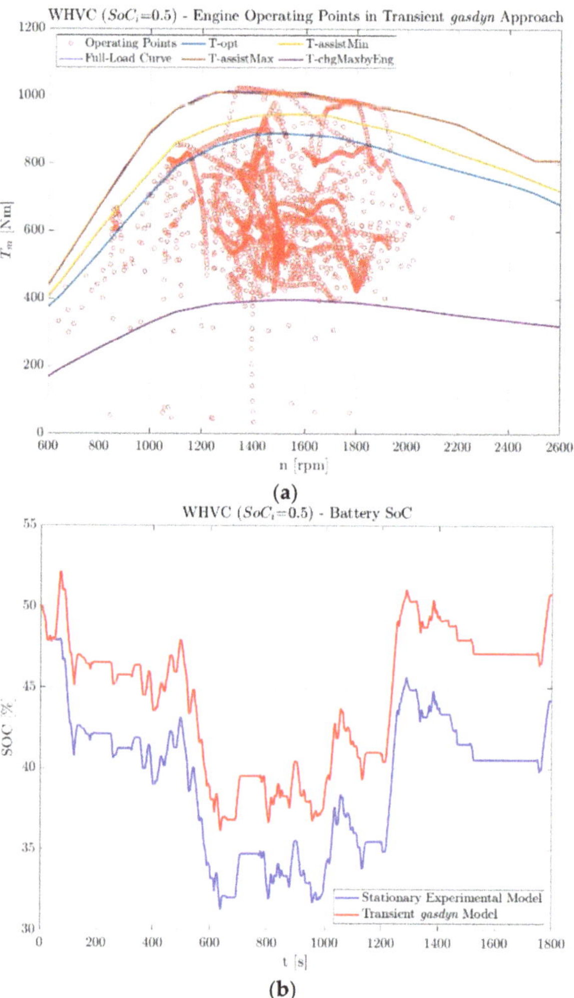

Figure 15. WHVC with SoC_i = 50%, simulation results: (**a**) cloud of engine operating points calculated by the transient *Gasdyn* approach; (**b**) battery SoC.

Moreover, the SoC at the end of simulation (SoC_i) reaches half of the battery storage energy. This convergence condition can be noticed for both the driving cycles and the three SoC_i values chosen, emphasizing the good design of the HCU logic.

Even if the WHVC is a mixed driving cycle, significant fuel savings can be achieved, probably mostly during its first phase: the urban drive. In fact, by comparing the Diesel and hybrid map-based calculations, the fuel economy improves by 15.96%, while for the hybrid transient simulation approach the reduction of fuel economy is equal to 16.16% (with respect to the Diesel transient model). Overall, a satisfactory agreement can be found, when comparing the two different powertrain architectures with the same modeling approach: independently of how the engine operation is calculated (through the map-based or transient method), the HCU is able to manage both, highlighting the advantages of the bus powertrain investigated in this work: the hybrid architecture.

Dealing with the simulation of RDE and WHTC test cycles, it is important to discuss the computational effort requested and the prediction accuracy achieved by the FSM (fast simulation method) methodology, compared to the standard solver applied to a 1D schematic with a refined mesh. This can be analyzed in terms of CPU time and of variation of engine output quantities. To this purpose, the simulation of the same WHTC cycle has been performed with both a refined mesh (1 cm) and a coarse one (10 cm with the FSM methodology). As shown in Figure 16a,b the reduction of computational effort is relevant, around 80% with respect to the same simulation in the refined mesh solution. To estimate if the accuracy has changed between the two approached, the cumulative fuel consumption and cylinder out emission are compared. As it can be seen, the two approaches differ by less than 1%, revealing that the overall accuracy has been preserved and that long-lasting driving cycle can be simulated by resorting to fast simulation strategies without introducing excessive approximations.

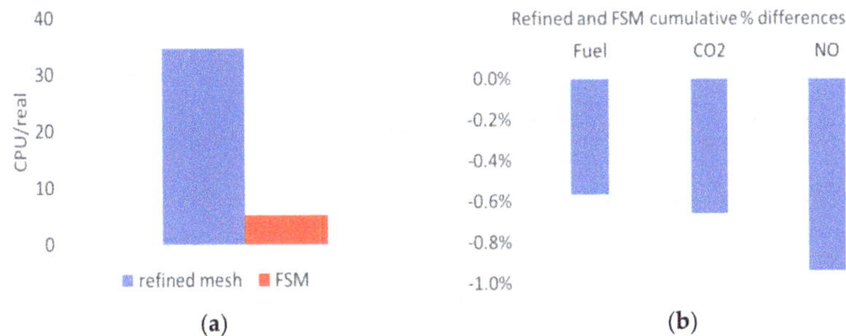

Figure 16. (a) Comparison between CPU time/real time of the driving cycle simulated by means of both refined mesh (1 cm) and coarse mesh (10 cm, FSM); (b) difference in the prediction of cumulative emissions of the engine emissions between the refined mesh case and FSM.

9. Conclusions

This research work focused on the prediction of fuel consumption of a Diesel heavy-duty engine in real driving cycles, by means of computer simulations. As investigated in previous works, the map-based approach (relying on tables derived through the engine steady-state analysis), to represent the IC engine characteristics within a vehicle model, can be very fast but it could also provide poor results in terms of fuel consumption and pollutants emissions, especially with new and demanding test cycles. In fact, this approach completely disregards the engine dynamics, which runs considering a simple sequence of steady-state points, with a sort of instantaneous transition from one to another.

To define a reliable simulation tool for the analysis of IC engines during transients, this work investigated the suitable coupling of a complete vehicle model to an IC engine model, interacting in real-time during an RDE.

A 0D/1D six-cylinder Diesel engine model has been developed and, with some calibration of its combustion model on the basis of the available experimental data, validated in steady-state (engine map) conditions. Hence, the study focused on the transient operation of the IC engine, which has been integrated into a complete vehicle model, by means of the Simulink® environment and coupled to vehicle models. A conventional vehicle architecture has been compared to a hybrid architecture, to highlight the performances of the two solutions and the prediction of fuel consumption by means of the usual map-based approach and the transient approach.

Future developments could focus on the prediction of pollutant emissions under transient operation. In fact, the complete transient model built could provide the input data, in terms of cylinder-out emissions, to the after-treatment simulation model coupled to the IC engine itself, which could highlight the tailpipe emissions during different RDE cycles and strategies.

Author Contributions: Conceptualization, T.C. and G.D.; Data curation, A.M. and G.M.; Formal analysis, A.M., G.M. and A.O.; Funding acquisition, A.O.; Investigation, M.T. and T.C.; Project administration, G.D. and A.O.; Resources, G.D. and E.P.; Supervision, A.M., T.C., G.M., G.D. and A.O.; Validation, E.P. and E.E.P.; Visualization, T.C.; Writing—original draft, A.M., E.P. and E.E.P.; Writing—review & editing, A.M., G.M. and E.E.P. All authors have read and agreed to the published version of the manuscript.

Funding: This research was funded by European Union, grant number 824314.

Institutional Review Board Statement: Not applicable.

Informed Consent Statement: Not applicable.

Acknowledgments: The authors wish to thank the European Union project VISION-xEV (virtual component and system integration for efficient electrified vehicle development), which received funding from the EU Horizon 2020 research and innovation program, under grant agreement No 824314. The authors are also grateful to the project partners involved the IC engine simulation activity, in particular to FPT Industrial for providing the experimental data required to validate the simulation models.

Conflicts of Interest: The authors declare no conflict of interest.

Abbreviations

RDE	Real Drive Emission
HCU	Hybrid Control Unit
BMEP	Brake Mean Effective Pressure
ATS	After Treatment System
LHV	Lower Heating Value
AHRR	Apparent Heat Released Rate
HRR	Heat Released Rate
SOC	Start of Combustion
EVO	Exhaust Valve Opening
BSFC	Break Specific Fuel Consumption
SoC	State of Charge
WHVC	World Harmonized Vehicle Cycle
WHTC	World Harmonized Transient Cycle
NEC	Net Energy Change
TFE	Total Fuel Energy

References

1. Kratzsch, M.; Wukisiewitsch, W.; Sens, M.; Brauer, M.; Tröger, R. The Path to CO_2-Neutral Mobility in 2050. In Proceedings of the 40th Internationales Wiener Motorensymposium, Wien, Austria, 15–17 May 2019.
2. Joshi, A. Review of Vehicle Engine Efficiency and Emissions. *SAE Int. J. Adv. Curr. Prac. Mobil.* **2020**, *5*, 2479–2507.
3. Onorati, A.; Montenegro, G. *1D and Multi-D Modeling Techniques for IC Engine Simulation*; SAE International: Warrendale, PA, USA, 2020.
4. Millo, F.; Giacominetto, P.F.; Bernardi, M.G. Analysis of different exhaust gas recirculation architectures for passenger car Diesel engines. *Appl. Energy* **2012**, *98*, 79–91. [CrossRef]
5. Wurzenberger, J.C.; Bardubitzki, S.; Bartsch, P.; Katrasnik, T. *Real Time Capable Pollutant Formation and Exhaust Aftertreatment Modeling-HSDI Diesel Engine Simulation*; SAE Technical Paper Series; SAE International: Warrendale, PA, USA, 2011.
6. Wahiduzzaman, S.; Wang, W.; Leonard, A.; Wenzel, S. Development of Integrated Engine/Catalyst Model for Real-Time Simulations of Control Design Options. In Proceedings of the FISITA 2010 World Automotive Congress, Budapest, Hungary, 30 May–4 June 2010.
7. Onorati, A.; Ferrari, G.; Montenegro, G.; Caraceni, A.; Pallotti, P. *Prediction of S.I. Engine Emissions During an ECE Driving Cycle via Integrated Thermo-Fluid Dynamic Simulation*; SAE Technical Paper Series; SAE International: Warrendale, PA, USA, 2004. [CrossRef]
8. Martínez, F.J.A.; Arnau, F.; Piqueras, P.; Auñon, A. *Development of an Integrated Virtual Engine Model to Simulate New Standard Testing Cycles*; SAE Technical Paper Series; SAE International: Warrendale, PA, USA, 2018.
9. Onorati, A.; Montenegro, G.; D'Errico, G.; Piscaglia, F. *Integrated 1D-3D Fluid Dynamic Simulation of a Turbocharged Diesel Engine with Complete Intake and Exhaust Systems*; SAE Technical Paper Series; SAE International: Warrendale, PA, USA, 2010. [CrossRef]
10. Montenegro, G.; Onorati, A.; D'Errico, G.; Cerri, T.; Marinoni, A.; Tziolas, V.; Zingopis, N. *Prediction of Driving Cycles by Means of a Co-Simulation Framework for the Evaluation of IC Engine Tailpipe Emissions*; SAE Technical Paper Series; SAE International: Warrendale, PA, USA, 2020.

11. Winterbone, D.; Pearson, R. Theory of Engine Manifold Design: Wave Action Methods for IC Engineers. *Appl. Mech. Rev.* **2001**, *54*, B109–B110. [CrossRef]
12. Onorati, A.; Ferrari, G.; D'Errico, G.; Montenegro, G. *The Prediction of 1D Unsteady Flows in the Exhaust System of a S.I. Engine Including Chemical Reactions in the Gas and Solid Phase*; SAE Technical Paper Series; SAE International: Warrendale, PA, USA, 2002. [CrossRef]
13. Serrano, J.R.; Arnau, F.J.; Piqueras, P.; Onorati, A.; Montenegro, G. 1D gas dynamic modelling of mass conservation in engine duct systems with thermal contact discontinuities. *Math. Comput. Model.* **2009**, *49*, 1078–1088. [CrossRef]
14. Montenegro, G.; Della Torre, A.; Onorati, A.; Fairbrother, R. A Nonlinear Quasi-3D Approach for the Modeling of Mufflers with Perforated Elements and Sound-Absorbing Material. *Adv. Acoust. Vib.* **2013**, *2013*, 546120. [CrossRef]
15. Corberan, J.M.; Gascon, M.L. Construction of Second Order TVD Schemes for Non-Homogeneous Hyperbolic Conservation Laws. *J. Comput. Phys.* **2001**, *172*, 261–297.
16. Benson, R.S. *The Thermodynamics and Gasdynamics of Internal Combustion Engines*; Clarendon Press: Wotton-under-Edge, UK, 1992; Volume 1.
17. Usai, V.; Marelli, S. Steady State Experimental Characterization of a Twin Entry Turbine under Different Admission Conditions. *Energies* **2021**, *14*, 2228. [CrossRef]
18. Hadef, J.E.; Colin, G.; Chamaillard, Y.; Talon, V. Physical-Based Algorithms for Interpolation and Extrapolation of Turbocharger Data Maps. *SAE Int. J. Engines* **2012**, *5*, 363–378. [CrossRef]
19. Montenegro, G.; Tamborski, M.; Della Torre, A.; Onorati, A.; Marelli, S. Unsteady Modeling of Turbochargers for Automotive Applications by Means of a Quasi-3D Approach. *J. Eng. Gas Turbines Power* **2020**, *143*, 071028. [CrossRef]
20. Bozza, F.; De Bellis, V.; Marelli, S.; Capobianco, M. 1D Simulation and Experimental Analysis of a Turbocharger Compressor for Automotive Engines under Unsteady Flow Conditions. *SAE Int. J. Engines* **2011**, *4*, 1365–1384. [CrossRef]
21. Eckert, P.; Henning, L.; Rezaei, R.; Seebode, J.; Kipping, S.; Behnk, K.; Traver, M.L. Management of Energy Flow in Complex Commercial Vehicle Powertrains. *SAE Int. J. Commer. Veh.* **2012**, *5*, 271–279. [CrossRef]
22. Millo, F.; Rolando, L.; Andreat, M. Numerical Simulation for Vehicle Powertrain Development. In *Numerical Analysis—Theory and Application*; IntechOpen: London, UK, 2011.
23. Woschni, G. *A Universally Applicable Equation for the Instantaneous Heat Transfer Coefficient in the Internal Combustion Engine*; SAE Technical Paper; SAE International: Warrendale, PA, USA, 1967. [CrossRef]
24. Žák, Z.; Emrich, M.; Takáts, M.; Macek, J. In-Cylinder Heat Transfer Modelling. *J. Middle Eur. Constr. Des. Cars* **2016**, *14*, 2–10. [CrossRef]
25. Soyhan, H.; Yasar, H.; Walmsley, H.; Head, B.; Kalghatgi, G.; Sorusbay, C. Evaluation of heat transfer correlations for HCCI engine modeling. *Appl. Therm. Eng.* **2009**, *29*, 541–549. [CrossRef]
26. Le Guen, S.; Maiboom, A.; Bougrine, S.; Tauzia, X. *Analysis of Systematic Calibration of Heat Transfer Models on a Turbocharged GDI Engine Operating Map*; SAE Technical Paper Series; SAE International: Warrendale, PA, USA,, 2018; Volume 1, p. 22018. [CrossRef]
27. Liu, J.; Dumitrescu, C.E. Single and double Wiebe function combustion model for a heavy-duty diesel engine retrofitted to natural-gas spark-ignition. *Appl. Energy* **2019**, *248*, 95–103. [CrossRef]
28. Liu, W. *Hybrid Electric Vehicle System Modeling and Control*; John Wiley & Sons: Hoboken, NJ, USA, 2017.
29. Test Procedure for Measuring Fuel Economy and Emissions of Low Carbon Buses Powered by Charge Sustaining Hybrid Powertrain. 2011. Available online: https://www.lowcvp.org.uk/ugc-1/1/2/0/annex_a2_-_test_procedure_for_charge.pdf (accessed on 10 February 2021).
30. Available online: https://dieselnet.com/standards/cycles/whvc.php (accessed on 10 February 2021).

MDPI

St. Alban-Anlage 66

4052 Basel

Switzerland

Tel. +41 61 683 77 34

Fax +41 61 302 89 18

www.mdpi.com

Applied Sciences Editorial Office

E-mail: applsci@mdpi.com

www.mdpi.com/journal/applsci